# Sustainability in Maritime Transport: Advances, Solutions and Pending Tasks

# Sustainability in Maritime Transport: Advances, Solutions and Pending Tasks

Editor

**José A. Orosa**

MDPI • Basel • Beijing • Wuhan • Barcelona • Belgrade • Manchester • Tokyo • Cluj • Tianjin

*Editor*
José A. Orosa
University of A Coruña
Coruña, Spain

*Editorial Office*
MDPI
St. Alban-Anlage 66
4052 Basel, Switzerland

This is a reprint of articles from the Special Issue published online in the open access journal *Applied Sciences* (ISSN 2076-3417) (available at: https://www.mdpi.com/journal/applsci/special_issues/Pending_Tasks).

For citation purposes, cite each article independently as indicated on the article page online and as indicated below:

LastName, A.A.; LastName, B.B.; LastName, C.C. Article Title. *Journal Name* **Year**, *Volume Number*, Page Range.

**ISBN 978-3-0365-8158-3 (Hbk)**
**ISBN 978-3-0365-8159-0 (PDF)**

# Contents

# About the Editor

**José A. Orosa**

José A. Orosa currently works at the University of A Coruña, E.T.S Náutica y Máquinas. He is a Marine Engineering, Naval Architect and Mechanical Engineer and Full Professor at the University of A Coruña, Spain. His research projects are centred on energy efficiency and ship design and efficiency. He was identified as the second-most-relevant researcher in the last ten years (2006-2016) in Sustainability and Energy Efficiency by the study entitled "Worldwide Research on Energy Efficiency and Sustainability in Public Buildings (Sustainability 2017, 9, 1294; doi:10.3390/su9081294)".

*applied sciences*

MDPI

*Editorial*

# Sustainability in Maritime Transport: Advances, Solutions and Pending Tasks

**José A. Orosa**

Department of Navigation Science and Marine Engineering, Universidade da Coruña, Paseo de Ronda, 51, 15011 A Coruña, Spain; jose.antonio.orosa@udc.es

## 1. Introduction

This Special Issue "Sustainability in Maritime Transport: Advances, Solutions and Pending Tasks", gives an up-to-date overview of the use new technologies to obtain more sustainable maritime transport based on new combustibles, working procedures and related industries, with all of them in agreement with the technical limitations indicated by safety on board.

## 2. New Procedures for Maritime Sustainability

Through their proposition of applied technical solutions to construct more sustainable maritime transport, the different perspectives of the international researchers can be deduced from the varying research topics. Some have focused on energy efficiency, others have attempted to synthesize new, less contaminating combustibles, and others have directed their attention to the operation of ships. Regarding the latter aspect, this operation was divided into two sections: the deck department in the form of navigational functions, and ship maintenance in the marine engineering department. Finally, all these points of view are derived from is the context of safety control, the most relevant factor to consider when changing technologies or working procedures on board.

Energy efficiency is associated with more sustainable transport, and the different works of this Special Issue showed how it can be implemented. When energy efficiency was investigated in a container terminal [1] it was concluded that there was a mutual interest for cooperation between the transport company and the terminal operator to mitigate truck congestion. When energy efficiency is achieved for onboard processes fuel reduction is obtained, but its emissions are still contaminated; more research papers aimed to reduce this effect using optimised scrubbers [2] to achieve a scrubber-silencer system. With the aim to reduce the emissions on board, different works aimed to analyse new combustibles like LNG or cold ironing [3,4] or electrical batteries [5], and other researchers propose the modification of ship [6] and port channels [7].

Once a ship is in service, few options to reduce its emissions are proposed, like crew attendance of navigational practices [8], the ship's interaction with tugboats [9] in different port manoeuvres, and daily maintenance [10]; these methods clearly ameliorate economic factors but do not do the same of rates of accidents. These accidents may have been a consequence of the previously proposed modifications to increase efficiency, so an analysis of the main risks on board was performed by another study [11], showing that machinery is the least-significant contributing cause of accidents.

## 3. Future Tasks

The collective results showed that a more sustainable maritime transport is possible, but it must be developed in collaboration with safety parameters. In consequence, future research works may be directed to parameters like resilience [12] and the maximum safe parameters [13] of ships in existing complex waterways [13]. What is more, the primary initial conclusion is that the study of human-related hazardous events [14] is the key to

**Citation:** Orosa, J.A. Sustainability in Maritime Transport: Advances, Solutions and Pending Tasks. *Appl. Sci.* **2023**, *13*, 7618. https://doi.org/10.3390/app13137618

Received: 26 June 2023
Accepted: 27 June 2023
Published: 28 June 2023

adequately move maritime transport improvement towards sustainability but within the safety limits recommended by international organizations.

**Conflicts of Interest:** The authors declare no conflict of interest.

## References

1. Im, H.; Yu, J.; Lee, C. Truck Appointment System for Cooperation between the Transport Companies and the Terminal Operator at Container Terminals. *Appl. Sci.* **2021**, *11*, 168. Available online: https://www.mdpi.com/2076-3417/11/1/168 (accessed on 27 June 2023). [CrossRef]
2. Kyaw Oo D'Amore, G.; Biot, M.; Mauro, F.; Kašpar, J. Green Shipping—Multifunctional Marine Scrubbers for Emission Control: Silencing Effect. *Appl. Sci.* **2021**, *11*, 9079. [CrossRef]
3. Martínez-López, A.; Romero, A.; Orosa, J. Assessment of Cold Ironing and LNG as Mitigation Tools of Short Sea Shipping Emissions in Port: A Spanish Case Study. *Appl. Sci.* **2021**, *11*, 2050. [CrossRef]
4. Butarbutar, R.; Gurning, R. Semin Prospect of LNG as Marine Fuel in Indonesia: An Economic Review for a Case Study of 600 TEU Container Vessel. *Appl. Sci.* **2023**, *13*, 2760. [CrossRef]
5. Mylonopoulos, F.; Boulougouris, E.; Trivyza, N.; Priftis, A.; Cheliotis, M.; Wang, H.; Shi, G. Hydrogen vs. Batteries: Comparative Safety Assessments for a High-Speed Passenger Ferry. *Appl. Sci.* **2022**, *12*, 2919. [CrossRef]
6. Hernández-Fontes, J.; Maia, H.; Chávez, V.; Silva, R. Toward More Sustainable River Transportation in Remote Regions of the Amazon, Brazil. *Appl. Sci.* **2021**, *11*, 2077. [CrossRef]
7. Paulauskas, V.; Paulauskas, D.; Paulauskas, V. Impact of Port Clearance on Ships Safety, Energy Consumption and Emissions. *Appl. Sci.* **2023**, *13*, 5582. [CrossRef]
8. Dorigatti, J.; Perić, T.; Mrčelić, G. Cruise Industry Trends and Cruise Ships' Navigational Practices in the Central and South Part of the Adriatic East Coast Affecting Navigational Safety and Sustainable Development. *Appl. Sci.* **2022**, *12*, 6884. [CrossRef]
9. Pérez-Canosa, J.; Orosa, J.; Pacheco, E. A New Understanding and Modelling of TSP and BP Indices Compared to Safety IMO Ship Requirements. *Appl. Sci.* **2021**, *11*, 7142. [CrossRef]
10. Stazić, L.; Račić, N.; Stanivuk, T.; Dobrota, Đ. Determination of Benefits of the Application of CMMS Database Improvement Proposals. *Appl. Sci.* **2023**, *13*, 2731. [CrossRef]
11. Kasyk, L.; Wolnowska, A.; Pleskacz, K.; Kapuściński, T. The Analysis of Social and Situational Systems as Components of Human Errors Resulting in Navigational Accidents. *Appl. Sci.* **2023**, *13*, 6780. [CrossRef]
12. Qiao, W.; Ma, X.; Liu, Y.; Lan, H. Resilience Assessment for the Northern Sea Route Based on a Fuzzy Bayesian Network. *Appl. Sci.* **2021**, *11*, 3619. [CrossRef]
13. Gucma, S.; Gucma, M.; Gralak, R.; Przywarty, M. Maximum Safe Parameters of Ships in Complex Systems of Port Waterways. *Appl. Sci.* **2022**, *12*, 7692. [CrossRef]
14. Qiao, W.; Guo, H.; Huang, E.; Deng, W.; Lian, C.; Chen, H. Human-Related Hazardous Events Assessment for Suffocation on Ships by Integrating Bayesian Network and Complex Network. *Appl. Sci.* **2022**, *12*, 6905. [CrossRef]

applied sciences

Article

# The Analysis of Social and Situational Systems as Components of Human Errors Resulting in Navigational Accidents

**Lech Kasyk [1], Anna Eliza Wolnowska [2], Krzysztof Pleskacz [3] and Tomasz Kapuściński [1,*]**

[1] Institute of Mathematics, Physics, and Chemistry, Maritime University of Szczecin, Wały Chrobrego 1-2, 70-500 Szczecin, Poland; l.kasyk@pm.szczecin.pl
[2] Faculty of Engineering and Economic of Transport, Maritime University of Szczecin, H. Pobożnego St. 11, 70-507 Szczecin, Poland; a.wolnowska@pm.szczecin.pl
[3] Faculty of Navigation, Maritime University of Szczecin, Wały Chrobrego 1-2, 70-500 Szczecin, Poland; k.pleskacz@pm.szczecin.pl
* Correspondence: t.kapuscinski@pm.szczecin.pl

**Abstract:** As in any industry exposed to risk, human and organizational factors are the main stakes of maritime safety. Understanding the causes and risks of maritime accidents is integral to the sustainability of shipping. The investigation of marine accidents is a crucial tool for their identification in areas related to operations and ships, including social and situational systems, their design, and technical systems. The authors conducted a cause–effect analysis of marine incidents. For this purpose, case-by-case analysis and an Ishikawa diagram were used, which is a tool that helps identify actual or potential causes of accidents. The study showed that by far the most significant cross-section of causes of accidents were elements of social and situational systems that affect the safety of the ship, crew, and environment. The least significant contribution came from the machinery area. Through the detailed descriptions, a picture emerges not so much of a lack of knowledge of the regulations as of a failure to comply with existing procedures or best practices. In the authors' opinion, more emphasis is needed on preventive measures, including safety culture, training, competence assessment, and increased awareness of the need for sustainability.

**Keywords:** safety; social systems; situational systems; navigation; sustainability; marine accidents; Ishikawa diagram

**Citation:** Kasyk, L.; Wolnowska, A.E.; Pleskacz, K.; Kapuściński, T. The Analysis of Social and Situational Systems as Components of Human Errors Resulting in Navigational Accidents. *Appl. Sci.* **2023**, *13*, 6780. https://doi.org/10.3390/app13116780

Academic Editors: José A. Orosa and Atsushi Mase

Received: 17 March 2023
Revised: 28 April 2023
Accepted: 26 May 2023
Published: 2 June 2023



## 1. Introduction

As in any risk-prone industry, human and organizational factors are a vital component of maritime safety. In most professional analyses carried out by authorized bodies, the human factor is important, if not the most important, out of the numerous factors leading to accidents. To reduce the number of adverse events, intensive research is carried out to identify their causes. Instructions, recommendations, regulations, and so-called checklists are developed. All these activities are intended to lead to the detailed development and strengthening of a safety culture [1,2], a term that first appeared in a preliminary report by the International Atomic Energy Agency (IAEA) after the Chernobyl disaster [3]. The early investigation into the accident initially focused on deficiencies in the power plant's design. However, more detailed analyses also pointed to problems with managerial and social and situational systems problems. Safety culture's primary objective is focused on creating and promoting certain habits among employees, following dedicated procedures in the work environment, paying special attention to any shortcomings, being cautious of any inconsistencies within equipment handling, and regarding other employees' labor. Similar aspects exist for shipping that are related to professions in general and, as per this article, its specific element, navigation. People working in the marine industry selected for interviews by Teperi [4] claim that safety culture has been shaped based on regulations and principles application.

Beyond safety culture, there are several formal international organizations that regulate the organization of maritime labor, with safety as one of the main goals. In order to improve safety across marine areas, The International Safety Management Code has been established, with hopes that an introduction of this ISM code will directly lead to the significant upgrade of safety culture [5].

On top of formal organizations, multiple ship safety procedures are subject to detailed regulations. Their lack or omission usually results in death, pollution of sea and land areas, and damage to the cargo or ship.

Current literature and use cases, e.g., official incident reports, treat the human component in a holistic way as one of the reasons for sea incidents and accidents, next to, for example, hydrometeorological conditions, mechanical failures, or force majeure [6]. They analyze human element subcomponents in a very general way, investigating so-called social and situational systems. These approaches, while having several advantages and being described in multiple sources, due to their general nature do not allow for the detailed analysis of specific root causes of sea incidents and are not suitable for improvement programs, neither for current control procedures nor for screening procedures performed on the ship. They can neither be used directly in marine educational units nor in training programs within STCW centers.

The objective of this research is to build a multilayer taxonomy of isolated and defined elements as components of the human error category, which is one of the main factors behind marine incidents. An investigation was performed that included a review of subject matter literature, analyses of individual cases, and a weighted Ishikawa diagram. The authors also performed a qualitative analysis of selected cases of several various sea accidents, followed by qualitative and quantitative analyses of incidents causes performed using a modified hierarchical Ishikawa diagram, comparison matrix, and stratification analysis. The analysis led to an isolation of key causes of human error which, as already mentioned, is a main factor in marine incidents. This work represents an innovation of how one can investigate the human component as a cause of marine incidents. Examining individual factors such as air quality or perceived pain allows educational and training institutions to modify their programs and focus on top key incident contributors.

## 2. Marine Safety in Relation to Human Errors and Their Root Causes: Background, Literature Review and the Concept of Social and Situational Systems

### 2.1. Marine Safety as a Goal of Formal Organizations

Safety is one of the primary goals of multiple organizations around marine space. The most prominent one—The International Maritime Organization (IMO)—has adopted Resolution A.850(20), defining its vision, principles, and goals for the human element [7]. It was updated by Resolution A.947(23), which was adopted in 2003 [8] and assumes that human factors are a complex, multidimensional issue that impacts maritime safety, security, and marine environment protection. Additionally, effective remediation following maritime casualties requires a solid understanding of the human factor's contribution to causing accidents; that adequate safeguards be put in place; that rules and regulations are simple, clear, and comprehensive; and that communication is flawless. It also recognizes that crew endurance is a function of many complex variables, including personal knowledge, management principles, cultural factors, experience, training, professional skills, and work environment. The resolution has several objectives:

- Implement a structured approach to the proper consideration of the human element.
- Conduct a comprehensive review of selected existing IMO instruments.
- Promote and communicate a maritime safety culture across the marine environment.
- Encourage the development of non-regulatory solutions.
- Implement a system to identify and disseminate human factor analysis and research.
- Provide educational material for seafarers on human factor issues.
- Provide a framework for understanding a highly complex system of interrelated human factors [7,8].

Thus, not only is the human factor being recognized as a key contributor to marine safety but there is also an understanding of how much culture can influence the practical application of formal rules. Another essential organization regulating maritime labor is the International Labour Organisation (ILO). Together with the IMO, they attempt to solve problems regarding the human factor in the marine sector. The result of this cooperation is the Seafarers' Employment Agreements from 2006, updated in 2016 [9,10].

Other organizations, such as the International Association of Ship Classification Societies (IASCS), are also contributing to improvements in safety on all ships by introducing their set of rules, regulations, and requirements. Also worth mentioning is the International Safety Management (ISM) Code, one of the most robust tools to prevent human errors in shipping, adopted by the IMO by Resolution A741(18), which became effective on 1 July 1998 as Chapter IX of the SOLAS Convention on the management of the safe operation of ships.

Every country has its own framework for ensuring that accidents are analyzed and learning applied. In Poland, for example, the body investigating the causes of incidents is the State Marine Investigation Commission (SMAIC), while in the UK, it is the Marine Accident Investigation Branch (MAIB). However, despite the ongoing detailed inspections of all ships sailing on the seas and oceans, and the detention of non-compliant ships, accidents happen almost every day due to the failure of safety mechanisms, and their number must be reduced. As systems improve, new and convenient technologies are introduced, resulting in excessive human reliance on them. [11]. Unfortunately, safety management systems are often over-regulated, too detailed, and too time-consuming [12]. Sometimes, such procedures are difficult to apply in real life, with multiple conflicting priorities around timeliness, cost, and efficiency [13]. While good practices exist to improve safety on every vessel, it is a common practice in the marine industry to purchase a standard, off-the-shelf safety management system designed and developed for a certain group of vessels or companies. This makes safety audits easier on the one hand, but these systems sometimes do not match the specific profile of a given ship [14,15]. For many seafarers, managing safety is an additional workload (documentation, procedures) requiring their concentration and distracting them from delivering their basic tasks [16,17]. Some researchers claim that managing safety takes time and may decrease seafarers' levels of concentration [18]. As displayed in multiple publications [19–21], safety procedures across many industries have been evaluated as too complex and requiring documentation that is too broad, which in itself could potentially lead to incidents. This leads to the conclusion that simply having formal organizations, regulations, and documented practices might not lead to an increase in safety. The marine industry must more actively adopt a safety culture by better analyzing past incidents and driving improvements to educational and training programs.

### *2.2. Human Errors as Causes of Incidents*

Every marine accident may have one or more causes. More than one accident may be associated with one incident. The five categorized causes are human action, system or equipment failure, other agent or vessel, hazardous material, and unknown. Each accident can also have one or more contributing factors, which are divided into three main categories: external environment, shore management, and ship operation. Studies have shown [21] that human error contributes to the following:

- 84–88% of tanker accidents [22].
- 79% of tugboat accidents [23].
- 89–96% of collisions [24].
- 75% of fires and explosions [24].

Contemporary sources [25] show the following breakdown of accidents and contributing factors:

- Cargo ships—in more than 80% of cases, the human factor was the leading cause or at least a component.

- Fishing vessels—56.5% of accidents were related to human factors.
- Passenger ships—65.4%.
- Specialist, service vessels—65.5%.
- Other ships—71.1%.

As shown in Figure 1, considering the 67.1% of human action in the basic breakdown of accident factors and the corresponding human–related percentages in the groups of: system/equipment failure, other agent or vessel, and hazardous material, there is in total more than 81% impact of the human factor on causes of accidents. It is evident, that the human factor is a direct or indirect cause of almost all marine accidents; thus, a proper analysis of the components that influence such errors can positively influence safety at sea.

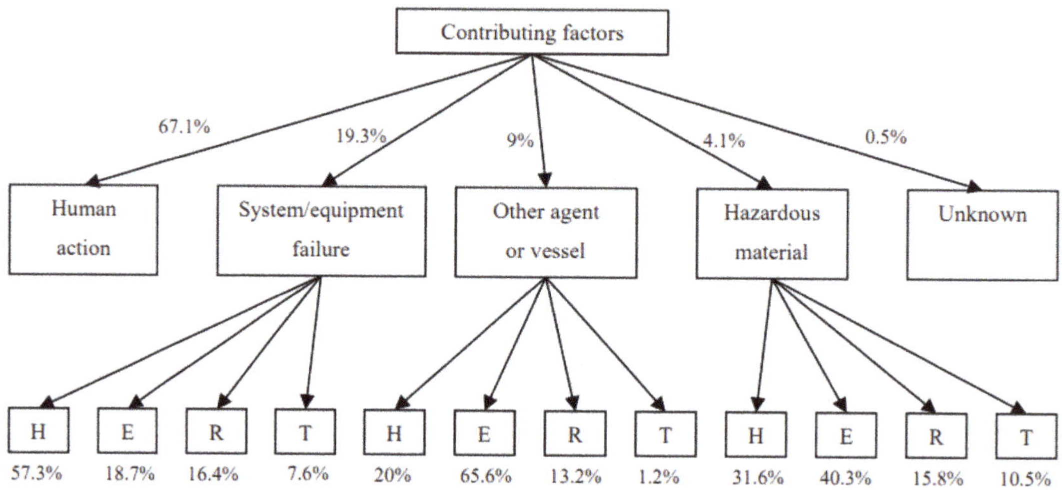


**Figure 1.** Tree of events of contributing factors to accidents. Source: own study based on [25]. H—Human behavior, E—Environment, R—Rules, procedures, and training, T—Tools and equipment.

### 2.3. *Review of the Literature*

A lot of documented research on the influence of the human factor on marine accidents is based on modelling techniques and focuses on the identification and quantification of the probability of human and organizational errors [26–29]. The HFACS method, being widely used, defines four cause categories: organizational influence, insufficient supervision, initial conditions for risky actions, and risky actions. In order to solve defined problems from a marine area, Chen et al. [30] proposed the HFACS—Maritime Accident (HFACS—MA) method. It uses a SHEL model, which stands for Software (S), Hardware (H), Environment (E), and Liveware (L), to describe initial conditions in a traditional HFACS environment. Another method based on a similar idea, HFACS for passenger vessels (HFACS—PV), has been proposed by Ugurlu et al. [31]. It treats operational conditions as a new HFACS category. The authors of this method believe that operational conditions are not a hidden fault but rather a result of a higher-level component that then leads to the accidental results of dangerous actions. Another modification of the original method is HFACS Fuzzy Cognitive Mapping (HFACS—FCM), proposed by Soner et al. [32], where a fuzzy cognitive map is used to identify and quantify definitions of causes that are initiated in HFACS. It is mainly used to strengthen organizational safety measures for fire accidents and collisions [33]. Other documented techniques that are a modification of HFACS include FAHP [34] and ANP [35], which use a process of fuzzy analytical hierarchy to identify causes contributing to HFACS; HFACS and Chi-square test [36]; HFACS and FTA [37]; and fuzzy FTA, ANN, and HFACS [38].

Another source [39] shows that human-related factors play a significant role in shipboard accidents. For example, in 2020, 63% of accidents were caused by human error and 37% of accidents had a technical cause. Therefore, this problem should still be presented and possible solutions proposed; all components of a ship's operation should be carefully analyzed, cataloged, and broken down into specific parts along with their source and a set of recommendations provided on how to counteract their negative effects on the safety of life at sea. An essential element is social and situational systems [40–42].

Looking further into significant causes of marine accidents, refraining from proper visual observation, over-reliance on GPS, fatigue, commercial pressures, and distraction are other significant causes of accidents. In an archive issue of The Navigator, David Patraiko, project director at The Nautical Institute, argued that new technologies and changing regulations can generate unknown direct causes of accidents [43]. A similar analysis can be found in another report from Acejo [44].

Nevertheless, shipping has maintained a long-term positive safety trend over the past year. Still, the recent COVID-19 pandemic and Russia's invasion of Ukraine are major impact factors on global supply chain routes and capacity that have placed enormous stress on the system, with potentially detrimental outcomes: loss of life, loss of ships, exacerbated crew crisis, trade disruption, sanctions burden, and increased cost and reduced availability of bunker fuel. The main places of incidents are Southern China, Indochina, Indonesia, and the Philippines. The increasing number of costly problems may be associated with manning larger ships, the challenges of port congestion due to the shipping boom, and managing ambitious decarbonization goals. Port congestion puts pressure on crews and facilities, meaning there is no room for complacency [43]. Additionally, the increased use of non-container ships to carry containers, despite bulk carriers not being designed to carry containers, can affect their maneuverability in bad weather, and crews may need assistance in responding appropriately to incidents.

All the mentioned components result in a situation where crew demand is high; however, many skilled and experienced seafarers are leaving the industry. A serious shortage of qualified staff is expected over the next five years. Among those who remain, morale is low as commercial pressures, cargo operations responsibilities, and workloads are high. This work situation is prone to error, with 75% of incidents involving human error, according to an AGCS analysis [45].

A Dutch study of 100 victims of navigation accidents [46] showed that the number of causes of accidents ranged from 7 to 58, with a median of 23. Therefore, half of the accidents had 7–23 causes and the other half had 23–58. Sometimes, small things go wrong or small mistakes may seem harmless. However, when these seemingly minor events come together, the result is a casualty. The study found that human error contributed to 96 of 100 accidents. In 93 accidents, there were multiple human errors, usually by two or more people, each making approximately two mistakes.

The key finding was that each human error was identified as a precondition for accidents. This means that if only one of these human errors had not occurred, the chain of events would have been broken, and the accident would not have happened.

There are many demanding aspects of shipping, such as the inability of employees to leave the workplace, extreme weather conditions, long periods away from home, and workplace traffic. Some of these are immutable and reflect the nature of the field. Sometimes, very ordinary situations, such as using the toilet, lead to a procedure breach (rest hours) when, for example, there is only a captain and an officer on duty on board the ship.

Nonetheless, it is possible to modify, supplement, and introduce new strategies or interventions to reduce the impact of these factors on the health and well-being of individual seafarers [47].

Maritime transport has a safety level that is comparable to rail transport and much higher than road transport. In the case of passenger transport in Europe, the risk of a fatal accident is estimated to be 1.1 for road transport and 0.33 for ferry transport [48]. In this context, accident risk and, more precisely, the place of the human factor in this risk, are central issues. Indeed, the human factor appears to be the leading cause of accidents at sea [49]. Among the factors that contribute to incidents are productivity loss (fatigue, stress, health problems), insufficient technical and cognitive skills, insufficient interpersonal competencies (communication difficulties, difficulty in mastering a common language), and organizational aspects (safety training, team management, safety culture) [34,49]. Following this, the article "On your watch automation on the bridge" took a closer look at issues of human–machine collaboration and the role of automation in marine accidents [50]. In the case of a collaborative crew or team, a shared mental representation is one of the key elements behind every safe action. Methods developed in cognitive psychology to analyze this mental structure can be used to assess its impact [51] on crew performance. A study of this type was conducted some years ago [51]. However, as presented in [52], this research remains marginal in maritime transport. Since human error (and usually multiple errors by multiple people) contributes to most marine accidents, preventing human error is essential to reduce the number and severity of maritime accidents. Many types of human error have been described, most of which are not the fault of the human operator. Instead, most of these errors occur due to technology, working environments, and organizational factors that do not account for the capabilities and limitations of the people interacting with them, thus setting up the operator for failure.

In general, there are ways to prevent some human errors or at least increase the chance that such errors will be noticed and corrected by improving the safety culture through better education and training for the better analysis of human factor causes of accidents. As such, we can achieve greater safety at sea and fewer casualties. Summarizing the available data, in the years 2014–2020 there were 6921 injuries, which corresponded to 6211 incidents, and crew members accounted for 81% of the victims [53]. These numbers are very high and should be a call to action for the marine industry.

### *2.4. Social and Situational Systems*

To fully describe social and situational systems, it is first necessary to understand what the human factor is—the interaction of humans with the environment and such human behavior that results in an error [54]. Attention must be paid to the cause effect relationship that contributes to an accident, and that the behavior of the people involved need to change to improve the whole system in the future and reduce the number of accidents [55]. People management must be constantly being improved to prevent errors.

Social systems can be defined as complex entities, i.e., interrelated elements linked by a relationship and interacting with each other. They are separated from the external environment by a clear boundary. The most important factor of such a system is people, without which, it cannot function, let alone exist [56]. Elements of the social system, such as social pressure, role, or life stress, are prevalent in a seafarer's work. Their common denominator is stress, i.e., a reaction responsible for the equilibration of the organism as a result of disturbing external stimuli. This manifests during unusual events or situations not previously encountered. Stress harms the functioning of a person in various spheres of their life and can also cause a deterioration in health. It is essential to realize that anxiety will never disappear from a person's life; it cannot be eliminated in any way. The average individual only focuses on its adverse effects, but what is usually overlooked is that stress, when controlled, can be a factor in self-improvement.

Social pressure exists in every society. In the case of a ship's crew, a certain attitude, behavior, or mindset is expected from the employee [57,58]. How they cope with this type of pressure depends on their mental state. It will be more stressful for some and for others less so. An example of social anxiety on a ship could be the shipowner's expectations of the captain, e.g., punctuality, or the chief officer's expectations of other seafarers, e.g., to complete a task quickly [59]. As competence increases, so does one's responsibility for the ship and less competent staff. The consequences of poor decisions made under stress can result in an accident or disaster. The essential skill, in this case, is to focus on solving the problem that has arisen and treating it as a challenge rather than on minimizing the stress caused by a problematic situation. In the case of life pressures, there are situations such as the death or illness of a family member, divorce, personal injury, loneliness, and risk of redundancy. These events usually cannot be influenced or changed; thus, adapting to a new situation is complex. With the intense emotional impact of such stress, a person can experience physical and psychological disorders [60] (Figure 2).

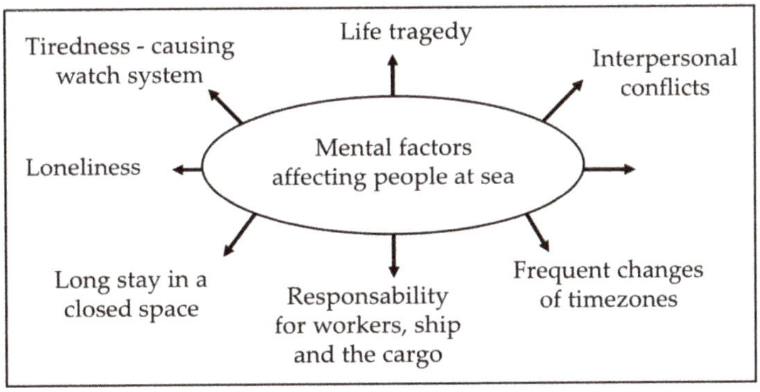


**Figure 2.** Mental factors that affect a person when working on a ship. Source: own study based on [61].

The situational system is the collection of individual factors that affect a person in a given environment and contribute to mistakes, such as employee fatigue during watch work, long physical work, perceived pain, and arduous hydrometeorological conditions. A tired worker is much more likely to make a mistake. These elements are interrelated to a greater or lesser extent [61].

Environmental stress occurs when the place where one works or lives is not organized correctly. Some external stimuli can cause this type of stress, such as air quality (dust, smells, allergic reactions); temperature (tropics or polar zone—high temperatures cause a decrease in employee efficiency and low temperatures increase drowsiness and decrease concentration, which affects one's ability to think quickly and logically and perform tasks flawlessly [62]); noise (operation of the main engine, navigation in ice); lighting (the polar night and tropical zones can cause eye pain and increase worker fatigue or feelings of drowsiness); and order in the room (personal space, the intrusion of others into one's living space when they are currently resting, constant intrusion, rocking of the ship, feeling physically threatened). The last example can be particularly stressful for people with seasickness. Even in the cabin, the rocking effect of the vessel can cause significant stress and discomfort. All this can lead to a deterioration in the quality of work and the possibility of mistakes [63].

There are also ergonomic aspects, such as repetitive tasks, hand strain, uncomfortable posture, vibration, and noise, which significantly affect the efficiency of the work performed and can cause adverse effects on human health. For example, repetition of the same activity can lead to monotony and fatigue. Continuous work with an uncomfortable posture can cause fatigue, discomfort, and pain. Vibrations from hand-held devices, such as electric and pneumatic hammers, grinders, and drills, can cause various types of diseases and damage to hand structures as well as fatigue, irritability, insomnia, and coordination problems [64,65]. These components that affect people at sea are pictured in Figure 3.

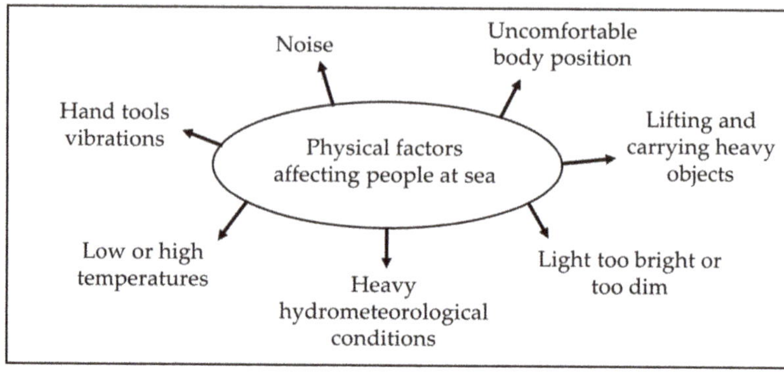


**Figure 3.** Physical factors that affect a person during work on a ship. Source: own study based on [61].

### 2.5. *Ship Safety: Training in Social and Situational Systems*

Training centers, universities, and schools related to maritime work teach the subject of Ship Safety, which is required by STCW and aims to provide knowledge of international and national regulations in which a ship's safety in various operating conditions has been addressed and to develop students' skills to apply them in hazardous conditions. The syllabus discusses the formal and process frameworks of the safety aspect, i.e., the regulations, organizations, processes, and procedures that are key to those in command of vessels and crews. However, a careful reading of post-accident reports shows that accidents are often caused not by a lack of knowledge of the regulations but by a lack of appropriate behavior. In one case, the commanding officer of the high-speed passenger ship, Express 1, undoubtedly knew the relevant rules, regulations, procedures, and laws. However, pressure from the shipowner—not explicitly stated, but hanging over the vessel in the form of a strict timetable—resulted in dealing with many issues simultaneously. This resulted in the commander being preoccupied with these activities instead of concentrating on navigation and steering in conditions of minimal visibility, leading to a collision with the Baltic Condor. The authors believe that both the captain and the second person at the helm of Express 1 would have been able to correctly apply radar noise reduction in calm conditions and consider the echo's potential position on the course dash. However, a flurry of tasks not directly related to steering in dense fog at very high speeds of more than 30 knots caused the echo of another craft on course to be overlooked, and a collision ensued [66].

When using any means of transport, accidents are inevitable and happen because of errors, with lasting consequences. Social and situational systems consider 12 types of effects of maritime accidents:

1. Collision and resulting oil spill. At dawn on 7 October 2018, approximately 15 nautical miles north of Cap Corse, in international waters, the ro-ro vessel, Ulysse, bound for Radès (Tunis), struck the starboard side of the unloaded Cypriot-flagged container ship, CSL Virginia, at anchor. Shortly after the collision, bunker fuel leaked from the damaged tank into the sea. The following conclusions are drawn from the report:
    - The collision resulted from a lack of watchkeeping on the vessel *m/v* Ulysse and an unreasonable anchoring position by CSL Virginia.
    - On the *m/v* Ulysse, the lack of proper observation was due to the lack of involvement of the officer on watch, who was occupied with his mobile phone. He was sitting in front of an unusable radar, depriving himself of the opportunity to make observations and correctly assess the situation. Using a second operational radar would have given him all the information he needed to evaluate the situation. Moreover, the place where he was sitting was lowered, which did not allow him to see the horizon line correctly.

- COLERG procedures and regulations were not carried out on both vessels.

The situational factor was the main component that caused the collision. There were other factors involved such as boredom, routine, and monotony associated with the length of periods at sea, which certainly affected the commitment of watch officers to their duties [67].

2. Cruise ship incidents A blackout and loss of main engine power occurred on the cruise ship, Viking Sky, as it sailed towards Stavanger, Norway, in strong winds and very rough seas. According to maritime lawyer Jim Walker, the focus of Viking Sky's accident report was whether the ship had any valid reason to sail during such bad weather conditions. In addition, Norwegian officials determined that low oil levels caused the engine to fail. Both social and situational factors led to the accident [68].
3. Fires on vessels lead to navigation difficulties or even vessel loss. According to statistics, almost 75% of such fires are caused by a simple error of the people working on the vessel. Twelve passengers went missing in a Greece ferry fire on Euroferry Olympia, near Corfu. A company representative stated that the fire started from a truck on the car deck. At the same time, relatives of the 12 missing were concerned and complained to the shipowner about the "miserable conditions" that prevailed on the Euroferry Olympia: there were not enough cabins and the ship was not designed for trucks, adding that the vessel was dirty and had bugs. Thus, again, a combination of both social and situational factors led to the accident [69].
4. Commercial fishing vessel accidents and navigation. The U.S. Bureau of Labour Statistics collects information on occupations and lists commercial fishing as the most dangerous of all professions. From rough weather to falling overboard to being struck by heavy equipment, working on a fishing boat can harm workers in many ways. Working on a fishing vessel means long, hard hours; exposure to cold weather; and the risk of accidents and injuries. Fatigue is common, especially if workers are not allowed breaks to which they are entitled. A lack of adequate protective equipment against cold and wet weather can also be problematic as fishermen can experience hypothermia or frostbite. The Seiner capsized, causing the deaths of four fishermen. An investigation revealed that the boat was unsuitable for the rough waters in which it was operating, which is considered negligent. The boat should never have gone into these waters, and someone made a bad judgment in deciding to send it out. Four workers paid for this mistake with their lives. Social and situational factors played a part in this unfortunate incident [70].
5. Accidents on tugboats related to the maneuvering of vessels. Sometimes, accidents occur due to tugs' visibility being blocked by larger ships. Additionally, human error on the skipper's part can lead to unwanted and unexpected incidents. Two crew members died after CMA CGM Simba's tugboat capsized [70]. The maneuver in which CMA CGM Simba's departure plan was discussed between the pilot and Captain Domingue before commencement was too general. During the maneuver, no one on board CMA CGM Simba monitored the tug's position. This accident resulted from a social system factor, the captain–pilot–tug skipper relationship, and a situational factor: failure to adapt the type of tug to the situation [71].
6. Explosions as a cause of accidents on tankers and cargo ships. The very nature of the materials carried on tankers is dangerous. Most ship fires and explosions are the result of human negligence. The fire on board the container ship, X-Press Pearl, which lasted for several days, appeared to be out of control after the explosion. According to Bloomberg, Sri Lankan authorities feared a severe oil spill. The vessel transported 1486 containers, including 25 tons of nitric acid and other chemicals. According to the Sri Lankan Navy, the fire was caused by chemicals on the Singapore-flagged container ship. The general manager of container ship operator MV X-Press Pearl confirmed that the vessel crew carrying the chemicals was aware of the spill. He added that Qatar and India had refused permission to unload the leaking container before a fire broke out on the vessel. The fire consumed most of the cargo, contaminating the

surrounding waters and a long stretch of beaches, and the ship sank. There were undisputed social and situational errors [72].

7. Stranding of a ship. This type of marine accident has a significant impact not only on the ship's hull but also on the entire ocean area around the incident. For the Shoei Kisena ship accident in the Suez Canal, the vessel's owner argued that the Suez Canal Authority (SCA) was at fault for allowing the ship to enter the waterway during poor weather. Ahmed Abu Ali, a member of the legal team, told Reuters that the authorities had failed to prove any fault of the vessel. The evidence that was presented to the court showed disagreements between SCA pilots and its control center over whether the ship should enter the canal. Shoei Kisen suggested that the vessel should have been accompanied by at least two tugs suitable for the size of the vessel [73]. Thus, again, an unfortunate set of situational and social errors occurred.
8. Maritime accidents due to drugs and alcohol. Drug abuse is a severe problem across the world and, in the marine industry, it can cause irreversible damage. If the crew members abuse psychoactive substances or alcohol, this can result in unpredictable behavior and lead to an accident. The Dutch-registered general cargo vessel, Ruyter, ran aground off the north coast of Rathlin Island in the UK. An investigation revealed that Ruyter ran aground because no action was taken to correct the deviation from the ship's intended track. According to the safety management system records, the shipowner should have arranged alcohol testing of the crew. However, the investigation found no evidence that alcohol testing was ever carried out on the vessel. Once again, we are confronted with problems inherent in social and situational systems as components of human error, resulting in navigational accidents [74].
9. Navigational accidents are indirectly linked to cargo-handling operations, which are integral to ship operations. One of the leading causes of crane accidents for loading and unloading marine cargoes is operator error. One known example happened when workers onshore were unloading aluminum bars from a ship using a crane. The workers were inexperienced in using a particular lifting system and needed to secure the load better. As the crane operator moved the load towards the quay, they struck the worker on the ground and killed him. A pattern of situational and social errors can be considered a cause of this accident [75].
10. Accidents in shipyards, where ship assembly and welding accidents are common, may spare workers' lives but limit their overall ability to work. Similarly, the continuous inhalation of toxic fumes also becomes another cause of shipyard accidents [76].
11. Marine accidents on dive support vessels. Suppose a dive support vessel is not fully operational and the crew is not adequately qualified. In this case, they cannot effectively supervise and direct the entire operation, which can lead to a severe accident. In one documented case, a large steel frame, known as a cursor, fell from a height of two meters, trapping the rigger and resulting in fatal crush injuries to his chest. Following an investigation, the company noted the following:
    - The control system was not fully commissioned before its use.
    - The company's internal project management systems were not fully utilized.
    - Communications and reporting lines within the project team and with offshore and onshore management were neither fully utilized nor understood.
    - Design intentions and pre-commissioning requirements for the safe operation of the new equipment had not been adequately communicated to the work team.
    - There were no secondary means of securing the active cursor.
    - A decision was taken to work under the cursor. The hazard of working under a suspended load was not recognized as it was not a typical load suspended from a crane [77].

This again describes an unfortunate set of errors triggered by situational and social systems.

12. Accidents on barges occur mainly because of their specific design and maneuvering capabilities. These problems can be caused by the inexperience of the person at the barge's helm or by the improper use of mooring lines. Other causes involving severe injuries while working on board a barge include the following:
    - Explosions, suffocation, and hypoxic-ischemic encephalopathy caused by hazardous and noxious fumes and gases.
    - Crushing while bunkering or mooring with a tug.
    - Trips and falls due to cluttered decks and workspaces [78].

It is evident from the above types of marine accidents that worker and operator errors play a significant role in causing accidents. Investigating a marine accident helps narrow down the actual cause of the accident, which helps those claiming damages assert their rights with absolute clarity. Some authors believe, however, that the complexity of causes of marine incidents and the lack of unified procedures for reporting such incidents makes it difficult to discover the real reasons and factors behind them [79].

## 3. Materials and Methods

The unique case method and a weighted Ishikawa diagram were used to address the complexity of the problem of the influence of social and situational systems on human error and, consequently, marine accidents considered in this paper.

The unique case method is dedicated to analyzing a specific single phenomenon, event, or person. For example, a special occasion or character can be studied in this way regarding the situation in which the object under study operates. This method is sometimes referred to as an individual case study. This method is often used to characterize an unusual case, often deviating from the norms commonly observed by society. It is widespread in the sciences, based on the analysis of specific issues, through which researchers can determine the causes of a situation. Thus, it has applications in medicine, law, environmental protection, and safety [80]. Using the unique case method, one can gain the necessary knowledge about a situation and obtain conclusions indicating whether a viable solution is possible. It is undoubtedly impossible to generalize based on research using the unique case method. Therefore, the results obtained in this way can only be applied to a select group. Given the above, the authors decided to use the following research methodology:

1. Analysis of selected incidents at sea.
2. Development of the Ishikawa diagram for the selected problem.
3. Determining the importance of each cause.
4. Diagram construction based on the stratification analysis and defining the main causes of incidents.

### *3.1. Analysis of the Causes of Maritime Accidents*

The first use case worth highlighting in a discussion about marine incidents is the case of fishing. In just one region, commercial fishermen sail across the Gulf of Mexico in almost all weather conditions as commercial pressure grows. It is a physically demanding and dangerous job, and, unfortunately, some fishing boat owners disregard job safety. Every year, commercial fishing boat crew members die in preventable accidents. Over 10 years, 116 commercial fishermen died in the Gulf of Mexico alone, many due to falling overboard and injuries sustained onboard. Commercial shrimpers were involved in an alarming number of fatalities as well, with many more injured. This is an evident effect of negative social and situational pressures [81].

The Neva accident was the next to be examined regarding the social and situational system elements involved. The vessel departed from the Szczecin ship repair yard with a pilot on board and the assistance of a tugboat. It was heading to the Szczecin-Świnoujście waterway to head for Riga. The tug was slowed down as it passed Huta quay. While attempting to increase the vessel's speed, it was noticed that the main engine was not working correctly. After some time, the engine stopped working, and the vessel lost speed. The ship was forced to leave the fairway and drop anchor. The senior officer on the bow supervised the anchoring maneuvers. The captain instructed him to throw one shackle of the chain. However, he executed this instruction incorrectly and threw more than two shackles. The additional 30 m of chain released contributed to the vessel partially running aground. At the same time, the ship's captain and pilot did not pay attention to the ship's position and the fact that it was not positioned with its bow to the wind. Their subsequent actions were based on poor judgment of the situation, related to the initial lack of communication between the navigation bridge and the bow position, leading to the ship running aground [68]. In analyzing the accident, it was found that the social system had an impact here. One of the two elements of this system that occurred in the described accident was social pressure; the master and the pilot did not obtain explicit information from the senior officer from the very beginning that he had made a mistake and the ship was not where it should be; thus, they made decisions chaotically and under pressure based on a misunderstanding of the situation. The second element was the role played by the senior officer on the bow. A person in such a position is expected to have the appropriate competencies, such as good situational awareness, quick responses, and proper communication skills to relay any relevant factors affecting maneuverability, including the immediate communication of an erroneous command.

The third accident investigated for the presence of social system elements was that of the ships Corvus J and Baltic Ace. Both ships had officers of Polish nationality on navigation watch. Corvus J was sailing from Scotland to Antwerp, Belgium, while Baltic Ace was leaving Belgium from the port of Zeebrugge to head to Finland. At some point during the voyage, the watch officer on the Baltic Ace vessel spotted the vessel Corvus J on the radar; it was about to pass within one nautical mile, and there were no vessels in the vicinity threatening safe passage. The ship Corvus J made a slight turn to the right, which prompted the officer of the watch on the vessel Baltic Ace to call the vessel on very high frequency (VHF) radio to understand its intentions and advised that it would alter course slightly to the left to increase the distance between the vessels. In the meantime, the vessel Corvus J again made a turn to the right and the Baltic Ace to the left. As the distance between the vessels decreased and the dangerous situation developed, it became increasingly difficult for the watch officers to communicate via VHF radio. The actions they agreed with each other during the conversation did not coincide with the actual steps, which resulted in the collision of the two vessels and the sinking of the vessel Baltic Ace [82]. Analyzing the accident studied, it was concluded that one element of the social system influenced the occurrence of the accident. Social pressure significantly impacted the course of events in this case. The watch officers of both ships attempted to communicate in English. However, their communication became increasingly incomprehensible to the other party under the influence of the developing dangerous situation. Actions that were agreed upon using VHF communication should have been followed. Under pressure and confusion, decisions to change course were taken chaotically and without analyzing the situation.

### 3.2. *Ishikawa Diagram*

One of the essential tools of quality management is the cause–effect diagram [83,84]. It was developed by Professor Kaoru Ishikawa and was first used by a Japanese company, Sumitomo Electric [85]. The diagram is known as the Ishikawa diagram after its creator [86]. Due to its structure and shape, this diagram is often called a herringbone or fishbone diagram. It visually represents causes and their interrelationships with a problem, error, or inconsistency in the area under investigation [87].

The Ishikawa diagram has a hierarchical structure in which the root causes are closest to the core, while the specific factors directly related to the root causes are their development. The principle of "from the general to the specific" applies in drawing the diagram. The leading causes are determined first, followed by the intermediate reasons: second-order causes and, if necessary, causes of subsequent orders [88]. Depending on the area under study, the diagram may use a layout appropriate to it:

- Object-oriented, used when the effect is analyzed, e.g., of poor product quality, related to technical and organizational aspects, divisible into sub-components;
- Technological, considered by the list of technical process operations;
- Participating factors, which can be used at any stage and level of the investigation of the quality of a process or event.

The primary 5M method can be used to develop a cause-and-effect diagram or it can be extended or modified as appropriate, depending on the area under analysis: 5Ms + E, 6Ms, or 8Ms:

1. Man—every aspect of a person's work, including routine, inexperience, monotony, and fatigue.
2. Machine—whether it is working correctly, whether every part is in working order, whether it needs calibration, etc.
3. Method (method/technology)—whether the process is as intended, whether the steps are performed correctly, whether the sequence of steps is optimal, etc.
4. Material—hidden or visible material defects, product dimensions, missing holes, problems at the supplier, etc.
5. Measurement—whether measurements have been carried out correctly, whether measuring instruments are operational and legalized, etc.
6. Mother nature (environment)—what effect does the environment have on the process/event, i.e., humidity, temperature, lighting, but also, e.g., noise, dust?
7. Management—whether the management of employees and crews is adequate and appropriate to the conditions, whether employees are given proper guidelines, etc.
8. Maintenance—whether the maintenance of the machine and its components, facility, tools, networks, pipelines, etc., is being neglected.

The Ishikawa diagram is also used outside its original production environment in the 8P design:

1. Product—physical aspects of the product.
2. Price.
3. Promotion—a type of promotion/advertisement.
4. Place—place/location/environment.
5. Process.
6. People.
7. Psychical evidence—physical aspects of customer interaction sites.
8. Performance—results compared to the competition.

In services, causes can be grouped according to the 4Ss:

1. Surroundings—surroundings, environment.
2. Suppliers—suppliers, sub-suppliers.
3. Systems—processes.
4. Skills—staff skills.

The classic Ishikawa diagram facilitates the analysis of a process in a cause–effect framework, but it does not contain quantitative information, only qualitative information. Gwiazda proposed supplementing the chart with the weights of individual causes. Once sets of primary and sub-causes for each leading cause have been determined, the next course of action is as follows:

- Each cause and sub-cause is assigned an appropriate weight.
- The absolute values of the sub-cause weights are determined.
- The Ishikawa diagram is completed with sub-weights [89].

The weights of the individual causal factors are determined using a matrix of pairwise comparisons using the following rule of thumb: if one of the comparable factors is considered more important, it is assigned a score of 1; the other element is given a score of 0. If both factors are considered equivalent, they are given a score of 0.5. A scale of 0.25 to 0.75

can be used to increase the precision of the assessment. These values denote being slightly less critical or slightly more critical, respectively [90].

## 4. Results

It is possible to analyze marine accidents using a weighted Ishikawa diagram when considering decision-making methods, including multi-criteria and multidimensional scaling techniques, which often support diagnosed issues and problems in various fields. The methodology above is divided into stages accordingly.

The determination of the problem category, i.e., causes of a marine casualty according to the selected 5Ms, 6Ms, 5Ms + E, or a combined idea is based on the previously conducted literature on the subject, with stages of the full process listed below:

1. Definition of first- and second-order causes.
2. Definition of the relative weights of the causes at each level.
3. Definition of the absolute value of the individual sub-cause weights.
4. Stratification analysis.
5. Definition of a set of critical causes.

Figure 4 shows the main categories of causes of marine accidents.

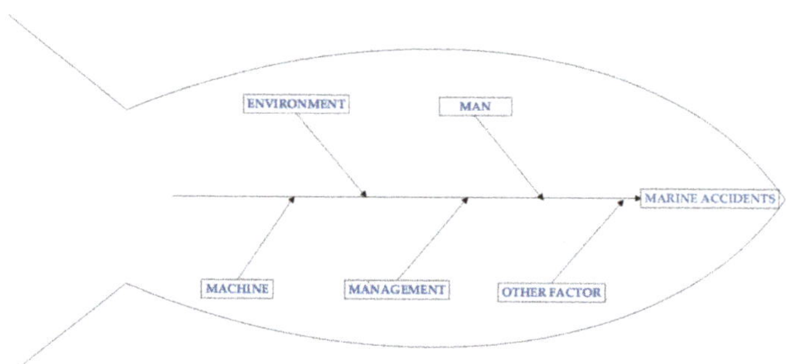


**Figure 4.** Ishikawa diagram leading causes.

The main cause weights were determined based on a comparison matrix, according to the 4Ms +A approach, i.e., Human, Machine, Management, and Other factor, see Table 1.

**Table 1.** Comparison matrix for leading causes.

| | Environment | Human | Machine | Management | Other Factor | Total | Ratio of Weight to Sum of Weights | Validity of Factors |
|---|---|---|---|---|---|---|---|---|
| Environment | X | 0.25 | 0.25 | 0.25 | 0.75 | 1.5 | 0.150 | 4B |
| Human | 0.75 | X | 0.1 | 0.5 | 0.75 | 3.0 | 0.300 | 1A |
| Machine | 0.75 | 0 | X | 0.25 | 0.75 | 1.75 | 0.175 | 3B |
| Management | 0.75 | 0.5 | 0.75 | X | 0.75 | 2.75 | 0.275 | 2A |
| Other factor | 0.25 | 0.25 | 0.25 | 0.25 | X | 1.0 | 0.100 | 5B |
| | | | | | Total | 10 | 1.000 | |

Components under analysis have been placed in the last column (Validity of Factors) looking into their respective weights vs. the sum of other weights. Letter A stands for important components, while B is assigned to less important ones. This was the start of the stratification analysis.

The standardized weights for the main factors are placed in circles on the Ishikawa diagram. The upper part of the circle contains the relative weight related to the factor in question, while the lower part contains the absolute weight associated with the whole group. The two weights are equal for the main factors (Figure 5).

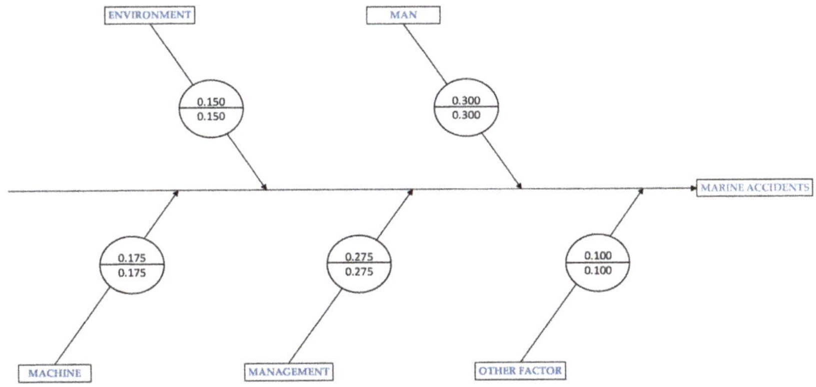


**Figure 5.** Weighted Ishikawa diagram with main cause weights.

Of the main reasons outlined, the group identified as "human" had the most significant weight, followed by "management", "machine", "environment", and "other". This preliminary assessment represents the beginning of inquiries in the area under investigation.

The next step is to identify the first- and subsequent-order sub-causes. A detailed set of causes with first-, second-, and third-order sub-causes is presented in Table 2.

**Table 2.** Marine accidents—leading and sub-causes of I level and sub-causes of II level.

| Main Causes | Sub-Causes of the First, Second, and Third Levels |
|---|---|
| 1. Environment | [1.1.] Adverse meteorological conditions |
| | [1.2.] Adverse hydrographical conditions |
| 2. Human | [2.1.] Adverse social system |
| | [2.1.1.] Social pressure |
| | [2.1.1.1.] Expectation of punctuality |
| | [2.1.1.2.] Expectation of speed |
| | [2.1.1.3.] Expectation of responsibility |
| | [2.1.2.] Life stresses |
| | [2.1.2.1.] Death of a loved one |
| | [2.1.2.2.] Illness of a close relative |
| | [2.1.2.3.] Harm (own) |
| | [2.1.2.4.] Loneliness |
| | [2.1.2.5.] Risk of dismissal |
| | [2.1.3.] Social role |
| | [2.2.] Inappropriate situational system |
| | [2.2.1.] Employee fatigue |
| | [2.2.1.1.] Long working hours |
| | [2.2.1.2.] Perceived pain |
| | [2.2.1.3.] Strenuous hydrometeorological conditions |
| | [2.2.2.] Stress caused by an external stimulus |

**Table 2.** *Cont.*

| Main Causes | Sub-Causes of the First, Second, and Third Levels |
| --- | --- |
| | [2.2.2.1.] Air quality |
| | [2.2.2.2.] Noise |
| | [2.2.2.3.] Lighting |
| | [2.2.2.4.] Personal space |
| | [2.2.3.] Ergonomic aspects |
| | [2.2.3.1.] Repetitiveness of activities |
| | [2.2.3.2.] Hand strain |
| | [2.2.3.3.] Uncomfortable posture |
| | [2.2.3.4.] Annoying vibrations |
| 3. Machine | [3.1.] Improper operation |
| | [3.1.1.] Non-compliance of operational activities with the requirements of the company and IMO regulations |
| | [3.1.2.] No regular inspections |
| | [3.2.] Ship failures |
| | [3.2.1.] Damage to the navigation system |
| | [3.2.2.] Loss of hull tightness |
| | [3.2.3.] Damage to the deck equipment |
| | [3.2.4.] Main engine failure |
| | [3.2.5.] Failure of automatic equipment |
| | [3.2.6.] Failure of auxiliary machinery in the engine room |
| | [3.2.7.] Damage to pipelines and fittings |
| 4. Management | [4.1.] Voyage planning errors |
| | [4.1.1.] Incomplete documents |
| | [4.1.2.] Non-compliance with certified system(s) |
| | [4.2.] Errors in the management of the ship's crew |
| | [4.2.1.] Errors in the delegation of tasks |
| | [4.2.2.] Inadequate staff appraisal system |
| | [4.2.3.] Inadequate monitoring |
| | [4.3.] Incomplete ship operating report |
| | [4.4.] Inaccurate inspection of the ship |
| | [4.4.1.] Internal audits by the company or external audits by the charterer |
| | [4.4.2.] Port inspections |
| | [4.4.3.] Flag State inspections |
| 5. Other factor | [5.1.] Cargo shifting |
| | [5.2.] Drifting wreck |
| | [5.3.] Terrorist attack |

As for the leading causes, i.e., based on the comparison matrix, relative weights were determined for the individual first-, second-, and third-order sub-causes. Subsequently, absolute weights were selected for the sub-causes based on the relative weights. The relative weights were multiplied by the value of the relative weight of the leading cause to which the sub-cause belonged. This can be written with a simple equation [91]:

$$W_{a.sn} = W_{r.i.c}\, W_{r.\ sn}$$

where

$W_{a.sn}$—Absolute weighting of sub-causes of n order,
$W_{r.i.c.}$—Relative importance of the main cause,
$W_{r.\ sn}$—Relative weighting of sub-causes of n order.

Based on the results obtained, which were ranked in descending order of importance, cumulative weights were determined for the defined sub-causes. A reference field was calculated, as shown in Table 3. The value of the reference field was obtained by multiplying the value of the cumulative weight by the value of N = (55 − n), where 55 is the number of all reasons, and n is the number of the next sub-reason [92,93]. The grey color indicates the causes with the most significant impact on marine casualties.

The cumulative weights presented in Table 3 are the coordinates of the Lorentz Curve and the basis for a stratification analysis based on the Pareto Rule. Thus, a group of influential and less essential factors influencing marine casualties was identified. The data obtained were transferred to the graph shown in Figure 6.

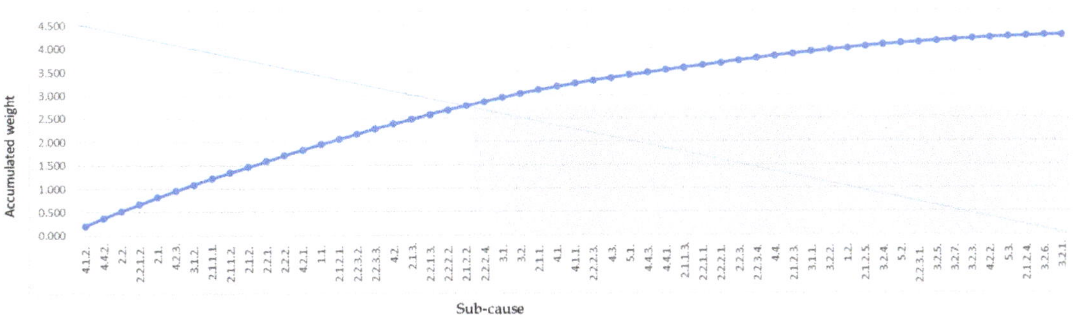


**Figure 6.** Lorenz diagram and stratification analysis.

The so-called "reference area" was taken as the index of division. This is the rectangle's area defined by the Lorenz Curve's inflection point. This area reached its maximum for the 23 sub-causes and amounted to 9.520, thus constituting the limit of stratification and definitively determining the division into the group of essential sub-causes A (marked in grey in Table 3) and the group of less important causes B. The most important and frequent causes included those from the "Management" and "Human" groups.

The stratification analysis made it possible to develop the simplified weighted Ishikawa diagram shown in Figure 7. The final version of the diagram presented contains the most critical causes of marine accidents.

**Table 3.** Results of the stratification analysis.

| Sub-Cause | Absolute Weight | Accumulated Weight | Reference Field |
|---|---|---|---|
| 4.1.2. Non-compliance with certified system(s) | 0.206 | 0.206 | 11.124 |
| 4.4.2. Port inspections | 0.165 | 0.371 | 19.663 |
| 2.2. Inappropriate situational system | 0.150 | 0.521 | 27.092 |
| 2.2.1.2. Perceived pain | 0.150 | 0.671 | 34.221 |
| 2.1. Adverse social system | 0.150 | 0.821 | 41.050 |
| 4.2.3. Inadequate monitoring | 0.138 | 0.959 | 46.991 |
| 3.1.2. No regular inspections | 0.131 | 1.090 | 52.320 |
| 2.1.1.1. Expectation of punctuality | 0.125 | 1.215 | 57.105 |
| 2.1.1.2. Expectation of speed | 0.125 | 1.340 | 61.640 |
| 2.1.2. Life stresses | 0.125 | 1.465 | 65.925 |
| 2.2.1. Employee fatigue | 0.125 | 1.590 | 69.960 |
| 2.2.2. Stress caused by an external stimulus | 0.125 | 1.715 | 73.745 |
| 4.2.1. Errors in the delegation of tasks | 0.115 | 1.830 | 76.860 |
| 1.1. Adverse meteorological conditions | 0.113 | 1.943 | 79.663 |
| 2.1.2.1. Death of a loved one | 0.113 | 2.056 | 82.240 |
| 2.2.3.2. Hand strain | 0.113 | 2.169 | 84.591 |
| 2.2.3.3. Uncomfortable posture | 0.113 | 2.282 | 86.716 |
| 4.2. Errors in the management of the ship's crew | 0.103 | 2.385 | 88.245 |
| 2.1.3. Social role | 0.099 | 2.484 | 89.424 |
| 2.2.1.3. Strenuous hydrometeorological conditions | 0.099 | 2.583 | 90.405 |
| 2.2.2.2. Noise | 0.099 | 2.682 | 91.188 |
| 2.1.2.2. Illness of a close relative | 0.090 | 2.772 | 91.476 |
| 2.2.2.4. Personal space | 0.088 | 2.860 | 91.520 |
| 3.1. Improper operation | 0.088 | 2.948 | 91.388 |
| 3.2. Ship failures | 0.088 | 3.036 | 91.080 |
| 2.1.1. Social pressure | 0.075 | 3.111 | 90.219 |
| 4.1. Voyage planning errors | 0.069 | 3.180 | 89.040 |
| 4.1.1. Incomplete documents | 0.069 | 3.249 | 87.723 |
| 2.2.2.3. Lighting | 0.062 | 3.311 | 86.086 |
| 4.3. Incomplete ship operating report | 0.058 | 3.369 | 84.225 |
| 5.1. Cargo shifting | 0.058 | 3.427 | 82.248 |
| 4.4.3. Flag state inspections | 0.055 | 3.482 | 80.086 |
| 4.4.1. Internal audits by the company or external audits by the charterer | 0.055 | 3.537 | 77.814 |
| 2.1.1.3. Expectation of responsibility | 0.050 | 3.587 | 75.327 |
| 2.2.1.1. Long working hours | 0.050 | 3.637 | 72.740 |
| 2.2.2.1. Air quality | 0.050 | 3.687 | 70.053 |
| 2.2.3. Ergonomic aspects | 0.050 | 3.737 | 67.266 |
| 2.2.3.4. Annoying vibrations | 0.050 | 3.787 | 64.379 |
| 4.4. Inaccurate inspection of the ship | 0.046 | 3.833 | 61.328 |
| 2.1.2.3. Harm (own) | 0.045 | 3.878 | 58.170 |
| 3.1.1. Non-compliance of operational activities with the requirements of the company and IMO regulations | 0.044 | 3.922 | 54.908 |
| 3.2.2. Loss of hull tightness | 0.044 | 3.966 | 51.558 |
| 1.2. Adverse hydrographical conditions | 0.038 | 4.004 | 48.048 |
| 2.1.2.5. Risk of dismissal | 0.038 | 4.042 | 44.462 |
| 3.2.4. Main engine failure | 0.035 | 4.077 | 40.770 |
| 5.2. Drifting wreck | 0.025 | 4.102 | 36.918 |
| 2.2.3.1. Repetitiveness of activities | 0.025 | 4.127 | 33.016 |
| 3.2.5. Failure of automatic equipment | 0.025 | 4.152 | 29.064 |
| 3.2.7. Damage to pipelines and fittings | 0.025 | 4.177 | 25.062 |
| 3.2.3. Damage to the deck equipment | 0.023 | 4.200 | 21.000 |
| 4.2.2. Inadequate staff appraisal system | 0.023 | 4.223 | 16.892 |
| 5.3. Terrorist attack | 0.017 | 4.240 | 12.720 |
| 2.1.2.4. Loneliness | 0.015 | 4.255 | 8.510 |
| 3.2.6. Failure of auxiliary machinery in the engine room | 0.012 | 4.267 | 4.267 |
| 3.2.1. Damage to the navigation system | 0.010 | 4.277 | 0.000 |

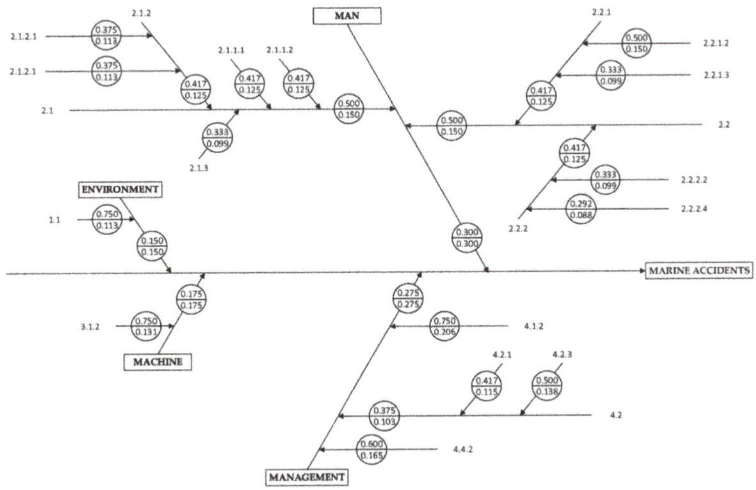


**Figure 7.** The final version of the Ishikawa diagram for marine accidents.

## 5. Discussion

The authors opened this article with a long review of the visions, objectives, and procedures in place of key industry organizations, aiming to improve safety across the maritime industry. With multiple checklists, procedures, audits, and inspections, one could think that shipping is as safe as aviation. Following this, there was a deeper dive into the literature on the subject of safety at sea, which contains a vast array of articles, publications, and reports, some from highly specialized agencies, many of which have found that incidents cause breaks down into various aspects and reasons [25]. Despite multiple safety measures and a large number of publications, the detailed descriptions of real-life sea incidents are striking regarding how trivial, simple, and obvious key root causes are. Hence, the objective of the authors was to go much deeper than an industry approach [6,25], whereby human errors are broken down into a few main categories, and, in reality, each category represents its own ecosystem of possible causes, triggers, and incident starters. This article goes as deep into the breakdown of components of human error as 'the death of a loved one or uncomfortable posture' components, analyzing how they potentially impact safety.

In this work, based on expert knowledge, 55 causes of incidents were included and, with the weighted Ishikawa diagram, 23 of causes were identified as playing the most significant role as contributors to accidents. Among them, as many as 16 were directly linked with social and situational systems, while an additional 5 were related to management, which is also a human factor element.

A causal analysis of the problem of the influence of social and situational systems on human error and, consequently, marine accidents based on the individual case method and weighted Ishikawa diagram provided positive evidence that this type of analysis can be applied to solve such complex issues.

A simplified version of the weighted Ishikawa diagram clearly showed the areas that require attention or even intervention by managers on ships and in ports. Decision makers have an impact on conditions. Among the most relevant issues are the following:

- Problems of non-compliance with certified systems and reliability on port inspection, with too few inspections and errors in delegation.
- A whole range of components of situational and social systems, for example, stress in life caused by an external stimulus, worker fatigue that can be a consequence of awkward posture, too much strain on the hands, etc. Strenuous hydrometeorological and meteorological conditions and noise can also cause perceived pain, affecting one's

ability and efficiency in activities performed, exacerbating frustration and errors due to the desire to meet deadlines and speed up action. Illness or death of a loved one, lack of personal space, and social expectations are also examples of this.

- A lack of regular reviews.

The authors were able to demonstrate the influence of social and situational systems on marine incidents, which also corresponds to the known direction of growing situational awareness of seamen by a bottom-up management approach [94]. There is also an opinion shared across several publications that seamen want to have the option of participation in decision-making processes [95]. Experienced crew members can deal with extraordinary situations, are flexible, and are able to adapt [96,97] as, sometimes, safety can be secured by not applying procedures [98–100].

Working in a maritime education space, authors now have evidence based on the results of this work, and can discuss with the management teams of learning programs at all levels potential changes and improvements in how they organize, run, and certify students, officers, and practitioners. The aspect of safety culture, going way beyond pure knowledge and awareness of safety procedures and regulations, can be discussed with the current analysis results, thereby reinforcing the need to also adapt the culture at the class level to provide stricter ambiance and approaches, offering students the feeling of real-life stress in controlled situations.

## 6. Conclusions

Issues concerning the influence of individual components of the entire category of causes classified as human error as sources of marine accidents were discussed. The authors have shown that social and situational systems interact and influence shipboard safety through a case study method. In each case, a detailed analysis of specific subcategories of direct incident sources was necessary to identify potential improvements. Detailed descriptions of the causes of incidents revealed ignorance of regulations and non-compliance with applicable procedures, standards, and good practices. Pure knowledge of regulations should not be the only criteria to ensure safety. The authors cannot find any protocols to be implemented under high pressure nor amid numerous stimuli distracting attention nor requiring decision-making in training programs conducted in simulators. Today, official reports from marine incidents are indeed analyzing step by step the chain of causes behind an incident, but are often not investigating the sub-categories of human factor elements and are not unified across investigations. This is not helping to plan specific improvements to procedures, systems, and legislative solutions. In conclusion, this work provides an honest basis for potential detailed corrective actions and shows the need for constant changes and modifications to prevent people from making the same mistakes.

Human error can be significantly reduced. There is often a time lag between developing safety culture weaknesses and an event with significant safety consequences. These vulnerabilities can interact to create a potentially unstable state that exposes the organization to safety incidents. The organization (which may be a particular vessel) and its regulators must be alert to signs of potential weaknesses in social and situational systems.

A marine accident investigation is now an essential tool for identifying human factor issues which, when investigated with care, can be one of the pillars of preventing accidents and improving marine safety. The long-standing positive trend in ship safety, with year-on-year improvements, has now been reversed [101]. This is worrying. More emphasis is needed on safety culture, safety training, and competence assessment.

Statistics show that the frequency of accidents in the maritime industry, including those related to navigation, has skyrocketed. Still, authors believe that technology, regulations, and compliance can achieve expected safety levels by emphasizing the human element of accidents.

Historically, safety in shipping has focused on technical improvements. Most shipping company personnel involved in shipping operations have a technical background. Audits and inspections pay great attention to technical compliance. This focus on technical issues has led to significant improvements in ship safety. However, the time has come to focus more on other topics that the authors of the article have raised, namely, the very thoroughly analyzed causes of accidents, so far quite commonly classified as human error and more precisely as social and situational systems, which have not yet been sufficiently explored. To improve safety at sea, therefore, a threefold approach must be adopted:

1. Improve safety culture.
2. Improve training programs.
3. Implement formal competence assessment programs.

This study's results indicate many areas for improvement in all parts of the traditional management system for the safe operation of ships, in which the human factor plays the most crucial role.

**Author Contributions:** Conceptualization, K.P., A.E.W. and T.K.; methodology, A.E.W. and L.K.; formal analysis, A.E.W. and L.K.; investigation, K.P. and T.K.; resources, K.P. and A.E.W.; data analysis, A.E.W. and L.K.; writing—original draft preparation, K.P. and T.K.; writing—review and editing, L.K. and A.E.W. All authors have read and agreed to the published version of the manuscript.

**Funding:** The APC was funded by the Maritime University of Szczecin.

**Institutional Review Board Statement:** Not applicable.

**Informed Consent Statement:** Not applicable.

**Data Availability Statement:** The data presented in this study are available upon request from the corresponding author.

**Conflicts of Interest:** The authors declare no conflict of interest.

## Abbreviations

| | | |
|---|---|---|
| IAEA | - | International Atomic Energy Agency |
| IMO | - | International Maritime Organization |
| ILO | - | International Labour Organisation |
| IASCS | - | International Association of Ship Classification Societies |
| ISM | - | International Safety Management Code |
| HFACS-MA | - | Classification System Maritime Accidents |
| SMAIC | - | Composition of the Commission and Commission Experts |
| MAIB | - | Marine Accident Investigation Branch |
| STCW | - | International Convention on Standards of Training, Certification and Watchkeeping |
| SCA | - | Suez Canal Authority |
| MLC | - | Maritime Labour Convention |

## References

1. Swift, A.J.; Bailey, T.J. *Bridge Team Management—A Practical Guide*; The Nautical Institute: London, UK, 2004.
2. *Bridge Procedures Guide*, 6th ed.; International Chamber of Shipping: London, UK, 2022.
3. International Atomic Energy Agency. *Safety Culture, A Report by the International Nuclear Safety Advisory Group*; Safety Series, No. 75-INSAG-4; IAEA: Vienna, Austria, 1991.
4. Teperi, A.M.; Lappalainen, J.; Puro, V. Assessing artefacts of maritime safety culture—current state and prerequisites for improvement. *WMU J. Marit. Aff.* **2019**, *18*, 79–102. [CrossRef]
5. Anderson, P. *Cracking the Code. The Relevance of the ISM Code and Its Impact on Shipping Practices*; Nautical Institute: London, UK, 2003.
6. Rothblum, A. Human error and marine safety. In Proceedings of the Maritime, Human Factors Conference, Linthicum, MD, USA, 13–14 March 2000.
7. International Maritime Organization. Human Element Vision, Principles and Goals for the Organization, Resolution A.850(20). 1997. Available online: https://wwwcdn.imo.org/localresources/en/KnowledgeCentre/IndexofIMOResolutions/AssemblyDocuments/A.850(20).pdf (accessed on 20 February 2023).
8. International Maritime Organization. Human Element Vision, Principles and Goals for the Organization, Resolution a.947(23). 2004. Available online: https://wwwcdn.imo.org/localresources/en/KnowledgeCentre/IndexofIMOResolutions/AssemblyDocuments/A.947(23).pdf (accessed on 20 February 2023).

9. IMO Website on Legal Affairs. IMO/ILO Work on Seafarer Issues. Available online: https://www.imo.org/en/OurWork/Legal/Pages/JointIMOILOWorkingGroupsOnSeafarerIssues.aspx (accessed on 3 August 2022).
10. International Labour Conference, Maritime Labour Convention, 2006, Consolidated Text Established by the International Labour Office, Including the Amendments of 2014 and 2016 to the Code of the Convention. 2016. Available online: https://www.ilo.org/wcmsp5/groups/public/{-}{-}-ed_norm/{-}{-}-normes/documents/normativeinstrument/wcms_554767.pdf (accessed on 10 September 2022).
11. Kaber, D.B.; Endsley, M.R. Out-of-the-loop performance problems and the use of intermediate levels of automation for improved control system functioning and safety. *Process Saf. Prog.* **1997**, *16*, 126–131. [CrossRef]
12. Oltedal, H.A.; McArthur, D.P. Reporting practices in merchant shipping, and the identification of influencing factors. *Saf. Sci.* **2011**, *49*, 331–338. [CrossRef]
13. Bye, R.J.; Aalberg, A.L. Why do they violate the procedures?—An exploratory study within the maritime transportation industry. *Saf. Sci.* **2020**, *123*, 104538. [CrossRef]
14. Størkersen, K.V.; Thorvaldsen, T.; Kongsvik, T.; Dekker, S. How deregulation can become overregulation: An empirical study into the growth of internal bureaucracy when governments take a step back. *Saf. Sci.* **2020**, *128*, 104772. [CrossRef]
15. Mišković, D.; Jelaska, I.; Ivče, R. Attitudes of experienced seafarers as predictor of ISM code. *Promet-Traffic Transp.* **2019**, *31*, 569–579. [CrossRef]
16. Österman, C.; Hult, C. Administrative burdens and over-exertion in Swedish short sea shipping. *Marit. Policy Manag.* **2016**, *43*, 569–579. [CrossRef]
17. Bailey, N. Risk perception and safety management systems in the global maritime industry. *Policy Pract. Health Saf.* **2006**, *4*, 59–75.
18. Rae, A.J.; Provan, D.; Weber, D.; Dekker, S. Safety clutter: The accumulation and persistence of 'safety' work that does not contribute to operational safety. *Policy Pract. Health Saf.* **2018**, *16*, 194–211. [CrossRef]
19. Bieder, C.; Bourrier, M. *Trapping Safety into Rules: How Desirable or Avoidable Is Proceduralization?* 1st ed.; Ashgate: Farnnham, UK, 2013.
20. Dekker, S. *Safety Differently: Human Factors for a New Era*, 2nd ed.; Taylor & Francis: Boca Raton, FL, USA, 2015.
21. Størkersen, K.V.; Antonsen, S.; Kongsvik, T.Ø. One size fits all? Safety management regulation of ship accidents and personal injuries. *J. Risk Res.* **2017**, *20*, 1154–1172. [CrossRef]
22. *Working Paper on Tankers Involved in Shipping Accidents 1975–1992*; Transportation Safety Board of Canada: Ottawa, ON, Canada, 1993.
23. Cormier, P.J. Towing Vessel Safety: Analysis of Congressional and Coast Guard Investigative Response to Operator Involvement in Casualties Where a Presumption of Negligence Exists. Master's Thesis, University of Rhode Island, Kingston, RI, USA, 1994.
24. Bryant, D.T. *The Human Element in Shipping Casualties*; Report; Department of Transport, Marine Directorate: London, UK, 1991.
25. Annual Overview of Marine Casualties and Incidents 2021. Available online: www.emsa.europa.eu/ (accessed on 9 September 2022).
26. Wu, B.; Yip, T.L.; Yan, X.; Guedes Soares, C. Review of techniques and challenges of human and organizational factors analysis in maritime transportation. *Reliab. Eng. Syst. Saf.* **2022**, *219*, 108249. [CrossRef]
27. Graziano, A.; Teixeira, A.P.; Guedes Soares, C. Classification of human errors in grounding and collision accidents using the TRACEr taxonomy. *Saf. Sci.* **2016**, *86*, 245–257. [CrossRef]
28. Isaac, A.; Shorrock, S.T.; Kirwan, B. Human error in European air traffic management: The HERA project. *Reliab. Eng. Syst. Saf.* **2002**, *75*, 257–272. [CrossRef]
29. Ung, S. Evaluation of human error contribution to oil tanker collision using fault tree analysis and modified fuzzy Bayesian Network based CREAM. *Ocean Eng.* **2019**, *179*, 159–172. [CrossRef]
30. Chen, S.; Wall, A.; Davies, P.; Yang, Z.; Wang, J.; Chou, Y. A Human and Organisational Factors (HOFS) Analysis Method for Marine Casualties Using HFACS-Maritime Accidents (HFACS-MA). *Saf. Sci.* **2013**, *60*, 105–114. [CrossRef]
31. Ugurlu, O.; Yildiz, S.; Loughney, S.; Wang, J. Modified human factor analysis and classification system for passenger vessel accidents (HFACS-PV). *Ocean Eng.* **2018**, *161*, 47–61. [CrossRef]
32. Soner, O.; Asan, U.; Celik, M. Use of HFACS–FCM in Fire Prevention Modelling on Board Ships. *Saf. Sci.* **2015**, *77*, 25–41. [CrossRef]
33. Chauvin, C.; Lardjane, S.; Morel, G.; Clostermann, J.; Langard, B. Human and organisational factors in maritime accidents: Analysis of collisions at sea using the HFACS. *Accid. Anal. Prev.* **2013**, *59*, 26–37. [CrossRef]
34. Celik, M.; Cebi, S. Analytical HFACS for investigating human errors in shipping accidents. *Accid. Anal. Prev.* **2009**, *41*, 66–75. [CrossRef]
35. Akyuz, E. A marine accident analysing model to evaluate potential operational causes in cargo ships. *Saf. Sci.* **2017**, *92*, 17–25. [CrossRef]
36. Yildirim, U.; Basar, E.; Ugurlu, O. Assessment of collisions and grounding accidents with human factors analysis and classification system (HFACS) and statistical methods. *Saf. Sci.* **2017**, *119*, 412–425. [CrossRef]
37. Zhang, M.; Zhang, D.; Goerlandt, F.; Yan, X.; Kujala, P. Use of HFACS and fault tree model for collision risk factors analysis of icebreaker assistance in ice-covered waters. *Saf. Sci.* **2019**, *111*, 128–143. [CrossRef]
38. Qiao, W.; Li, Y.; Ma, X.; Liu, Y. A methodology to evaluate human factors contributed to maritime accident by mapping fuzzy FT into ANN based on HFACS. *Ocean Eng.* **2020**, *197*, 106892. [CrossRef]

39. Galieriková, A. The human factor and maritime safety. *Transp. Res. Procedia* **2019**, *40*, 1319–1326. [CrossRef]
40. Porathe, T.; Hoem, Å.S.; Rødseth, Ø.J.; Fjørtoft, K.; Johnsen, S.O. At Least as Safe as Manned Shipping? Autonomous shipping, safety and "human error". In *Safety and Reliability—Safe Societies in a Changing World*; Haugen, S., Barros, A., Gulijk, C.V., Kongsvik, T., Vinnem, J., Eds.; CRC Press: London, UK, 2018.
41. Endsley, M.R. Toward a theory of situation awareness in dynamic systems. *Hum. Factors* **1995**, *37*, 32–64. [CrossRef]
42. Hollnagel, E. *Safety-I and Safety-II: The Past and Future of Safety Management*; Ashgate: Farnham, UK, 2014.
43. Patraiko, D. Into the Future, the Technology of Tomorrow, The Navigator, Issue 24. The Nautical Institute. 1 June 2020. Available online: https://www.nautinst.org/resources-page/into-the-future-the-technology-of-tomorrow.html (accessed on 21 September 2022).
44. Acejo, I.; Sampson, H.; Turgo, N.; Ellis, N.; Tang, L. *The Causes of Maritime Accidents in the Period 2002–2016*; Seafarers International Research Centre (SIRC), Cardiff University: Cardiff, UK, 2018.
45. Covid-19 losses settle, but pandemic effects linger. In *Global Claims Review 2022*; Available online: www.agcs.allianz.com/ (accessed on 8 July 2022).
46. Wagenaar, W.A.; Groeneweg, J. Accidents at sea: Multiple causes and impossible consequences. *Int. J. Man-Mach. Stud.* **1987**, *27*, 587–598. [CrossRef]
47. Parker, A.W.; Huhinger, L.M.; Sargent, L.; Boyd, R. *Health Stress and Fatigue in Shipping*; Australian Maritime Safety Agency: Brisbane, Australia, 2002.
48. Maekay, M. *Safer Transport in Europe: Tools for Decision-Making*; European Transport Safety Council Lecture; The European Transport Safety Council: Etterbeek, Belgium, 2000.
49. Hetherington, C.; Flin, R.; Mearns, K. Safety in shipping: The human element. *J. Saf. Res.* **2006**, *37*, 401–411. [CrossRef]
50. Lülhoft, M.; Dekker, S. On your watch automation on the bridge. *J. Navig.* **2002**, *55*, 83–96.
51. Brun, W.; Johnsen, B.; Iabertg, J.C.; Ekomas, B.; Kobbeltvedt, T. Bridge resource management training enhancing shared mental models and task performance. In *How Professionals Make Decisions*; CRC Press: Boca Raton, FL, USA, 2004; pp. 183–193.
52. Salas, E.; Wilson, C.A.; Burke, C.S.; Wighlman, D.C. Does crew resource management training work? An update, an extension, and some critical needs. *Hum. Factors* **2006**, *48*, 392–412. [CrossRef]
53. Kobylinski, L. Risk analysis and human factor in prevention of CRG casualties. *Int. J. Mar. Navig. Saf. Sea Transp.* **2009**, *3*, 443–448.
54. Makarowski, R. Czynnik ludzki w lotnictwie. *Przegląd Psychol.* **2012**, *3*, 305–326.
55. Uzarski, M.; Abramowicz-Gerigk, T. Human factor in air and sea traffic management. *Pr. Nauk. Politech. Warsz. -Transp.* **2014**, *102*, 159–169.
56. Encyklopedia Zarządzania. Available online: www.mfiles.pl (accessed on 12 October 2021).
57. Sheridan, T.B. Risk, human error, and system resilience: Fundamental ideas. *Hum. Factors* **2008**, *50*, 418–426. [CrossRef] [PubMed]
58. Pauksztat, B. 'Only work and sleep': Seafarers' perceptions of job demands of short sea cargo shipping lines and their effects on work and life on board. *Marit. Policy Manag.* **2017**, *44*, 899–915. [CrossRef]
59. Dawidziuk, K.; Lishchynskyy, Y.; Wojciechowska, M.; Kopański, Z.; Marczewska, S.; Uracz, W. Stres jako źródło wypalenia zawodowego. *J. Public Health Nurs. Med. Rescue* **2011**, *4*, 26–29.
60. Jasielska, A.; Molińska, M.; Cejrowska, J. Profil zawodowy oficera floty handlowej—udział cech osobowości w radzeniu sobie ze stresem. Humanizacja Pracy. *Gaz. Inst. Psychol.* **2016**, *2*, 101–118.
61. Krystosik-Gromadzińska, A. Safety and ergonomics on board. *Logistyka* **2015**, *5*, 1035–1042.
62. Piechowicz, B.; Stawarczyk, K.; Stawarczyk, M. Temperature as an important factor in the work environment—selected aspects. *Bezpieczeństwo Pr. Nauka Prakt.* **2013**, *7*, 8–11.
63. Przewidywanie Stresu. Available online: www.szybkanauka.net (accessed on 11 July 2021).
64. Lubaś, P. *Diagnoza Ergonomicznych Czynników Ryzyka*; Okręgowy Inspektorat Pracy w Szczecinie: Szczecin, Poland, 2010; pp. 4–14.
65. Wróblewska, M. *Ergonomia, Wydaw*; Politechniki Opolskiej: Opole, Poland, 2004; pp. 160–172.
66. Państwowa Komisja Badania Wypadków Morskich, EXPRESS 1 & BALTIC CONDOR, Summary Report on Collision, 10 May 2019. Available online: https://pkbwm.gov.pl/wp-content/uploads/images/Reports/enFinal-marine-accident-report{-}{-}-EXPRESS-1-og-BALTIC-CONDOR{-}{-}-Collision-on-10-May-2019.pdf (accessed on 15 September 2022).
67. CSL Virginia. Available online: https://wwz.cedre.fr/Ressources/Accidentologie/Accidents/CSL-Virginia (accessed on 11 March 2023).
68. Walker, J. Cruise Law News. Everything Cruise Lines Don't Want You to Know. Available online: www.cruiselawnews.com/2019/ (accessed on 23 October 2022).
69. Nautilus International. IMO Raises Concern over Euroferry Olympia Ferry Fire, 24 February 2022. Available online: https://www.nautilusint.org/en/news-insight/news/imo-raises-concern-over-euroferry-olympia-ferry-fire/ (accessed on 10 January 2023).
70. Seiner Fishing Boat Accidents and Injuries. Available online: www.maritimeinjurycenter.com/accidents-and-injuries/seiner-fishing-boat/ (accessed on 10 January 2023).
71. Marine Accident Investigation Branch. Accident Report, Capsize of Tug Domingue while Assisting CMA CGM Simba Resulting in Two Fatalities, July 2017. Available online: https://assets.publishing.service.gov.uk/media/596dd157e5274a0a690001ba/MAIBInvReport16_2017.pdf (accessed on 10 January 2023).

72. UN Environment Programme, X-Press Pearl Maritime Disaster Sri Lanka—Report of the UN Environmental Advisory Mission, 6 August 2021. Available online: https://www.unep.org/resources/report/x-press-pearl-maritime-disaster-sri-lanka-report-un-environmental-advisory-mission (accessed on 10 January 2023).
73. WION Web Team, Ship Owner Says Suez Canal was at Fault over Ever Given Grouding, Wionews.com, 23 May 2021. Available online: https://www.wionews.com/world/ship-owner-says-suez-canal-was-at-fault-over-ever-given-grounding-lawyer-386811 (accessed on 1 September 2022).
74. Accident Report. Marine Accident Investigation Branch. 2018, 11. Available online: https://assets.publishing.service.gov.uk/media/5b30b379ed915d587f69eee6/MAIBInvReport11_2018.pdf (accessed on 20 February 2023).
75. Cargo and Crane Accidents in the Maritime Industry. Available online: www.maritimeinjurycenter.com/accidents-and-injuries/cargo-and-crane/ (accessed on 1 December 2022).
76. Examining Fatal Shipyard Accidents. Volume 2. Available online: www.osha.gov/video/shipyard-accidents/transcripts#v2_4 (accessed on 12 December 2022).
77. Fatal Incident Onboard a Dive Support Vessel. IMCA Safety Flash, 2009, 18. Available online: www.imca-int.com/safety-events/fatal-incident-onboard-a-dive-support-vessel/ (accessed on 1 December 2022).
78. Barge Accidents. Available online: www.simmonsandfletcher.com/maritime-law/offshore-accident/barge-injury/ (accessed on 20 February 2023).
79. Grech, M.R.; Horberry, T.; Smith, A. Human error in maritime operations: Analyses of accident reports using the Leximancer tool. *Proc. Hum. Factors Ergon. Soc. Annu. Meet.* **2002**, *46*, 1718–1721. [CrossRef]
80. Zainal, Z. Case Study as a Research Method. *J. Kemanus.* **2017**, *5*. Available online: https://jurnalkemanusiaan.utm.my/index.php/kemanusiaan/article/view/165 (accessed on 23 November 2022).
81. Arnold and Itkin. Available online: www.offshoreinjuryfirm.com/ (accessed on 12 December 2022).
82. Department of Merchant Shipping, Ministry of Communications and Works, Republic of Cyprus, MARINE ACCIDENT REPORT Collision between "CORVUS J" and "BALTIC ACE" in the North Sea on 5 December 2012. Available online: https://pkbwm.gov.pl/wp-content/uploads/images/uchwaly_raporty/raporty/raporty_obce/Final_report_Corvus_J_WIM_07_12_DMS_Cyprus.pdf (accessed on 15 October 2021).
83. Swift, J.A.; Ross, J.E.; Omachonu, V.K. *Principles of Total Quality*, 2nd ed.; St. Lucie Press: Boca Raton, FL, USA, 1998; pp. 256–262.
84. Rios, N.; Spínola, R.O.; Mendonça, M.; Seaman, C. Supporting Analysis of Technical Debt Causes and Effects with Cross-Company Probabilistic Cause-Effect Diagram. In Proceedings of the 2019 IEEE/ACM International Conference on Technical Debt (TechDebt), Montreal, QC, Canada, 26–26 May 2019; pp. 3–12.
85. Wong, K.C.; Woo, K.Z.; Woo, K.H. *Ishikawa Diagram in Quality Improvement in Behavioral Health*; Springer: Cham, Switzerland, 2016; pp. 119–132.
86. Ishikawa, K.; Loftus, J.H. *Introduction to Quality Control*, 3rd ed.; 3A Corporation: Tokyo, Japan, 1990.
87. Doggett, A.M. Root Cause Analysis: A Framework for Tool Selection. *Qual. Manag. J.* **2005**, *12*, 34–45. [CrossRef]
88. Lira, L.H.; Hirai, F.E.; Oliveira, M.; Portellinha, W.; Nakano, E.M. Use of the Ishikawa diagram in a case-control analysis to assess the causes of a diffuse lamellar keratitis outbreak. *Arq. Bras. Oftalmol.* **2017**, *80*, 281–284. [CrossRef]
89. Gwiazda, A. Koncepcja ważonego wykresu Ishikawy. *Probl. Jakości* **2005**, *4*, 13–17.
90. Wiśniewska, M. Jasiak-Kujawska, Cause and Effect Analysis of Hospital Infections with the Use of Weighted Ishikawa Diagram. *Zarządzanie i Finanse* **2012**, *3*, 328–343.
91. Leśniak, A.; Górka, M.; Skrzypczak, I. Barriers to BIM Implementation in Architecture, Construction and Engineering Projects—The Polish Study. *Energies* **2021**, *14*, 2090. [CrossRef]
92. Jasińska, J.; Świderski, A. Risk assessment methodology in quality assurance of logistic systems. *Pr. Nauk. Politech. Warsz. Transp.* **2008**, *64*, 59–66.
93. Łybacki, W.; Zawadzka, K. Zastosowanie koncepcji ważonego wykresu Ishikawy do analizy wad odlewów. *Przegląd Odlew.* **2009**, *59*, 430–433.
94. Kongsvik, T.Ø.; Størkersen, K.V.; Antonsen, S. The relationship between regulation, safety management systems and safety culture in the maritime industry. In *Safety, Reliability and Risk Analysis: Beyond the Horizon*; Steenbergen, R.D.J.M., va Gelder, P.H.A.J.M., Miraglia, S., Vrouwenvelder, A.C.W.M., Eds.; Taylor and Francis Group: London, UK, 2014; pp. 467–473.
95. Røyrvik, J.O.D.; Skarholt, K.; Lamvik, G.M.; Jonassen, J.R. Risk Management in Anchor-handling Operations: The Balance Between Control and Autonomy. In *Safety and Reliability*; Nowakoski, T., Mlyńczak, M., Jodejko-Pietrucuzuk, A., Werbińska-Wojciechowska, S., Eds.; Taylor & Francis: London, UK, 2015.
96. Schröder-Hinrichs, J.U.; Praetorius, G.; Graziano, A.; Kataria, A.; Baldauf, M. Introducing the Concept of Resilience into Maritime Safety. In Proceedings of the 6th Symposium on Resilience Engineering, Lisbon, Portugal, 22–25 June 2015; Sophia Antipolis Cedex. 2016. Available online: http://www.resilience-engineering-association.org/wp-content/uploads/2016/11/Proceedings-REA6SYM-2016--230916-1.pdf (accessed on 4 April 2021).
97. Størkersen, K.V.; Johansen, J.P.K. No Swans in Sight. Analyzing the Resilience in Norwegian Water Passenger Transport. In *Safety, Reliability and Risk Analysis: Beyond the Horizon*; Steenbergen, R.D.J.M., van Gelder, P.H.A.J.M., Miraglia, S., Vrouwenvelder, A.C.V.M., Eds.; Taylor & Francis: London, UK, 2014; pp. 1619–1626.
98. Bhattacharya, S. The effectiveness of the ISM Code: A qualitative enquiry. *Mar. Policy* **2012**, *36*, 528–535. [CrossRef]

99. Oltedal, H.A. The Use of Safety Management Systems Within the Norwegian Tanker Industry: Do They Really Improve Safety? In *Reliability, Risk, and Safety: Theory and Applications*; Bris, R., Soares, C.G., Martorell, S., Eds.; Taylor & Francis: London, UK, 2010; pp. 2355–2362.
100. Vandeskog, B. The legitimacy of safety management systems in the minds of norwegian seafarers. *TransNav Int. J. Mar. Navig. Saf. Sea Transp.* **2015**, *9*, 101–106. [CrossRef]
101. Madsen, O.M. *A New Look at Safely Culture*; DNV Forum: Bærum, Norway, 2011; p. 2.

*applied sciences*

*Article*

# Impact of Port Clearance on Ships Safety, Energy Consumption and Emissions

**Vytautas Paulauskas [1,*], Donatas Paulauskas [2] and Vytas Paulauskas [3]**

1 Marine Engineering Department, Klaipeda University, H. Manto g. 84, LT-92219 Klaipeda, Lithuania
2 Maritime Engineering Department, Klaipeda University, H. Manto g. 84, LT-92294 Klaipeda, Lithuania; paulauskasd75@gmail.com
3 Klaipeda Shipping Research Centre, V. Berbomo str. 7-5, LT-92219 Klaipeda, Lithuania; vytaspaulauskas1@gmail.com
* Correspondence: vytautaskltc@gmail.com

**Abstract:** The safety of shipping, energy consumption and environmental impact in ports and port channels is very critical. One of the most important elements in the provision of safe navigation, energy consumption and emissions generation is the depth of ports so that under all conditions the hull of a ship does not touch the bottom of the channels or the bottom of the basin, as well as optimizing energy consumption and minimizing the environmental impact. The very high depth reserves in ports make it possible to ensure the safety of shipping, but at the same time require huge investments in the dredging and maintenance of a port's channels and basins, which can have a negative impact on a port's economic results. Optimizing the depth of port channels and basins is very important from an economic, maritime safety, energy saving and environmental point of view, as vessels navigating port channels and basins must not only keep their hulls off the bottom of the channel or basin, but also have good controllability, use minimal energy consumption and minimize their environmental impact. With good maneuverability, the number of and need for auxiliary vehicles (tugs) can be minimized. This article analyses the relationship between ships' draught and port channels and basins depths, which influences the aspects of a ship's controllability, in order to optimize the depths of port channels and basins and, at the same time, minimize energy consumption and environmental impact while preserving the necessary navigational safety.

**Keywords:** ship draught; depth of port channels and basins; ship maneuverability at low depths; energy consumption; emissions generated by ships

**Citation:** Paulauskas, V.; Paulauskas, D.; Paulauskas, V. Impact of Port Clearance on Ships Safety, Energy Consumption and Emissions. *Appl. Sci.* **2023**, *13*, 5582. https://doi.org/10.3390/app13095582

Academic Editor: José A. Orosa

Received: 14 March 2023
Revised: 26 April 2023
Accepted: 27 April 2023
Published: 30 April 2023



## 1. Introduction

Depth in port channels and basins is one of the most critical elements in ensuring navigation safety in ports, which is why ports carry out maintenance and dredging operations to reach and maintain the necessary depths [1]. The depth of navigational channels is important to ensure not only that vessels navigating the channels do not touch the bottom of the channel with their hulls, but also that the vessel has good controllability [2]. The clearance (the distance between a ship's hull and the navigational channel bottom) in a port's channels and basins has an influence on energy consumption due to the ship's increased resistance while maintaining speed, as well as an environmental impact [3,4].

Today, the depth of channels and basins is calculated by mainly taking into account the potential maximum draught of the vessel; the accuracy of the depth measurement; the water level and its possible change (accuracy of the measurement); the variance in the ship's draft depending on the ship's speed and clearance [1,4,5]; the potential increase in the vessel's draft due to the vessel's corner angle; and the navigational margin, which takes into account the potential changes in the bottom of the channel (accumulations of soil sediments) [4,5].

The depth of the channel must be sufficient enough to ensure the safe passage of vessels, including in and out of the ports under difficult conditions, i.e., when the speed of the biggest possible vessel entering the port may be increased by up to 1.3–1.5 times [1,5,6].

The maneuvering of vessels in port is subject to external forces such as wind, current, swell and the effects of tugboats, as well as internal influences such as the ship's own speed and the ship's inclination during maneuvering. Thus, all possible influences have to be taken into account when planning the depths of port channels and basins [2,4,5,7].

When planning the depths of port channels and basins, it is also necessary to take into account the processes of sediment accumulation, movement and the possibility of periodic dredging in order to guarantee optimum depths in port channels and basins at all times [4,8].

When ships are sailing in port approaches and port internal channels at shallow depths, the resistance of the ship's hull increases significantly [3,5]. To maintain the proper speed of the ship, it is necessary to increase the power of the main engine(s) of the ship [3,9,10]. Increasing the power of a ship's main engine(s) significantly increases fuel consumption [11,12] and, accordingly, generates more emissions [13].

Ports are trying to become "green ports"; therefore, the amount of emissions generated from shipping in ports is a very important goal, so it is necessary to follow international and national requirements to reduce emissions from shipping [14–16].

Air quality in regions and especially in large ports has a significant impact on human health; therefore, the issues of environmental protection and decarbonization of shipping are very important for the quality of life and the economy of countries [17–19].

People play a very important role in reducing emissions in ports, i.e., the human factor, without which significant results cannot be achieved in any field, including the development of "green port" ideas [16,20].

The main objective of this article is to develop a methodology to determine the required optimal or minimum possible ship clearance in port channels and basins (the space between the ship's hull and the bottom of the port channel) while maintaining the appropriate ship speed and guaranteeing the safety of shipping in port channels (shipping safety is a priority) and to assess the relationship between the ship's clearance and energy (fuel) consumption and emissions. In this way, the main aims of this article are as follows: to assess ships' safe navigation capabilities in port channels, i.e., the ability of a ship to navigate independently in the bends of a port channel with the existing relationship between the draft of the ship and the bottom of the navigation channel; to determine the change in the power of a ship's main engine(s) at shallow depths to maintain the specified speed of the ship in the navigation channel, i.e., at the existing draft of the ship and the depth of the navigation channel to maintain the prescribed safe cruising speed; to determine the change in energy (fuel) of a ship's main engine(s) and the emission levels generated based on the current ratio of the ship's draft to the depth of the port (navigation) channels.

In this way, the structure of the research and the article is oriented as follows: first of all, it must be assessed whether the ship can enter the port, which is related to the depths of the navigation channels (port entrance and internal navigation channels compared to the ship's berth). In the event that the ship has the opportunity to enter the port independently, then the following task is solved, i.e., the power of the ship's main engine(s), depending on the ratio of the ship's draft to the depth of the navigation channels, respectively, energy (fuel) consumption and the amount of generated emissions. If the depth of the port channels is satisfactory, the ship can enter the port independently or, if necessary, port tugs or additional special maneuvers can be used. In the event that it is possible for the ship to enter the port on its own, then the task is solved, i.e., the power of the ship's main engine(s), depending on the ratio of the ship's draft to the depth of the navigation channels, energy (fuel) consumption and generated emissions, respectively, is sufficient.

## 2. Analysis and Literature Review of Clearance in Port Channels and Basins

Many ports in the world have different requirements for minimum depths in port channels and basins and generally use guidelines such as PIANC [21] and EAU [22]. The distance between a ship's hull and the bottom of a navigation channel is very important in order to avoid the ship's hull coming into contact with the bottom of the navigation channel and to maintain sufficient controllability of the ship. Studies have been conducted on this topic [1,4,5,7,8]. At the same time, it should be noted that ship handling hearings require additional research, as ports try to attract larger and larger ships, and sometimes emergency situations occur due to too little clearance when ships touch the bottom of the navigation channel with their hulls due to ship heeling caused by external forces, or ships sink due to the ship's speed in the shipping channel or become uncontrollable and float outside the shipping channel [3,6,7].

At the same time, other very important aspects, such as the energy consumption of ships entering and maneuvering in ports and the environmental impact of low clearance under the ship's hull, have not yet been taken into account when assessing depths in port approaches and internal navigational channels and basins. The above aspects of energy consumption and environmental impacts from ships have been analyzed individually and without considering the potential impact of clearance on these aspects [16,23,24].

Vessel speeds are restricted in many ports and are linked to maritime safety and environmental impacts (the effect of waves generated by the vessel on moored vessels and the coastline) (Figure 1) [25–27].

**Figure 1.** LNG tanker sailing in port (Klaipeda port), permitted speed up to 8 knots.

Tugboats are widely used for assisting with maneuvering large ships in port because they are equipped with powerful engines and can use high engine power; however, they consume a lot of energy (fuel), generating large amounts of emissions, especially when turning ships and approaching or leaving the quay (Figure 2) [8].

**Figure 2.** LNG tanker turns in ships turning basin using port tugboats.

The assessment of sustainable transport systems in ports and their environmental impact is very important for the ports themselves and for individual regions, so research in this area is very essential [28,29].

Maritime transport plays an important role for many countries and creates a significant part of their gross domestic product (GDP). At the same time, maritime transport, especially in port regions, has a significant impact on the environment, so many countries are conducting research on that issue and are looking for optimal solutions to minimize the impact on the environment while improving the economic situation of the countries [29–33].

The emission control regions adopted by the International Maritime Organization (IMO) such as the $SO_x$ emission control areas—the Baltic and North Seas, the English Channel and the east and west coasts of North America—have allowed a significant reduction of $SO_x$ emissions from ships in these regions [13,34,35].

Ports, as the most important maritime transport points, also try to maximize their influence on the reduction of emissions from ships and other vehicles [15,16,36,37].

The navigational safety of ports and the necessary minimum depths of ports are studied in many countries, since dredging works are expensive and in the development of sustainable ports it is necessary to combine navigation safety, economic, environmental and other possible factors [16,17,24–26].

For the provision of navigation safety, when sailing ships in shallow depths, i.e., the minimum depth below the ship's hull, methods for its determination, such as Bernoulli's flow formulas ideas [4], estimates of the connected liquid mass and others [1,5,12,21,22], are used in many countries, but the results obtained are often similar.

As follows from the analysis of the situation and the literature, the necessary depths in ports and the minimum clearance under the ship's hull in ports and navigation channels, as well as emissions from ships, have been studied by many researchers. At the same time, the change in the ship's controllability at low depths (clearance under the ship's hull) and the clear connection between the clearance under the ship's hull and the ship's energy consumption and emissions are missed.

## 3. Basis for the Theoretical Calculations of Possibilities for the Ships Sail in Port Channels and Basins, Energy Consumption and Emissions Generation

### *3.1. Steps of Research Methodology*

The following steps of research methodology were used to conduct the research (Figure 3). After conducting the literature review, the mathematical model was developed.

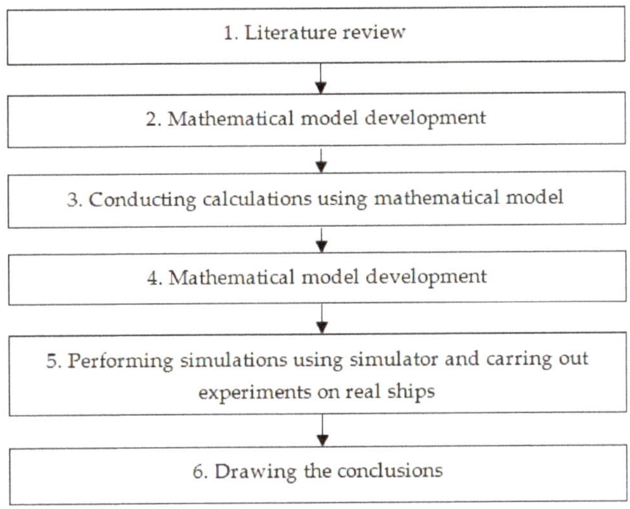


**Figure 3.** The steps of research methodology.

Based on the presented methodology, theoretical calculations of possibilities for ships sailing in port channels and basins, energy consumption and emissions created by ship were performed. The variation of hull resistance to longitudinal movement with clearance, i.e., the ratio *T/H* (draught to depth), was used to estimate the ship's sailing speed in ports [1,5]. For the assessment of the ship's controllability, the methods of calculation of the ship's circulation elements and trajectory at low depths were used [3,8,27]. The method of calculating the ship's circulation trajectory in shallow depth was applied as given in sources [4,5]. The method is based on the calculation of the basic ship movement parameters (ship speed, ship angular velocity and ship drift angle) and the ship's trajectory in the case of low clearance, which often occurs in port navigation channels and basins. Estimation of the ship's sailing trajectory in the presence of small clearances is necessary when navigating port approach and internal navigation channels so that the ship can safely pass through channel bends. Due to the variation of the ship's draft and the power of the main engine(s) when sailing in shallow waters (port navigation channels) and maintaining the set speed, at low depths (clearance), additional resistance forces of the ship hull are formed which are related to the change of the added water mass [4,5].

For the calculation of the ship's energy consumption, fuel consumption was calculated as a function of the engine power at a given speed, and for the assessment of emissions, the parameters of the ship's fuel consumption, engine power and operating time were used [15,38]. For this calculation, the maximal distribution method was used utilizing data achieved by conducting experiments on simulators and real ships [39]. The maximal distribution method could be applied in case at least 5 measurements were carried out.

In order to verify theoretical calculations and practical application of the presented methodology, experiments were performed with the assistance of a simulator and on real ships. Simulations were carried out using the full mission simulator "SimFlex Navigator" (Force Technology product) [40], which analyzed similar maneuvers as the real ships, considering the set forces acting on ships sailing in port channels and ships turning in turning basins. Experiments performed on real ships covered two port areas with specific navigational conditions (port approach and internal navigational channels).

Then, the results were analyzed, discussions initiated, conclusions were drawn and suggestions for future research were outlined.

*3.2. Mathematical Model*

The results of the literature review, the results of simulators and the results of the real ship tests when different clearances were used between the ship's hull and the bottom of the channels were used to develop mathematical models of shipping navigational safety, energy (fuel) consumption and emission generation at shallow depths [1,9,15,37,41–44]. When conducting research and creating mathematical models, it was assumed that ships in navigational channels sail independently, and the controllability of the ship is ensured by the ship's own steering equipment. In channel bends and turning basins, ships turn with the help of their control equipment (propulsion complex), and if necessary, they can use the help of ship steering devices and tugboats [5,7]. It is also assumed that when ships move straight through channels due to the effect of low depth, only the longitudinal resistance of the ship changes, and when ships sail in bends of navigation channels and when turning in turning basins, the resistance of the ship's lateral movement additionally increases [43,45].

The safe depth of the port channels and port water areas, so that the largest ship in the port does not touch the bottom of the navigation channels and basins with its hull, can be calculated with the help of the following formula [1,4]:

$$H_{\min} = T + \Delta T_v + \Delta T_\theta + \Delta T_\Psi + \Delta H_m + \Delta H_{V.L} + \Delta H_{\Delta V.L} + \Delta H_n, \quad (1)$$

where $T$—the maximum draught of the calculated vessel; $\Delta T_V$—the increase in draught due to settlement (speed), sometimes named squat [4,5]; $\Delta T_\theta$—the increase in draught due to heeling [4,5]; $\Delta T_\psi$—the increase in draught due to the effect of swell (change in the difference) [5]; $\Delta H_m$—the accuracy of the depth measurement, depending on the port depth measurement technique used; $\Delta H_{V.L.}$—the level of the water in the particular port; $\Delta H_{\Delta.V.L.}$—the accuracy of the measurement of the water level, depending on the port water level measurement technique used; $\Delta H_{n.}$—navigational margin, which can be decomposed into a direct navigational margin, which is assumed to be about 2–3% of the ship's draught, by means of accurate bottom depth measurements (using modern depth measurement techniques), and a layer of sediment, which has to be periodically removed (cleaning). The above elements of Formula (1) can be calculated using the methodology presented in [4,5].

The increase in draught ($\Delta T_v$) due to low-depth effects and the ship's speed can be calculated using the following formula and graph (Figure 4) to estimate the added water mass coefficients [1,4,5]:

$$\Delta T_V = \frac{\rho L B}{\delta}\sqrt{\frac{1 + k'_{11}}{1 + k_{11}}} \quad (2)$$

where $\rho$—water density; $L$—ship's length between perpendiculars; $B$—ship's width; $\delta$—overall hull fullness factor; $k_{11}$ and $k'_{11}$—added water mass coefficients at high depth and in shallow water, depending on the ship's speed and $T/H$ ratio (Figure 4) [5].

The added water mass depends on the speed of the ship and the draft of the ship, as well as the depths of the navigation channels and the port water area ratio (T/H). Added water mass coefficients, which are presented in Figure 4 when the ship is moving in the longitudinal direction, are usually used for calculation [4,5].

In the article, the water mass coefficients presented in Figure 4 are used to calculate the change in the ship's draft and the increase in the ship's draft, as well as the correspondence of the specified coefficients, verified by experiments by measuring clearance on real ships [4,5].

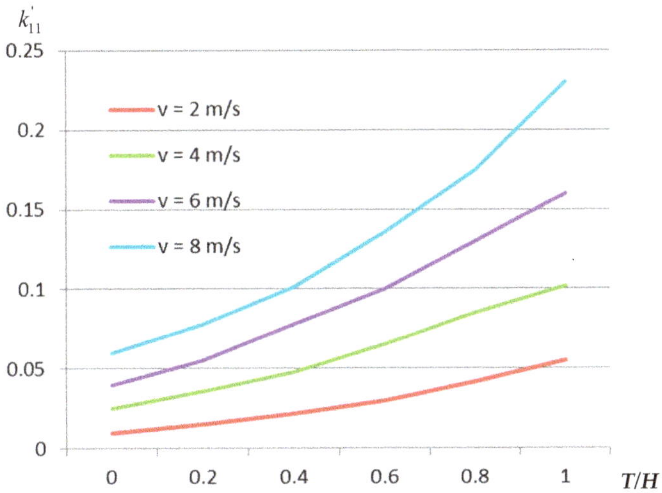


**Figure 4.** Dependence of the added water mass coefficients ($k_{11}$—deep water, $T/H$ ratio is zero); $k'_{11}$—shallow water) on the $T/H$ ratio and the vessel's sailing speed $v$.

Formulas can be used to calculate the trajectory of a ship at shallow depth and in the presence of wind and current, which is a characteristic of ports [4,40,43,45]:

$$X_{0i(s)} = \int v_{i(s)} \cdot \cos(\int (\omega_{i(s)}dt - \beta_{i(s)}))dt + \int v_{cr}dt \cdot \cos q_{cr} + \int v_{d}dt \cos q_{a}; \quad (3)$$

$$Y_{0i(s)} = \int v_{i(s)} \cdot \sin(\int (\omega_{i(s)}dt - \beta_{i(s)}))dt + \int v_{cr}dt \cdot \sin q_{cr} + \int v_{d}dt \cdot \sin q_{a}, \quad (4)$$

where $v_{i(s)}$—ship's speed at low depths; $\omega_{i(s)}$—turning velocity at low depths; $\beta_{i(s)}$—drift angle at low depths; $v_{cr}$—current velocity; $q_{sr}$—current course angle during the start of the maneuverer; $v_d$—ship's drift speed; $q_a$—wind course angle during the start of the maneuver. The above elements of Formulas (2) and (3) can be calculated using the methodology presented in [5].

When examining the passage of ships in ports, it is assumed that the movement of ships in the port approach and internal navigation channels lead to the fact that ships independently sail on a straight or almost straight trajectory (turning up to 30–40 degrees). The research also assumes that the change in the power used by the ship's main engine due to the effect of shallow water is basically related to the change in the resistance of the ship's longitudinal movement [3,10,11]. Experiments with real ships were carried out in the open sea and in ports (shallow waters) for the power factor, maintaining a set constant speed and accurately recording the main engine power, the speed of the ship with the help of electronic lag and DGPS and the depth with the help of echo sounder. On the basis of studies carried out on real ships, a dependence of the calculation of the power factor of the ship's main propulsion system ($\Delta N$) on the $T/H$ ratio (ship's draught/depth ratio) and the overall hull fullness factor is obtained as follows:

$$\Delta N = (1 + 1.25(\frac{T}{H})^{2})\sqrt{\frac{\delta}{0.65}}, \quad (5)$$

Calculation of the power factor of the ship's main engine due to the influence of shallow water according to Formula (5) and experimental tests with real ships showed that the difference between the calculations and the results of experimental tests is no more than 10%. Thus, Formula (5) can be successfully used in practice. The limitations of Formula (5) evaluated during the experiments are as follows: the speed of the ships should be between

6 knots and 10 knots, and the overall fullness factor of the ship's hull should be between 0.65 and 0.90. If the overall fullness ratio of the ship's hull is less than 0.65, Formula (5) can be used when the ratio of the ship's draft to the depth of the navigation channel (*T/H*) is greater than 0.3.

As hull resistance increases, the power of the ship's engine must be increased to maintain the ship's target or planned speed. The relative power of the ship's engine ($N'$) and the relative speed of the ship ($v'$) can be expressed in the following formula:

$$N' = \frac{N}{N_0}; \quad (6)$$

$$v' = \frac{v}{v_0}, \quad (7)$$

where $N_0$—nominal power of the main engine(s) of the vessel; $v$—speed of the vessel at the power of the vessel's engine(s) $N$; $v_0$—speed of the vessel at the nominal power of the vessel's engine(s) $N_0$.

The relationship between the ship's relative speed and the relative power of the ship's main engine(s), presented in the literature [3,10,11], was verified by the authors through experiments on real ships of various types and sizes. Some of the results of the experiments carried out on real ships are presented in Table 1. Comparing the results of real ship experiments regarding the relative power of the ships' main engine(s) and the ships' relative speed using the maximum distribution method, the maximum changes, expressed as a percentage, were obtained (presented in Table 1). Experiments were carried out for many years on various ships, in which at least one of the authors of the article participated. The navigation equipment available on the ship was used for the experiments (ship speed was measured with an accuracy of at least 0.1 knots) as were the power measurement indicators of the main engine(s) on board the ships, the measurement accuracy of which was at least 2% of the instantaneous power.

**Table 1.** Experiments results of relative main engine power and relative speed of the real ships.

| Ship Type | Displacement, t | $N_0$, MW | $v_0$ | $N'$ | $v'$ | Max Difference between Calculation and Experiments Results, % |
|---|---|---|---|---|---|---|
| Container | 220,000 | 75 | 25 | 0.15 | 0.28 | 5.2 |
| Container | 130,000 | 50 | 24 | 0.17 | 0.32 | 4.8 |
| LNG tanker | 91,500 | 24 | 21 | 0.44 | 0.69 | 7.6 |
| Bulk | 80,000 | 10 | 14.6 | 0.33 | 0.54 | 8.5 |
| Oil tanker | 72,900 | 10 | 14 | 0.34 | 0.58 | 6.5 |
| Container | 70,230 | 25 | 23 | 0.33 | 0.52 | 4.6 |
| Oil tanker | 34,300 | 11 | 15.5 | 0.31 | 0.51 | 5.4 |
| General | 15,200 | 7 | 14.3 | 0.24 | 0.41 | 3.5 |
| Ro-Pax | 10,200 | 12 | 18.2 | 0.41 | 0.65 | 4.4 |
| Coaster | 5200 | 3.5 | 12.5 | 0.45 | 0.70 | 8.9 |

The power of the engine and the speed of the ship are related by a quadratic relationship [3,10,38]. In most cases, the relative power of the ship's engine(s) and the ship's speed can be used. For this purpose, a graph based on the experimental results from more than 1000 ship passages can be used [3,10,11] (Figure 5).

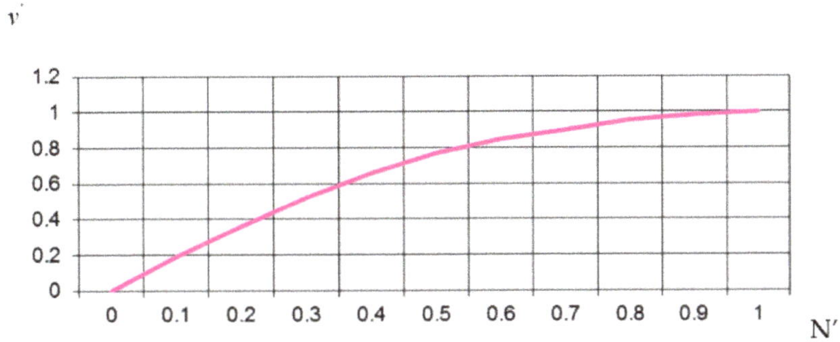


**Figure 5.** Relative vessel speed ($v'$) versus relative engine power ($N'$).

A limitation of the graph (Figure 5) with a very high overall hull fullness factor (δ) is that with the overall fullness factor of the ship's hull greater than 0.9, the form resistance parameters of the ship's hull shape change significantly and the accuracy of the graph is not good enough (error size can reach more than 10 percent).

Engine power can be calculated by taking into account the amount of fuel consumed over a given period of time, e.g., an hour, and the relative fuel consumption, i.e., [9,11,12]:

$$N = \frac{q_k}{q_k' \cdot t}, \tag{8}$$

where $N$—engine power, kW; $q_k$—fuel consumption, kg; $q_k'$—relative fuel consumption, kg/kWh; $t$—engine running time, h.

Due to their powerful engines, ships consume a lot of fuel while sailing, especially when performing additional, not always justified, maneuvers in ports. A ship's fuel consumption is often calculated over a voyage or other period. In a general case, the fuel consumption of a ship on a voyage ($q_{LP}$) or other sailing places and times can be calculated according to the following formula [9,11,38,42]:

$$q_{LP} = \int_0^t q_k' \cdot N_{av} \cdot dt, \tag{9}$$

where $N_{av}$—ship main engine average power during time $t$.

The relative fuel consumption of the main and auxiliary engines of most ships ranges from 0.13 to 0.25 kg/kWh (for more precise data, please refer to the engine specification of the individual ship) [11]. Depending on the type of fuel, the amount of fuel used can be different, so when using LNG, its calorific value is on average about 15 percent higher than other petroleum products, which means that about 15 percent less fuel mass is consumed [15,46].

Emissions from ships and other transport vehicles directly depend on the quantity and quality of fuel used, engine power and engine running time [11,17,42]. The main emissions from ships constitute carbon dioxide ($CO_2$), nitrogen oxides ($NO_x$), carbon monoxide ($CO$), sulfur oxides ($SO_x$) and particulate matter ($PM$) [11].

Emissions are calculated according to the formula that includes fuel consumption, actual engine power used and the relative magnitude of specific emissions. Thus, the carbon dioxide emissions are calculated according to the following formula [15,47]:

$$CO_2 = q_{LP} \cdot \Delta CO_2, \tag{10}$$

where $\Delta CO_2$—carbon dioxide coefficient, which for petroleum products (diesel, fuel oil) is between 3.0 and 3.5 and for LNG between 2.5 and 2.9.

The Sulphur oxide content can be calculated using the following formula [14,15]:

$$SO_x = q_{LP} \cdot \Delta SO_x, \tag{11}$$

where $\Delta SO_x$—Sulphur oxide coefficient, which depends on the type of fuel; for petroleum products, it ranges from 0.001 to 0.035 and for LNG it is around zero.

The carbon monoxide content can be calculated using the following formula [41]:

$$CO = \int_0^t N_{av} \cdot \Delta CO \cdot dt, \tag{12}$$

where $\Delta CO$—carbon monoxide coefficient, which depends on the type of engine [42].

The amount of nitrogen oxides generated is calculated using the following formula [41]:

$$NO_x = \int_0^t N_{av} \cdot \Delta NO_x \cdot dt, \tag{13}$$

where $\Delta NO_X$—nitrogen oxide coefficient, depending on the engine type.

The particulate matter generation is calculated using the following formula [48]:

$$PM = \int_0^t N_{av} \cdot \Delta PM \cdot dt, \tag{14}$$

where $\Delta PM$—the particulate matter coefficient, which depends on the type of engine and the type of fuel, and is up to 10 g/kWh for petroleum products and close to zero for LNG fuels [48].

The emission factors and sizes of marine engines depend on the type of engine and the type of fuel used. For marine engines, average emission factors and relative values are given in Table 2 (as an example) [11].

**Table 2.** Average emission factors for marine engines by fuel type.

| Types of Emission | Petroleum Products | LNG |
|---|---|---|
| $CO_2$—depending on the amount of fuel | 3.2 | 2.5 |
| $SO_x$, %—depending on the amount of fuel | 0.1 | 0.0–0.1 |
| $NO_x$, g/kWh | 10 | 4 |
| $CO$, g/kWh | 5 | 3 |
| $PM$, g/kWh | 0.5 | 0.0–0.1 |

Thus, the reduction of emissions from engines depends on the type and design of the engine, the type of fuel used and the engine's operating conditions (maneuvering mode). Fuel consumption depends on the operating mode of the engine, especially the modes of transitional mechanisms.

The qualifications and experience of ship crews and port pilots greatly influence the amount of emissions from ships [15,38,48,49].

The power of engines used in ships has a significant influence on the generation of individual emissions. Emission factors and relative magnitudes are shown in Table 3 (as an example) [11,49,50].

**Table 3.** The relative amount of emissions based on engine power.

| N, kW | $CO$, g/kWh | $NO_x$, g/kWh | $PM$, g/kWh |
|---|---|---|---|
| 30 | 5.0 | 7.5 | 0.40 |
| 100 | 5.0 | 7.2 | 0.30 |
| 250 | 5.0 | 7.2 | 0.20 |
| 1000 | 5.0 | 7.5 | 0.25 |
| 3000 | 5.0 | 9.8 | 0.50 |
| 10,000 | 5.0 | 10.5 | 0.50 |
| ≥10,000 | 5.0 | 11.0 | 0.50 |

Thus, shipping emissions depend on the type of fuel, the quality of the engine and its power. Knowing how much fuel is consumed in shipping, what type of emissions are emitted and what their quantities are, it is possible and necessary to look for opportunities and methods to reduce the impact on the environment.

## 4. Case study and Results

This case study has analyzed a few types of ships: PANAMAX type bulk ships (length about 206 m, width about 36.0 m, draft about 12.5 m, deadweight about 80,000 t), POST PANAMAX container vessels (length about 300 m, width about 46 m, draft about 13.0 m, container capacity about 8500 TEU, deadweight about 130,000 t), G class container vessels (length about 400 m, width about 61 m, draft about 13.5 m, container capacity about 19,500 TEU, deadweight about 220,000 t) and LNG standard tankers (length about 290 m, width about 49 m, draft about 12 m, capacity about 150,000 m$^3$ LNG).

Real ships and the full mission simulator SimFlex Navigator were used for the experiments. The following methodology of conducting experiments was used: first, an experiment plan was drawn up to achieve specific goals, then possible real ships were selected and the possibility of using the simulator was checked. Experiments with real ships were carried out both at sea and while entering and leaving ports.

For the purposes of this article, previously mentioned conducted experiments and targeted experiments were used when sailing ships at sea at great depths and when sailing in ports or other navigational channels where there were limited depths, such as the Oresund, Belt and other straits where there are limited depths, in which one or all authors participated. During the experiment, the ship's speed was recorded using the ship's navigation equipment (DGPS or GPS and others on the ship's bridge), as well as a port pilot RTK (real-time kinematic) system, which was implemented in Klaipeda port, clearance (distance between ship's hull and navigational channels bottom) was measured by the ship's navigational equipment (ship's echo sounder on the ship's bridge), the load (power) of the ship's main engine(s) and the propeller rotation frequency (on the ship's bridge and in the ship's engine room) were measured and in separate cases, when there was a real possibility, fuel consumption during a particular voyage (in the ship's engine room) was measured.

Experiments were carried out to measure the advance of the ship during circulation (during the ship's waiting for entry into ports or special trials of ships after their construction or repair, in which at least one of the authors participated) as well as during navigation during large turns.

According to the obtained results of real ship experiments, a suitable ship was selected in the simulator and the relevant ship movement and other parameters obtained in real ships and in the simulator under identical conditions were compared. During the matching process between the real ship and the ship in the simulator, the calibration coefficients were calculated, and then the experiments were continued with the help of the simulator using the calibration coefficients.

For the checking ship's advance [5,51] (Figure 6), all tested ships sailed with an initial speed of 8 knots and a rudder turn angle of 25° to starboard. All tested ships had conventional propulsion, i.e., one propeller and one rudder [52]. The vessel's advance in circulation analysis is important to assess whether the vessel is able to turn on bends in the channel on its own, whether it needs the assistance of tugboats or whether additional maneuvers by the vessel are necessary [1,7,43].

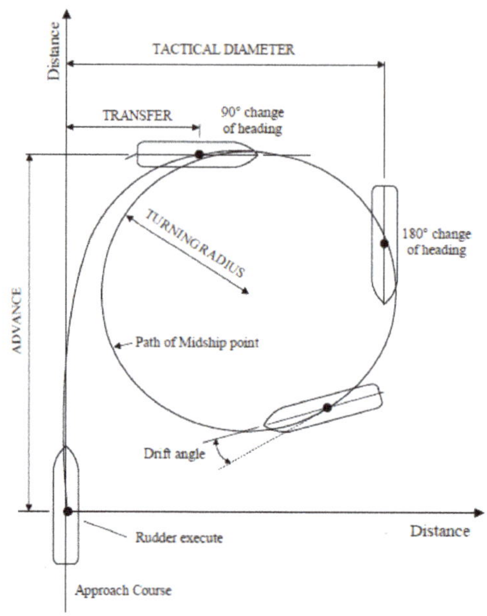


**Figure 6.** Ship's circular trajectory details.

Ships advance testing was made in good navigational conditions, i.e., wind velocity less than 10 m/s, wave height less than 1 m and current less than 0.5 kn. The mentioned ship's advance testing results are presented in Figure 7. In all cases, a rudder angle of 25 degrees was adopted (to leave a margin of maneuverability). Tests of real ships for simulator calibration were performed as follows: bulk cargo ship—three ships; container ship (9000 TEU)—two ships; container ship (19,500 TEU)—one ship. The "Tables of maneuvering elements" available on the ship's bridge were also used. With the help of a calibrated simulator, at least 10 tests (for G class container vessel—10 tests; for 9000 TEU container vessel—12 tests; for 80,000 t bulk cargo ship—12 tests) of each mentioned ship were carried out. In the simulator, we selected ships from the simulator library which were analogous to real ships with which real experimental tests were performed in terms of type and parameters. The differences, although minor, were mostly due to differences in draft and displacement. For example, the simulator library contained a G-class container ship with an average draft of 14.0 m, while the real ship had a draft of 13.7 m. In order to unify the obtained results, it was necessary to calibrate the simulator, i.e., coefficients of the relationship of the received simulator data with the real ship. In practice, for such simulator calibration, one or two datapoints of a specific parameter of a real ship were sufficient. In the article, there were at least 2–3 real results for specific parameters of specified ships, and up to 7–12 such real ship test results were used for individual ships.

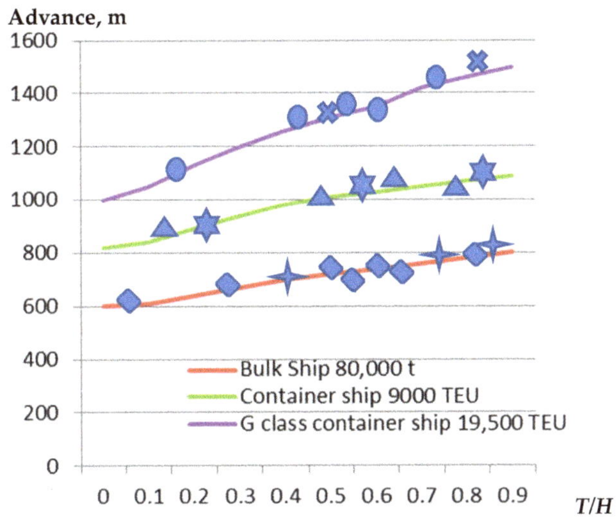


**Figure 7.** Ships advance on circular trajectory in deep and shallow waters (depending on *T/H*), received by calculation (lines), real ship experiments (G class container ship, container ship 9000 TEU, bulk ship 80,000 t) and calibrated simulator (G class container ship, container ship 9000 TEU, bulk ship 80,000 t).

The accuracy of the real ship test results obtained, necessary for research and simulator calibration using the RTK system, was: ship location—up to 0.1 m; ship speed—up to 0.1 knots; ship angular rotation speed—0.2 degrees per minute. When using the DGPS system, the accuracy of the received data consisted of: the location of the ship—up to 0.5–1.0 m (depending on the distance to the base station); the speed of the ship—up to 0.2 knots; the angular speed of the ship—up to 0.5 degrees per minute.

The study was conducted as presented in Section 3 using the full mission simulator SimFlex Navigator [40] and using real similar ships for experiments. All received data were filtrated by a Kalman filter [53] and the differences between calculated, simulated and real ships' experimental data were analyzed.

The results obtained from the advance calculation and experiments (Figure 7) show that the calculation results using the methodology presented in Section 3 are in high compliance with the results of the real ship experiments and the simulator results.

The results of calculation and experiments (obtained on real ships and with the help of a calibrated simulator) of ships' advance in circulation at a rudder angle of 25 degrees are shown in Figure 7, and the differences (accuracy) at different *T/H* ratios in meters and percentages from the experimental results are shown in Table 4.

The power changes in the main engines of the ships, due to hull longitudinal speed, and additional resistance in shallow waters were calculated according to the methodology presented in Section 3 and verified using a calibrated simulator and the results of real ship experiments under similar conditions (Figure 8).

At least eight types of ships were used to determine the power factor of the ship's main engine, depending on the ship's draft and the depth of the navigation channels, for conducting research and calibrating the simulator. During the experiments, equipment on the ship was used: ship echo sounders were used for depth measurement, the accuracy of which was up to 0.1 m, and the accuracy of the power of the ship's main engine(s) was up to 2% of the engine(s) power. During experiments, the fuel consumption of the real ships was measured by existing sensors in the ship's engine room, the accuracy of which was up to 2–3% of the amount of fuel consumed, and instantaneous fuel consumption sensors, the accuracy of which was about 4%. All data were recorded automatically.

**Table 4.** The difference (accuracy) between the results of calculation and experiments (obtained on real ships and with the help of a calibrated simulator) of ships' advance in circulation in meters and percentages from experimental results.

| Ship | T/H | Calculated Advance, m | Difference Advance between Experimental and Calculation, m | Percentages from Experimental Results |
|---|---|---|---|---|
| G class Container | 0.76 | 1430 | 30 | 2.1 |
| G class Container | 0.61 | 1350 | 48 | 3.6 |
| G class Container | 0.55 | 1330 | 32 | 2.4 |
| G class Container | 0.37 | 1260 | 22 | 1.7 |
| G class Container | 0.14 | 1120 | 18 | 1.6 |
| Container 9000 TEU | 0.73 | 1050 | 20 | 1.9 |
| Container 9000 TEU | 0.62 | 1030 | 27 | 2.6 |
| Container 9000 TEU | 0.47 | 1005 | 31 | 3.1 |
| Container 9000 TEU | 0.13 | 850 | 41 | 4.8 |
| Bulk 80,000 t | 0.83 | 790 | 22 | 2.8 |
| Bulk 80,000 t | 0.61 | 740 | 30 | 4.1 |
| Bulk 80,000 t | 0.31 | 680 | 28 | 4.1 |
| Bulk 8000 t | 0.06 | 603 | 18 | 3.0 |

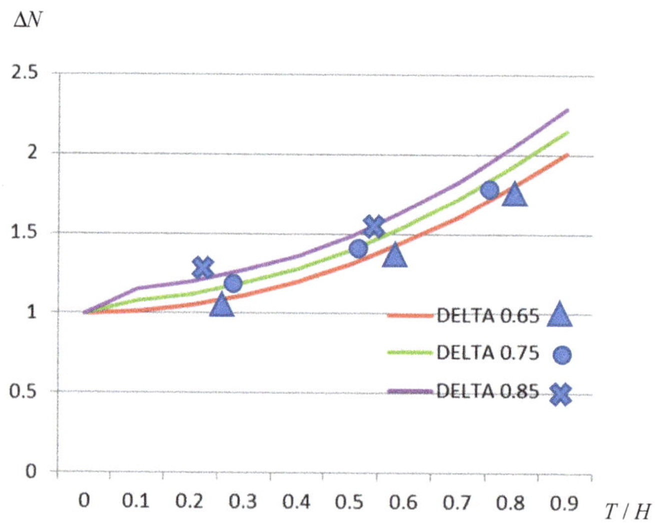


**Figure 8.** Ship's main engine power coefficient depending on ship's draft and depth ratio (*T/H*) and ship's overall hull fullness factor (DELTA) (calculation and experimental results).

The compliance between the results of the calculations and the experiments is quite high (the difference does not exceed 10 percent (Table 1)) (Figure 8), and therefore it can be concluded that the methodology presented in Section 3 for the calculation of the ship's engine power increasing during sailing at low depths can be used for practical purposes for the assessment of the performance of ships when sailing through channels and other similar locations with low depths.

The port of Klaipeda was chosen for the case analysis. The passage of a standard LNG tanker from the entrance channel to the southern turning basin of the port was analyzed (Figure 9).

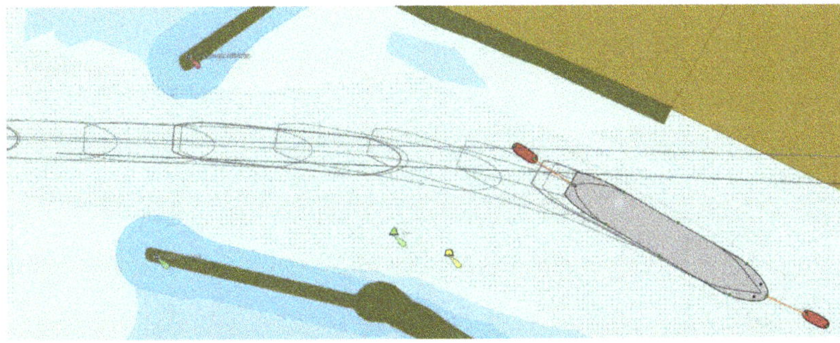

**Figure 9.** LNG standard tanker sailing trajectory in Klaipeda port.

The main engine power, fuel consumption and emissions of the ships were calculated according to the methodology presented in Section 3. The SimFlex Navigator simulator was used to change the clearance under the hull. Experiments were carried out on real ships (LNG standard tankers, length approx. 290 m, beam approx. 49 m, draft approx. 12 m, overall fullness coefficient approx. 0.75). The speeds adopted and used in the calculations, simulator and actual LNG carriers for the majority of the passage (up to the turning basin) were between 7 and 8 knots.

The calibration of the simulator was carried out by comparing the results of a real ship and a ship in the simulator. The simulator calibration was produced by calibration coefficients for the ship's main engine power, fuel consumption and the ship's speed at high and low depths. Following the simulator calibration, experiments were carried out on the simulator at various depths and speeds in the range from 6 to 10 knots and at characteristic depths in harbor approaches and ports, i.e., a *T/H* between 0.6 and 0.92.

The main engine power, speed and fuel consumption of the LNG standard tanker in the approach and internal navigational channels of the port of Klaipeda, obtained in a calibrated simulator, are shown in Figure 10.

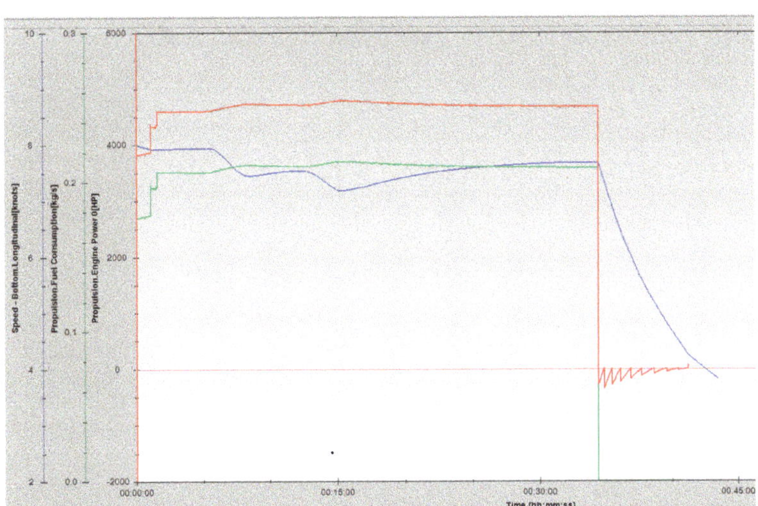


**Figure 10.** The LNG standard tanker engine power, ship's speed and fuel consumption obtained in a calibrated simulator.

The variation of the main engine power and fuel consumption factors for the LNG standard tanker sailing at a constant speed as a function of the ship's draught/depth ratio, which is characteristic of harbor approaches and internal navigation channels, using the methodology presented in Section 3, and the results obtained on the real ship, are presented in Figure 11.

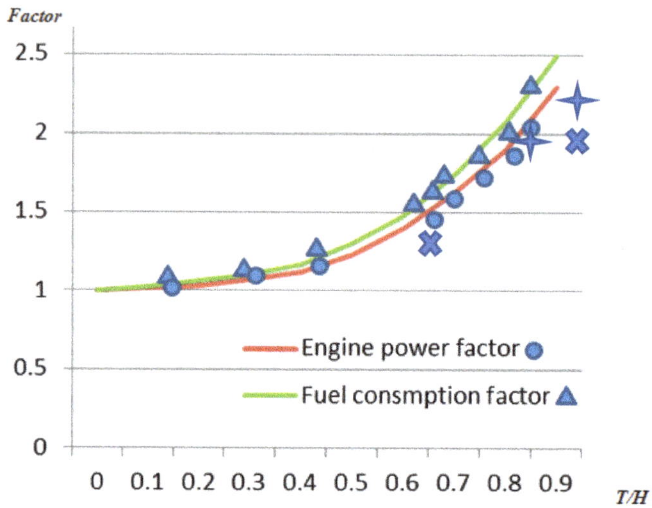


**Figure 11.** Standard LNG tanker main engine power and fuel consumption factors depending on the ratio of the ship's draft and depth (*T/H*) received by the theoretical method (lines) presented in Section 3, and experiments' results of the real ship (fuel consumption factor, engine power factor) and calibrated simulator.

As can be seen from the results obtained, the methodology presented in this article for the estimation of the usable power of a ship's main engine for ships navigating in the port approach and the internal channels can be applied for practical purposes.

The comparative studies of the calculation and experimental results of the received engine power and fuel consumption factors of the ship showed (LNG standard tanker) that the maximum difference between the calculation and experimental results (real ship and calibrated simulator) was up to 0.23, or, as a percentage, up to 9.3 percent.

On the basis of the results obtained, it can be concluded that the methodology presented in Section 3 for the estimation of the fuel consumption of ships in port approach and internal navigation channels can be successfully used for practical purposes and further calculations, for example, for the estimation of generated emissions.

Fuel consumption of the standard LNG tankers while sailing (Figures 9 and 10) at the specified sailing distance using LNG and diesel fuel depending on the ratio of the ship's draft and depth while the ship is sailing at a speed of 7–8 knots obtained by calculation and experimentally are presented in Figure 12.

As can be seen from the obtained calculation results (ship's circulation advance at low depths and ship's main engine(s) factors using the methodology presented in Section 3) and the results of experiments on real ships in corresponding conditions, there is a good correlation between the calculation and experimental results, which allows the use of the methodology developed and presented in this paper for practical purposes. At the same time, it is necessary to appreciate the fact that fuel consumption for ships sailing and maneuvering in ports depends up to 10–12 percent on the qualifications of ship crews and port pilots [15].

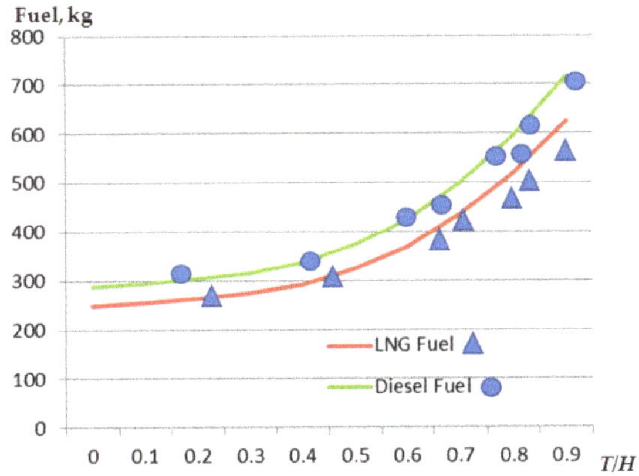


**Figure 12.** The fuel consumption of a standard LNG tanker sailing at a speed of 7–8 knots at sea and when entering the port of Klaipeda, obtained by calculation and experiments (sailing distance 5 n. miles).

The methodology presented in Section 3 is used to calculate the emissions. Emission values were calculated using petroleum products (diesel) and LNG fuel. The values of $CO_2$ and $SO_x$ emissions, depending on the amount of fuel used, when sailing from the beginning of the port entrance channel to the turning basin (taking into account the differences in the energy capacity of LNG and diesel) were calculated. When the ship sails at a speed of 7–8 knots and different T/H ratios, the amount of $CO$, $NO_x$ and $PM$ emissions of a standard LNG tanker, depending on the T/H ratio, was calculated according to the methodology presented in Section 3, based on engine power, engine operating time and the corresponding emission generation factors presented in Tables 1 and 2. The results of generated emissions are presented in Table 5.

**Table 5.** Standard-LNG-tanker-generated emissions during entrance to Klaipeda port, as shown in Figures 9 and 10.

| T/H | 0 | 0.1 | 0.2 | 0.3 | 0.4 | 0.5 | 0.6 | 0.7 | 0.8 | 0.9 |
|---|---|---|---|---|---|---|---|---|---|---|
| N, kW | 2220 | 2253 | 2287 | 2375 | 2486 | 2731 | 3108 | 3619 | 4040 | 5106 |
| $CO_2$ (LNG), kg | 675 | 691 | 716 | 725 | 788 | 872 | 1000 | 1175 | 1399 | 1688 |
| $CO_2$ (diesel), kg | 922 | 945 | 976 | 1011 | 1075 | 1197 | 1360 | 1600 | 1907 | 2298 |
| $SO_x$ (diesel), kg | 0.28 | 0.29 | 0.31 | 0.32 | 0.34 | 0.37 | 0.43 | 0.50 | 0.60 | 0.72 |
| CO (LNG), kg | 3.80 | 3.85 | 3.92 | 4.00 | 4.21 | 4.62 | 5.31 | 6.14 | 6.91 | 8.7 |
| CO (diesel), kg | 6.40 | 6.50 | 6.62 | 6.90 | 7.22 | 7.93 | 9.01 | 10.5 | 11.7 | 14.8 |
| $NO_x$ (LNG), kg | 5.1 | 5.2 | 5.3 | 5.5 | 5.7 | 6.3 | 7.2 | 8.3 | 9.3 | 11.7 |
| $NO_x$ (diesel), kg | 12.9 | 13.1 | 13.3 | 13.8 | 14.2 | 15.8 | 18.0 | 21.0 | 23.4 | 29.6 |
| PM (diesel), kg | 0.69 | 0.70 | 0.71 | 0.72 | 0.74 | 0.82 | 0.93 | 1.10 | 1.21 | 1.53 |

The fuel consumption and emissions of the vessels to maintain the same speed in channels and other similar locations at low depths were calculated using the methodology presented in Section 3 and verified with simulators and real vessels under similar sailing conditions. As can be seen from the obtained results, the difference between the calculation and experimental results is not significant (maximum difference of 8 percent) and therefore the calculation methodology presented in Section 3 can be applied to the estimation of fuel consumption and emissions of ships sailing in harbors and other channels at low depths.

## 5. Discussion

For further studies related to fuel consumption and emissions, it is important to study the power of the ship's main engine when the ship moves at a constant speed, depending on the ratio of the ship's draft to the depth of the channel, which is typical for ships sailing in ports. Research on the variation of the power of the ship's main engine at shallow depths in the evaluation of the fullness factor of the ship's hull should cover a wider range of ship types and designs. In addition, additional aspects of further research such as ships turning in turning basins and ship towing to and from quays, including the performance of tugs at low clearances, are important for finding methods to reduce environmental impact. These could be further directions of research.

The research results presented in the paper are critical because they clearly showed the importance of finding optimal methods for ships to enter and leave ports safely, primarily to ensure the safety of shipping (safety first) and at the same time reduce energy (fuel) demand and emissions. In this way, further complex studies are very important for the safety of shipping in the approaches to ports while constituting as low a possible impact on the environment.

The results of the scientific literature review showed that a specific methodology for assessing the trajectory of a ship's movement in shallow depths is important for predetermining the ship's maneuverability and safe navigation in port entrances and internal port navigation channels. At the same time, in order to reduce non-standard situations as much as possible, especially when ships pass near port infrastructure and ships moored at the quays, further studies of the controllability of ships in difficult conditions, especially regarding the effect of tides on the trajectories of ships, are important.

The change in the power of the ship's engines at shallow depths is important; therefore, the developed methodology for estimating the power of a ship's engines when the ship is sailing at a shallow depth is extremely important in ensuring the safety of shipping in ports. At the same time, for ships with relatively low-power engines, such as some bulkers, a preliminary assessment of the capabilities of such ships and further research is very important. Vessels with relatively weak main engines often have to use maximum or near-maximum main engine power when navigating port navigation channels in bad weather conditions where there is very little clearance, which requires high fuel consumption and generates high emissions. Therefore, according to the authors, similar studies are very important and may be another direction for future research.

The methodology developed to more accurately estimate the fuel consumption and emissions of vessels operating in shallow waters is very important, but further research is needed to determine the optimal safe speed of ships in ports while minimizing fuel consumption and emissions. This is especially important for ports that are within the boundaries of large cities.

## 6. Conclusions

This article examines the possibility of maneuvering ships sailing in port navigation channels and in the presence of small turns in the channels, and the obtained results allow for increasing the safety of navigation in port approaches and ports. Carrying out studies of the necessary energy (fuel) consumption while maintaining a planned speed is important from the point of view of shipping safety and environmental impact minimization. The results obtained in the article can be applied in the planning of port infrastructure (depths and turns of port navigation channels).

Developed and verified by experiments on real ships and with the help of a simulator, the methodology allows for estimating the power changes of a ship's main engine(s) at low clearances and can be successfully used in port planning and assessing possible emissions changes depending on the clearance. In this way, the methodology developed can help in planning the shipping channels of ports and the environmental impact of ships sailing in them (regarding the amount of pollutants emitted) depending on the depths of the existing or planned shipping channels.

The developed methodologies for evaluating the influence of shallow water on the required power of a ship's main engine, fuel consumption and emissions are important both for ensuring the safety of shipping in ports and for optimizing fuel consumption and reducing emissions in ports.

**Author Contributions:** Conceptualization, V.P. (Vytautas Paulauskas) and D.P.; methodology, V.P. (Vytautas Paulauskas); software, D.P. and V.P. (Vytas Paulauskas); validation, V.P. (Vytautas Paulauskas) and D.P.; formal analysis, V.P. (Vytautas Paulauskas); investigation, V.P. (Vytautas Paulauskas) and D.P.; resources, V.P. (Vytautas Paulauskas) and D.P.; data curation, V.P. (Vytautas Paulauskas), V.P. (Vytas Paulauskas) and D.P.; writing—original draft preparation, V.P. (Vytautas Paulauskas) and D.P.; writing—review and editing, V.P. (Vytautas Paulauskas) and V.P. (Vytas Paulauskas); visualization, V.P. (Vytautas Paulauskas) and D.P.; supervision, V.P. (Vytautas Paulauskas); project administration, V.P. (Vytautas Paulauskas); funding acquisition, V.P. (Vytautas Paulauskas). All authors have read and agreed to the published version of the manuscript.

**Funding:** This research received no external funding.

**Institutional Review Board Statement:** Not applicable.

**Informed Consent Statement:** Informed consent was obtained from all subjects involved in the study.

**Data Availability Statement:** Information about authors reporting results could be find in ResearchGate platform

**Conflicts of Interest:** The authors declares no conflict of interest.

## References

1. Sharma, A.; Nazir, S.; Ernstsen, J. Situation awareness information requirements for maritime navigation: A goal directed task analysis. *Saf. Sci.* **2019**, *120*, 745–752. [CrossRef]
2. Tomczak, A. Safety evaluation of ship's maneuvers carried out on the basis of integrated navigational system (INS) indications. *J. Konbin* **2008**, *4*, 247–266. [CrossRef]
3. Molland, A.F.; Turnock, S.R.; Hudson, D.A. *Ship Resistance and Propulsion. Practical Estimation of Propulsive Power*; Cambridge University Press: Cambridge, UK, 2011; 537p. [CrossRef]
4. Nakamura, S. Study on Maneuvering Criteria for Safety Assessment in Shallow Water. *TransNav Int. J. Mar. Navig. Saf. Sea Transp.* **2017**, *11*, 401–407. [CrossRef]
5. Paulauskas, V. *Ships Entering the Ports*; N.I.M.S publish house: Riga, Latvia, 2013; 240p, ISBN 9984-679-71-3.
6. Mou, J.M.; Chen, P.F.; He, Y.X.; Yip, T.L.; Li, W.H.; Tang, J.; Zhang, H.Z. Vessel traffic safety in busy waterways: A case study of accidents in western Shenzhen port. *Accid. Anal. Prev.* **2019**, *123*, 461–468. [CrossRef]
7. Lisowski, J. Analysis of Methods of Determining the Safe Ship Trajectory. *TransNav Int. J. Mar. Navig. Saf. Sea Transp.* **2016**, *10*, 223–228. [CrossRef]
8. Paulauskas, V.; Simutis, M.; Placiene, B.; Barzdžiukas, R.; Jonkus, M.; Paulauskas, D. The Influence of Port Tugs on Improving the Navigational Safety of the Port. *J. Mar. Sci. Eng.* **2021**, *9*, 342. [CrossRef]
9. Le, L.T.; Lee, G.; Kim, H.; Woo, S.-H. Voyage-based statistical fuel consumption models of ocean-going container ships in Korea. *Marit. Policy Manag.* **2020**, *47*, 304–331. [CrossRef]
10. Lack, D.A.; Corbett, J.J. Black carbon from ships: A review of the effects of ship speed, fuel quality and exhaust gas scrubbing. *Atmos. Chem. Phys.* **2012**, *12*, 3985–4000. [CrossRef]
11. IMO. *Marine Engine Regulations*; IMO: London, UK, 2015.
12. Lang, X.; Zhang, C.; Jonasson, L.; Mao, W.; Eriksson, L.; Zhang, D. Comparison between full-scale measurements and theoretical fuel consumption model in a real arctic ship navigation. *Proc. Int. Offshore Polar Eng. Conf.* **2019**, *1*, 886–892.
13. Endresen, Ø.; Sørgård, E., Sundet; Gravir, G. Emission from international sea transportation and environmental impact. *J. Geophys. Res. Atmos.* **2003**, *108*, D17. [CrossRef]
14. Pallis, A.A.; Vaggelas, G.K. Cruise Shipping and Green Ports: A Strategic Challenge. In *Green Ports: Inland and Seaside Sustainable Transportation Strategies*; Elsevier: Amsterdam, The Netherlands, 2018; pp. 255–273. [CrossRef]
15. Paulauskas, V.; Filina-Dawidowicz, L.; Paulauskas, D. The Method to Decrease Emissions from Ships in Port Areas. *Sustainability* **2020**, *12*, 4374. [CrossRef]
16. Di Vaio, A.; Varriale, L. Management innovation for environmental sustainability in seaports: Managerial accounting instruments and training for competitive green ports beyond the regulations. *Sustainability* **2018**, *10*, 783. [CrossRef]
17. Czermański, E.; Pawłowska, B.; Oniszczuk-Jastrząbek, A.; Cirella, G.T. Decarbonization of maritime transport: Analysis of external costs. *Front. Energy Res.* **2020**, *8*, 28. [CrossRef]
18. UNCTAD. Review of Maritime Transport. 2019. Available online: https://unctad.org/en/PublicationsLibrary/rmt2019_en.pdf (accessed on 25 February 2023).
19. Eyring, V.; Köhler, H.W.; Lauer, A.; Lemper, B. Emissions from international shipping: 2. Impact of future technologies on scenarios until 2050. *J. Geophys. Res. Atmos.* **2005**, *110*, D17306. [CrossRef]
20. Corrigan, S.; Kay, A.; Ryan, M.; Ward, M.E.; Brazil, B. Human factors and safety culture: Challenges and opportunities for the port environment. *Saf. Sci.* **2019**, *119*, 252–265. [CrossRef]

21. PIANC. Harbour Approach Channels and Design Guidelines—Report No. 121-2014, s.l.: The World Association for Waterborne Transport Infrastructure. 2014. Available online: http://marineman.ir/wp-content/uploads/2015/04/NAVIGATION-PIANC-Harbour-Approach-Channels-Design-Guidelines-2014.pdf (accessed on 25 February 2023).
22. Wilhelm Ernst & Sohn. *Recommendations of the Committee or Waterfront Structures Harbours and Waterways EAU 2012*, 9th ed.; Wiley: Hoboken, NJ, USA, 2015; 676p, ISBN 978-3-433-03110-0.
23. IMO. Amendments to the Annex of the Protocol of 1997 to Amend the International Convention for the Prevention of Pollution from Ships, 1973, as Modified by the Protocol of 1978 Relating thereto (MARPOL Annex VI). 2008. Available online: http://www.imo.org/includes/blastDataOnly.asp/data_id%3D23760/176 (accessed on 25 May 2020).
24. Sorte, S.; Rodrigues, V.; Borrego, C.; Monteiro, A. Impact of harbour activities on local air quality: A review. *Environ. Pollut.* **2020**, *257*, 113542. [CrossRef] [PubMed]
25. Langenus, M.; Dooms, M. Creating an industry-level business model for sustainability: The case of the European ports industry. *J. Clean. Prod.* **2018**, *195*, 949–962. [CrossRef]
26. Lozano, R.; Fobbe, L.; Carpenter, A.; Sammalisto, K. Analysing sustainability changes in seaports: Experiences from the Gävle Port Authority. *Sustain. Dev.* **2019**, *27*, 409–418. [CrossRef]
27. Paulauskas, V.; Filina-Dawidowicz, L.; Paulauskas, D. Ships speed limitations for reliable maintenance of the quay walls of navigation channels in ports. *Eksploat. Niezawodn.–Maint. Reliab.* **2020**, *22*, 306–315. [CrossRef]
28. Ypsilantis, P.; Zuidwijk, R. Collaborative fleet deployment and routing for sustainable transport. *Sustainability* **2019**, *11*, 5666. [CrossRef]
29. Gnap, J.; Varjan, P.; Ďurana, P.; Kostrzewski, M. Research on relationship between freight transport and transport infrastructure in selected European countries. *Transp. Probl.* **2019**, *14*, 63–74. [CrossRef]
30. Soto, C.G. The potential impacts of global climate change on marine protected areas. *Rev. Fish. Biol. Fish.* **2002**, *11*, 181–195. [CrossRef]
31. Bagoulla, C.; Guillotreau, P. Maritime transport in the French economy and its impact on air pollution: An input-output analysis. *Mar. Policy* **2020**, *116*, 103818. [CrossRef]
32. Bouman, E.A.; Lindstad, E.; Rialland, A.I.; Strømman, A.H. State-of-the-art technologies, measures, and potential for reducing GHG emissions from shipping—A review. *Transp. Res. Part D Transp. Environ.* **2017**, *52*, 408–421. [CrossRef]
33. Novac, V.; Rusu, E. Air emissions from ships—Western Black sea case study. *Int. Multidiscip. Sci. Geoconference Surv. Geol. Min. Ecol. Manag. Sgem.* **2019**, *19*, 813–819. [CrossRef]
34. Cullinane, K.; Bergqvist, R. Emission control areas and their impact on maritime transport. *Transp. Res. Part D Transp. Environ.* **2014**, *28*, 1–5. [CrossRef]
35. Pinchasik, D.R.; Hovi, I.B.; Mjøsund, C.S.; Grønland, S.E.; Fridell, E.; Jerksjö, M. Crossing borders and expanding modal shift measures: Effects on mode choice and emissions from freight transport in the Nordics. *Sustainability* **2020**, *12*, 894. [CrossRef]
36. Di Vaio, A.; Varriale, L.; Alvino, F. Key performance indicators for developing environmentally sustainable and energy efficient ports: Evidence from Italy. *Energy Policy* **2018**, *122*, 229–240. [CrossRef]
37. Aulinger, A.; Matthias, V.; Zeretzke, M.; Bieser, J.; Quante, M.; Backes, A. The impact of shipping emissions on air pollution in the greater North Sea region—Part 1: Current emissions and concentrations. *Atmos. Chem. Phys.* **2016**, *16*, 739–758. [CrossRef]
38. Heinrich, L.; Koschinsky, A.; Markus, T.; Singh, P. Quantifying the fuel consumption, greenhouse gas emissions and air pollution of a potential commercial manganese nodule mining operation. *Mar. Policy* **2020**, *114*, 103678. [CrossRef]
39. Jin, H.; Peng, S. Optimal unbiased estimation for maximal distribution. *Am. Inst. Math. Sci.* **2021**, *6*, 189–198. [CrossRef]
40. "SimFlex Navigator" Simulator, 2016. Available online: https://forcetechnology.com/en/about-force-technology/news/news-before-2017/simflex-updates (accessed on 26 April 2023).
41. Grosso, M.; Marques dos Santos, F.L.; Gkoumas, K.; Stępniak, M.; Pekár, F. The Role of Research and Innovation in Europe for the Decarbonisation of Waterborne Transport. *Sustainability* **2021**, *13*, 10447. [CrossRef]
42. Lion, S.; Vlaskos, I.; Taccani, R. A review of emissions reduction technologies for low and medium speed marine Diesel engines and their potential for waste heat recovery. *Energy Convers. Manag.* **2020**, *207*, 112553. [CrossRef]
43. Paulauskas, V.; Filina-Dawidowicz, L.; Paulauskas, D. Navigation of Ships in Channel Bends under Special Conditions Using Sensors Systems. *Sensors* **2022**, *22*, 8783. [CrossRef]
44. Quy, M.N.; Łazuga, K.; Gucma, L.; Vrijling, J.K.; van Gelder, P.H.A.J.M. Towards generalized ship's maneuver models based on real time simulation results in port approach areas. *Ocean Eng.* **2020**, *209*, 107476. [CrossRef]
45. Ortolani, F.; Mauro, S.; Dubbioso, G. Investigation of the radial bearing force developed during actual ship operations. Part 1: Straight ahead sailing and turning maneuvers. *Ocean Eng.* **2015**, *94*, 67–87. [CrossRef]
46. González-Cancelas, N.; Serrano, B.M.; Soler-Flores, F. Seaport sustainable: Use of artificial intelligence to evaluate liquid natural gas utilization in short sea shipping. *Transp. J.* **2019**, *58*, 197–221. [CrossRef]
47. Cariou, P. Is slow steaming a sustainable means of reducing $CO_2$ emissions from container shipping? *Transp. Res. Part D Transp. Environ.* **2011**, *16*, 260–264. [CrossRef]
48. Anderson, J.O.; Thundiyil, J.G.; Stolbach, A. Clearing the air: A review of the effects of particulate matter air pollution on human health. *J. Med. Toxicol.* **2012**, *8*, 166–175. [CrossRef]
49. Agrawal, H.; Welch, W.A.; Miller, J.W.; Cocker, D.R. Emission measurements from a crude oil tanker at sea. *Environ. Sci. Technol.* **2008**, *42*, 7098–7103. [CrossRef]

50. Bermúdez, F.M.; Laxe, F.G.; Aguayo-Lorenzo, E. Assessment of the tools to monitor air pollution in the Spanish ports system. *Air Qual. Atmos. Health* **2019**, *12*, 651–659. [CrossRef]
51. Broglia, R.; Dubbioso, G.; Durante, D.; Di Mascio, A. Simulation of turning circle by CFD: Analysis of different propeller models and their effect on maneuvering prediction. *Appl. Ocean Res.* **2012**, *39*, 1–10. [CrossRef]
52. Bruzzone, D.; Gaggero, S.; Podenzana Bonvino, C.; Villa, D.; Viviani Diten, M. Rudder-propeller interaction: Analysis of differentapproximation techniques. In Proceedings of the 11th International Conference on Hydrodynamics (ICHD 2014), Singapore, 19–24 October 2014; pp. 1–10.
53. Chauhan, S.; Patil, C.; Sinha, M.; Halder, A. Fuzzy state noise-driven Kalman filter for sensor fusion. *Proc. Inst. Mech. Eng. Part G J. Aerosp. Eng.* **2009**, *223*, 1091–1097. [CrossRef]

*applied sciences*

MDPI

*Article*

# Prospect of LNG as Marine Fuel in Indonesia: An Economic Review for a Case Study of 600 TEU Container Vessel

**Riko Butarbutar, Raja Oloan Saut Gurning * and Semin**

Department of Marine Engineering, Faculty of Marine Technology, Institut Teknologi Sepuluh Nopember, Surabaya 60111, Indonesia
* Correspondence: sautg@its.ac.id; Tel.: +62-(0)82140060563

**Abstract:** The alternative use of environmentally friendly marine fuel by Indonesian vessel owners complies with IMO regulations. Marine fuels with low carbon and sulfur are alternative fuels to the current fossil fuels used by the shipping industry. Some alternative marine fuels are being used or developed such as LNG, hydrogen, and methanol. LNG is one alternative fuel that is used significantly as a marine fuel in the shipping industry. As one of the LNG producers, Indonesia is still behind in using LNG as an alternative marine fuel. One of the main reasons is the use of conventional marine fuels such as HFO, MDO, MGO and the understanding of LNG as an expensive and high-risk commodity. However, vessel owners face various challenges when selecting alternative fuel, which is associated with price and technology. This study aims to analyze a 600 TEU container vessel by calculating its net present value, the capital recovery factor and life cycle analysis (LCA) to determine whether owners carry out the investment. The result of the economic analysis for the 600 TEU vessel showed that the investment of retrofit for LNG as a marine fuel will be a good choice for owners due to the challenge of capital cost for financing a new vessel.

**Keywords:** fuel gas supply system; life cycle analysis; LNG

**Citation:** Butarbutar, R.; Gurning, R.O.S.; Semin Prospect of LNG as Marine Fuel in Indonesia: An Economic Review for a Case Study of 600 TEU Container Vessel. *Appl. Sci.* **2023**, *13*, 2760. https://doi.org/10.3390/app13052760

Academic Editor: José A. Orosa

Received: 13 January 2023
Revised: 14 February 2023
Accepted: 16 February 2023
Published: 21 February 2023



## 1. Introduction

The shipping sector is an important player in the Indonesian economy because sea transportation is cost-effective. Its growth is impacted by indigenous and international regulatory bodies such as IMO. However, the current regulatory standard adopted by IMO is emission control from the vessel's exhaust. Ref. [1] Arefin et al. stated that the increased demand for energy triggers the production of greenhouse gases (GHGs) in enormous quantities. GHGs are obtained from burning fossil fuels, which ultimately cause global warming. Since the implementation of emission control by IMO, several studies have been carried out on alternative fuels, and presently, various types are available in the market. Vessel operators have no choice but to select advanced alternative fuel technology as a management strategy. In terms of sulfur emission, traditional marine fuel is influenced by component and hydrocarbon composition and the structure of asphaltenes [2]. The characteristics, both physical and chemical, of asphaltenes will also impact the sulfur content [3]. The major alternative marine fuels in development are hydrogen, LNG, methanol and batteries [4]. Jack Sharples stated that transportation modes are significant sources of carbon emission (CO2) [5]. Air pollution containing SO, NOx and particulate emissions significantly impacts human health. In 2015, approximately 32.3 billion tons of CO2 emissions were recorded globally, of which 7.7 billion was obtained from the transportation sector with 5.8 billion tons on land. This is followed by sea transportation and aviation, with approximately 657 million tons and 530 million tons of emissions, respectively. While land transportation contributes a huge amount of CO2, NOx, and particulate emissions, the sea contributes approximately 90% of SOx emissions and impacts the local port [5]. One factor that causes high emissions from vessels is the cheap price and filtering technology of fuel.

The use of LNG as alternative marine fuel for decarbonization is implemented in various types of vessels. In South Korea, there was a study on LNG fuel application to new bulk carriers [6]. LNG as a marine fuel is not only implemented in deep-sea shipping, but it has implemented for short-sea shipping or domestic shipping. Fishing vessels are another type of vessel that increase the impact on the environment due to vessel emission, and these types of vessels have started to use alternative marine fuels. The study shows that LNG fuel is a good option for fishing vessels to reduce environmental impact [7]. Another type of vessel that started to use LNG as a marine fuel is the Ro-ro ferry vessel. The conversion of Ro-ro vessels to LNG-fueled vessels will be technically feasible and a good option for local ship operators [8]. In Indonesia, vessel owners are usually more comfortable with conventional fuels such as HFO, MDO or MGO rather than engaging in environmentally friendly vessel operations. According to one study, heavy fossil hydrocarbons are transformed into natural gas, and their constituents can reduce emissions [9]. This flexibility is significant to the termination of dependency on conventional fuel; many countries are yet to develop renewable energy [10]. Yun et al. stated that energy derived from fossil fuels is expensive, impacts the economy, social life protection, welfare, and the educational sector and triggers air pollution [11]. The government needs to create awareness of the importance of environmentally friendly fuel, specifically in the shipping sector. Its realization is bound to impact the economy and social environment for decades significantly. Vessel owners can utilize several fuel alternatives to comply with the IMO regulation. Natural gas is the preferred fuel in this sector due to its innumerable advantages. These include GHGs reduction, better combustion efficiency, attractive cost, and renewability through biomass production [12]. Irrespective of the fact that natural gas is majorly used in the transportation sector due to its availability and environmental benignity, it is still limited to small engines, specifically spark-ignition (SI) and is rarely found in large diesel engines [12]. In the next decade, the number of LNG-fueled vessels is forecasted to increase immensely, even though certain segments are bound to experience massive expansion [13]. The possibility of vessels switching from using fossil fuel to LNG is because this has gained significant concern among the currently evaluated technologies [14]. In this present study, LNG is used as an alternative fuel due to its numerous advantages. These include advanced LNG vessel technology and the market price established across major ports globally. Another journal has studied the LNG fuel and diesel engine based on the Energy Storage System (ESS) using the NGSA-II algorithm and discovered the optimal scheme to reduce pollutants and cost [15].

Previous studies stated that LNG is the most advanced energy technology implemented on board vessels. Its source is readily available in terms of emission reduction compared to other alternative fuels. Natural gas is a promising alternative fuel source in transportation because of its remarkable advantages [1]. A survey was conducted in some shipping companies that render several services related to the container, offshore, general cargo vessels, and crew boats. The objective of the survey was to understand the emission control requirement mandated by IMO and vessel owners' plans for using environmentally friendly marine fuel. Eight companies were selected randomly as respondents, and due to confidentiality, the company name is classified. Their responses were used as sampling representatives with respect to marine fuel transition. These eight companies represented the types of vessels operating in Indonesia such as container vessels, general cargo vessels, oil tanker vessels, LCT and offshore support vessels.

In accordance with the distributed questionnaires, most respondents (70.6%) stated that they knew about IMO regulation to reduce sulfur content and even felt the impact on their businesses. The graphic representation in Figure 1 shows that the rest of the respondents do not understand the IMO regulation on emissions. The challenge is to ensure vessel owners in the country realize that IMO regulation should be implemented in target regions to obtain zero emissions by 2050.

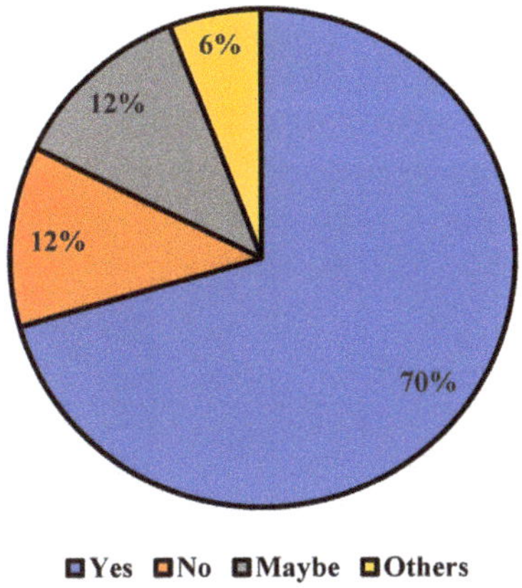


**Figure 1.** Respondents' understanding of IMO emission control and impact.

The next questionnaire about the shipping company plan is on its use as an alternative fuel and interestingly only 35.3%, 17.6% and 29.4% plan to use, not use and consider using it, respectively. Figure 2 below illustrate the respond for ship owner plan on use of alternative marine fuel.

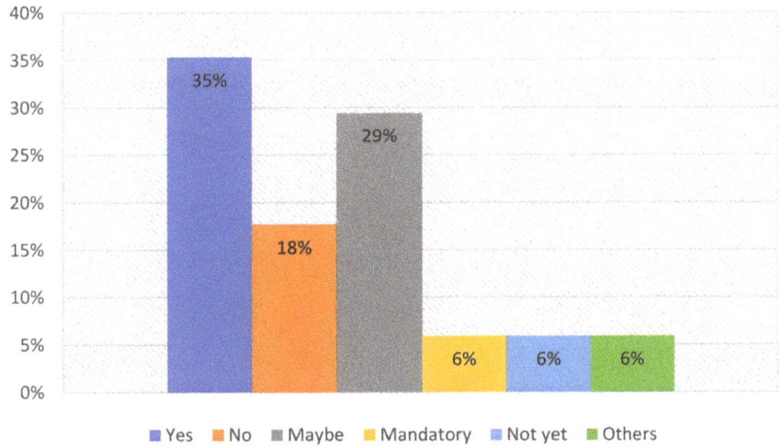


**Figure 2.** Plan for using alternative fuel.

Furthermore, approximately 45% of the respondents intend to use LNG, while 5% used other alternatives such as ethanol, methanol, and hydrogen. The remaining 14% and 9%, intend to use electricity and LPG, respectively. With respect to this question, respondents can use more than one energy alternative as planned. Figure 3 below illustrate the respond to alternate fuel for Indonesia shipping sector.

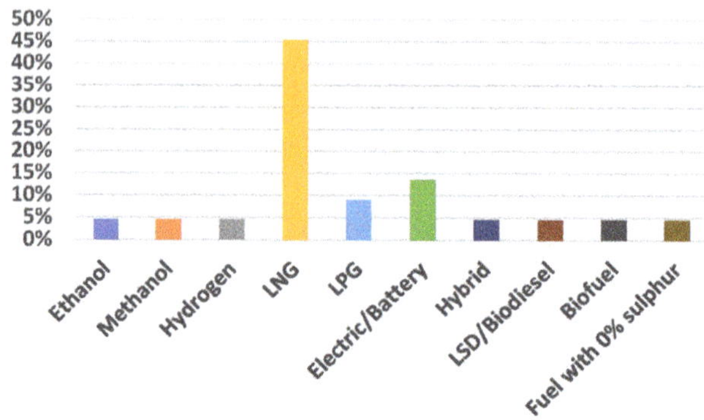


**Figure 3.** Alternative fuel for Indonesia shipping sector.

The majority of the shipping companies stated that the selection of alternative fuel was based on the considerations of low price investment (30%), government regulation or authorization (25%), advanced technology (25%), energy content (10%) and resource (10%). From this mapping, most of the companies tend to consider low fuel prices, which is the primary selection factor, followed by advanced technology. Figure 4 below illustrate the reason for select the alternative fuel.

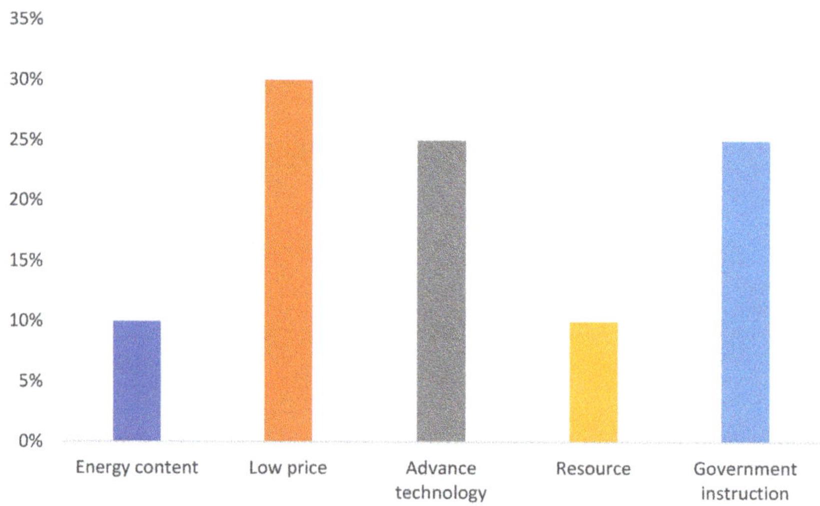


**Figure 4.** Reason for selecting alternative fuel.

Based on the earlier mentioned survey, several factors need to be taken into account by Indonesian vessel owners when investing in alternative energy. The most challenging factors to be considered are freight rate versus investment. Currently, some owners use conventional fossil fuels, such as MGO and HFO in their comfort zone. However, when IMO strengthened its position to reduce emissions from vessels, it was supported by environmental energy, which is the main objective, alongside domestic and international trade. The study of LNG-fueled vessel investment in recent years is increasing, although none analyzed the capability of Indonesian vessel owners to invest in this alternative energy. Therefore, this present study focuses on the challenges that vessel owners face in the country, specifically in understanding the investment strategy concerning the use of LNG as a marine fuel. It seeks to economically analyze this strategy by considering the potential retrofit for owners' existing fleets.

## 2. Literature Review

### 2.1. LNG Technology

The choice of alternative fuel by vessel owners is mainly driven by investment costs and advanced technology. Elkafas et al. stated that both natural gas and hydrogen are already used [16]. However, compared to natural gas, hydrogen has safety issues. The advanced technology depends on the availability of a bunker and related infrastructure. In 2019, DNV identified some alternative fuels used in shipping companies, such as LNG, LPG, methanol, biofuel and hydrogen. LNG is the most popular and promising alternative fuel because its technology is developed correctly [17]. It has been developed through significant innovation, hence, its ability to reduce the high content of fuel emissions. They also stated that the capability to reduce sulfur and nitrogen levels is due to the use of marine fuel, such as LNG, in a diesel engine [18]. Clean and renewable energies are ideal, although, in practice, LNG is usually selected by owners [19]. LNG is categorized as the leading alternative fuel, followed by methanol and biofuel [4].

LNG technology on board vessels depends on the fuel gas supply system (FGSS). Wang et al. stated that the fuel tank needs to be kept in the liquid phase at $-163$ °C. Furthermore, it is designed to supply gas to dual-fuel engines under the required temperature and pressure. It also needs to avoid being over-pressurized due to its ability to improve fuel efficiency [20]. Most vessel technology uses dual fuel systems, while the boil-off gas produced in the LNG tank is used for steam turbines [21]. In 2022, the Maritime Executive stated that the retrofit concept reduces the cost of LNG conversion operations [22].

### 2.2. Investment Analysis Outlook

Generally, vessel owners need a reference for their investment because they usually encounter difficulties, such as changing the current fuel to an alternative one that is environmentally friendly. Some previous studies stated that as a marine fuel, LNG would positively impact the future; its technologies are bound to pay off in a matter of years (DNV-GL, 2015). Some methods can help owners adopt an ideal investment strategy, for example, the cash flow. The uncertain price of LNG is also a huge drawback for transitioning to alternative fuel. Chen et al. stated that no international market is currently dealing with natural gas. Furthermore, the common economic analysis approach that considers time value, namely present, final, and annual worth methods, is employed in selecting these alternatives [14]. Some literature stated that most shipping investment evaluations use Real Options Analysis (ROA). ROA is used because it incorporates the uncertain prices of both LNG and conventional fossil fuels [14]. Previous studies stated that the shipping investment decision is based on the relation vessel between the current freight and trigger rates from ROA and Net Present Value (NPV). Kou et al. stated that it impacts the mean freight rate [23]. Figure 5 illustrates the challenges Indonesian vessel owners face regarding the regulation requiring them to comply with emission control. Another economic assessment method is using life cycle cost assessment (LCCA) which this method used to investigate the total cost including the sum of investment, maintenance and operations costs [24]. LCCA is a method for analyzing cost throughout the lifecycle of a product or service and it is a preferred method for the decision-making process. The LCCA method has been demonstrated to be effective when it is used for assessing the yacht cost model [25]. In the shipping industry, it is difficult to assess fuel prices for certain periods and the result from one study showed that the sensitivity of lifecycle cost for uncertain fuel prices can be observed [26]. In terms of LNG fuel options, a study from Alvestad which compares MGO, LNG and scrubbers has concluded that for new build vessels, LNG fuel might be the most economical marine fuel alternative [27].

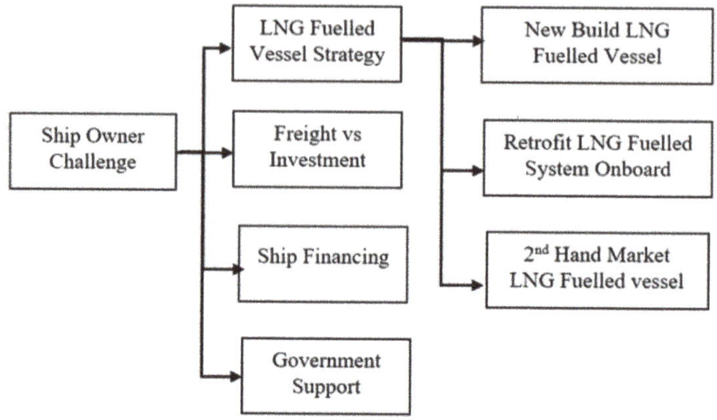


**Figure 5.** Indonesian vessel owners challenging condition.

As mentioned earlier, vessel owners need to consider the capital cost of the initial investment whenever they want to use alternative fuels such as LNG. The owners were exposed to three options, namely building new LNG-fueled vessels, retrofitting and purchasing from the second-hand market. New build and second-hand purchase markets depend on sales, while their characteristics are centered on the vessel type [28]. According to Rivieramm News, DNV reported that 240 LNG-fueled vessels were ordered in 2021, consisting of the container ship, tanker and bulk carrier sectors. Snyder further stated that based on DNV data, 251 LNG-fueled vessels are presently in operation globally, and 403 fleets are under construction [29]. This implies that the development of LNG-fueled vessels has progressed significantly. Some studies were carried out to analyze the investment in fuel transition. [20] Wang et al. calculated the low-cost analysis (LCC) for boil-off gas management and discovered that the universal solution is not applicable in all situations. It was further stated that the fuel gas supply system depends on the vessel's scale, operation, and LNG fuel price [20]. Yoo conducted an economic assessment of LNG as marine fuel for CO2 carriers and compared it to MGO. It was found that LNG is more cost-effective compared to MGO. He also used the discount rate, and the project lifetime functions to calculate the annual cost index on LNG and MGO [30]. According to studies on Discount Cash Flow Method (DCFM) LNG fuel container vessels with low-speed diesel attract economic investment compared to the Tier III complied oil-fueled container vessels [26].

## 3. Methodology

### *3.1. Selection for Vessel*

Vessel owners who invested in fuel transition are demanding to know when they can benefit from vessels in the market. The container vessel is extremely important in the Indonesian shipping industry and has an impact on emission control regulation. Other factors that need to be considered during selection are tankers, offshore supply vessels, tug boats, and fuels paid for or provided by the charterer. This restricts the vessel owners' flexibility to change to another alternative fuel. This study selected a container vessel with a capacity of 600 TEU because it was considered suitable and the capacity size is commonly available in the Indonesian shipping market compared to other container capacities. The container vessel with 600 TEU capacity is also the feeder size container that plays an important role in short sea shipping within Indonesia and the nearest regional countries such as Singapore and Malaysia. Furthermore, it has a company schedule and voyage, which simply means that assuming owners change to LNG, the maintenance program can be predicted and managed quickly. For this analysis, the vessel route is from the Port of Tanjung Priok, Indonesia, to the Port of Singapore, with a distance and economical speed of approximately 591 nm and 11 knots, respectively.

### 3.2. Vessel Design

In this study, the existing container vessel was used to carry out certain analyses. This is intended to provide an overview of the vessel owners' perspective on the fuel transition strategy. Assuming this is not a new build vessel, the ideal methodology that needs to be adopted is retrofit. The availability of technology reduces the cost of the vessel and improves efficiency. Furthermore, it was stated that zero-emission fuel impacts the already-built vessel [31]. It provides retrofit, which most vessel owners usually consider. The 600 TEU container vessel serves as a retrofit to dual fuel. According to the Retrofit Series (2020), the three vessels subjected to retrofit have significant potential savings, such as lube oil cleaning and other attributes that are often overlooked [32]. This includes potential savings from machine learning. Some other studies carried out on a mini-cape size bulk carrier stated that the payback period for LNG-fueled vessel retrofit is 4.5 years compared to a 0.5% compliant fuel vessel [33]. A retrofit vessel that uses LNG fuel is an attractive option to meet the new regulation. Another study stated that Hapag-Lloyd investigated a 15,000 TEU Sajir retrofitted for LNG fuel. This concept has LNG cylinders contained in open frames with 40-foot containers. The venting system and LNG piping, including the fire-fighting technique, are integrated into the container cell guide structures handling the gas adjacent to the storage. It feeds the low and high-pressure fuel gas system to the current four-stroke dual-fuel engines [22]. In this study, the 600 TEU container vessel has a similar concept with retrofit, as stated by previous studies on 15,000 TEU Sarji by Hapag-Lloyd. Wang et al. designed a three-configuration fuel gas supply system, and as mentioned earlier, FGSS is a critical factor in the LNG fuel system. The three configurations of FGSS are GCU, AE, and reliquefaction schemes with the combustion of boil-off gas-by-gas combustion unit (GCU), supply boil-off gas using auxiliary engine (AE) and reliquefaction boil-off gas by reverse Brayton cycle (RBC) system, respectively. In line with a previous study [20], this present study selected a suitable retrofit configuration for a 600 TEU container vessel dependent on a GCU scheme because the system is reliable, simple, and compact. Figure 6 illustrates the configured FGSS with boil-off gas handled by GCU. The configured FGSS is adapted from Wang et al.'s scheme [19], and the LNG Tank is fitted into a deck with a similar arrangement as a refrigerator container tank. It uses the plug-in system on the LNG tank and container cell, thereby reducing the cost of the conversion vessel [22].

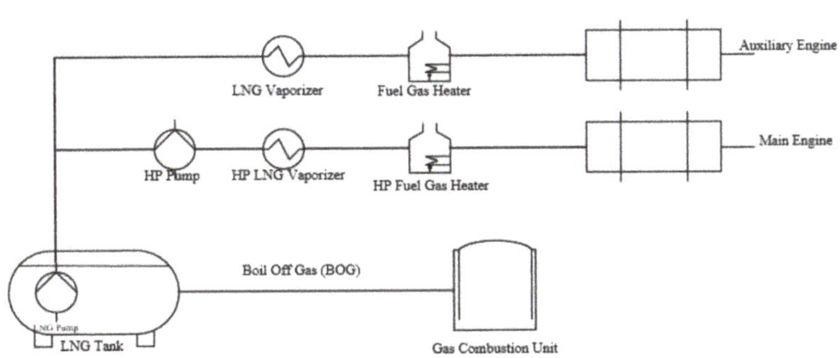


**Figure 6.** Configuration of FGSS with BOG handled by GCU.

The retrofitted designed vessel has employed a promising strategy to avoid uncertainty. They defined the retrofit cost using a Pareto-optimal solution, and interestingly, it depends on different alternative fuel types. The retrofit cost was calculated by analyzing certain aspects, namely machinery, tank, piping, shipyard and lost income. Figure 6 shows an illustration formulated by Lagemann et al. [34] as a reference. Another study that proposed the use of the calibrated method for the fuel substitution ratio, economy and particulate matter emission proved brake-specific consumption for the dual fuel model is higher than the diesel [35]. The generated boil-off gas tends to have certain advantages, such as energy efficiency [36]. The calculated lost income and the time needed during retrofit at the shipyard are perceived as a challenge to Indonesian vessel owners because it requires opportunity costs to compensate for the time lost.

### 3.3. Maintenance and Crew Cost

This study defined and considered three maintenance scenario assumptions. These consisted of high, medium, and low scenarios dependent on the Moore Maritime Index. For a high scenario, the assumption of all maintenance and crew costs is increased by 10%. Meanwhile, for the medium scenario, there is no difference between existing and retrofit vessels, and for the low medium, the lubricating oil and spare parts are reduced by 50%, as opposed to maintenance costs by 33% less than the initial. The maintenance and crew costs are the most significant operational expenditure, besides fuel oil prices.

Table 1 shows the information of these three scenarios with operational cost in accordance with Moore Maritime Index 2021 [37] on an average level.

**Table 1.** Maintenance and crew cost scenario.

| | High Scenario | Medium Scenario | Low Scenario |
|---|---|---|---|
| LNG price | USD 1100/MTon | USD 1100/MTon | USD 1100/MTon |
| Maintenance cost and crewing | 10% increase | No difference between existing and retrofitted vessels | Lubricating Oil reduce by 50%, Spare parts reduce by 50%, and maintenance becomes 33% lesser than the initial cost |
| Operational cost based on Moore Maritime Index | Average | Average | Average |

### 3.4. LNG Fuel Prices

LNG fuel prices are the dominant factor in determining the economic analysis of its transition. The fuel cost depicts approximately 60 to 80% of the total operating cost, while the rising oil price poses a huge challenge [38]. From the beginning of the fuel transition plan, the vessel design is not altered since the vessel will be retrofitted, and the changing prices tend to impact the LNG fuel system [26]. The increasing LNG fuel cost affects the overall system as well. Further, Wang et al. stated that the preference for appropriate FGSS configuration depends on the LNG price [20]. Some other studies stated that the LNG investment option depends on three parameters. These include the price differentials between LNG and conventional fossil fuels, new build LNG fueled vessels compared to the conventional type that entails burning traditional maritime fuels, and the shared operations within ECAs. They also observed the cost change in different bunker locations, such as Japan [14]. The major source of LNG fuel prices referenced in the market is Henry Hub for the east coast US and TTG or NBP for North West Europe and Asian markets. Japanese prices are perceived as an option. Figure 7 shows the illustration of the marine fuel price differential [39], which is cheaper based on a negative differential.

Lagemann et al. also described the fuel prices for some alternative fuels within a certain period [34]. This group of fuel types was sorted based on prices and divided into fossil, bio, and e-fuel.

### 3.5. Flow Analysis

Another challenge encountered is that the Indonesian government has yet to implement green environmental fuel regulations to support vessel owners to change from conventional to alternative fuel. The government must provide some incentives to attract these individuals to use alternative fuels such as LNG. Figure 8 illustrates Indonesian vessel owners' challenging situation before changing their fuel management to an alternative type, such as LNG.

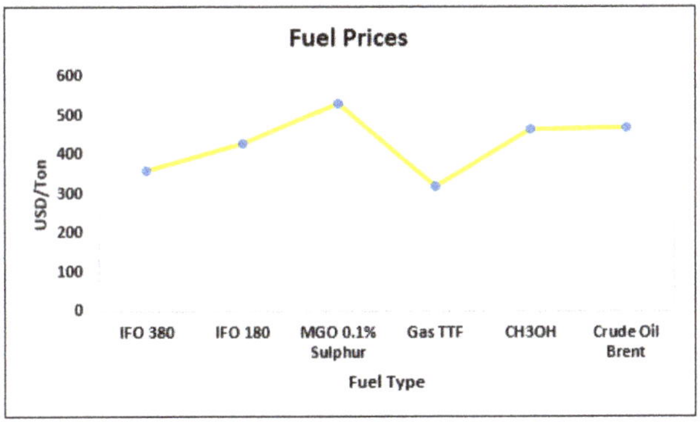


**Figure 7.** Marine fuel price differentials.

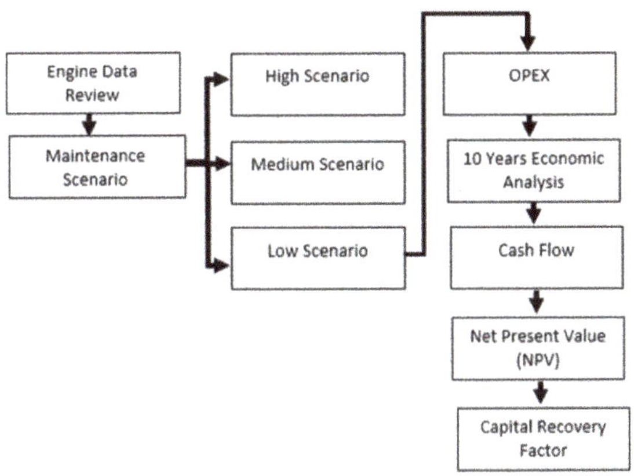


**Figure 8.** Flow Diagram for data analysis.

The data were analyzed using the lower scenario, assuming that the maintenance cost and spare parts were reduced due to LNG usage. This study employed some processes to obtain the cost recovery factor for investing in LNG fuel existing 600 TEU vessels. The first step for the analysis is gaining engine information from vessel owners. The expected data are engine power, specific fuel consumption, and type. The next step is to select the maintenance and crew cost scenarios. In addition, this study compared the high and low scenario investments. A particular study on Niigata's engine manufacturing proved customer satisfaction with gas engine series with low running cost and required maintenance at 4000 h running intervals [40]. Wartsila stated that switching to LNG as a marine fuel, whether new build or converting existing technology, will generate significant savings in fuel cost, thereby increasing profitability [41].

Operating Expenditure (OPEX) consists of the spares, lubricant, repair and maintenance. Retrofit investment for container vessel 600 TEU constitutes modifying the fuel and gas supply system. This is realized by installing the LNG tank, piping, changing the main engine and installing a gas combustion unit. As discussed earlier, vessel owners usually consider these, including vessel modification and additional equipment installation. The economic analysis requires a 10-year scheme because it is sufficient to review the potential payback from the vessel owners' view and offers future plans for vessel acquisition. Net Present Value (NPV), with respect to a 10-year investment scheme, shows differences between the present cash inflows and outflows.

Furthermore, the vessel owner requires an analyzed loan payment in a different scenario. It is calculated based on loan principal, interest, payment, and remaining amount.

Some formulas used to calculate NPV and CRF [20] are as follows:

$$\text{NPV} = \sum_{t=1}^{t} \frac{C_t}{(1+r)^t} - C_0 \tag{1}$$

where:

$C_t$ = cash flow for time (t)

$r$ = interest rate

$C_0$ = initial investment on year 0

$t$ = time

$$\text{CRF} = \frac{i(1+i)^n}{(1+i)^n - 1} \tag{2}$$

An economic analysis of the investment in trans-ocean LNG-fueled container ships, 9300 TEU sailing between Asia and Europe, showed that the LNG low-speed diesel vessels compared to oil-fueled SCR is more attractive [27]. Furthermore, Adachi et al. discovered that the NPV with a lifetime of 20 years is larger, while the refund time to payback is shorter for LNG vessels. Wang et al. also economically analyzed the lowest CAPEX for retrofit fuel gas supply systems and discovered a lower LCC on the auxiliary scheme [18]. LCC analysis allows the assessment of some shipping costs. These include acquisition (capital cost), operation or running costs, fuel consumption, operational services, maintenance, and ship disposal costs [25]. Some of the important elements of this economic assessment can be defined as follows.

#### 3.5.1. CAPEX

Capital expenditure or cost is an essential element that owners consider when making investment decisions. Since the retrofit approach was employed, investing in conversion vessels has become a fundamental option. Wang et al. stated that there are three fuel gas supply systems, and the one with a gas combustion unit has a lower cost. However, this study used an FGSS with the GCU approach as the capital cost includes direct and indirect prices. Direct cost is related to purchase, installation and other related labor expenditures. The indirect costs are related to transportation, insurance, tax, construction overhead, and engineering expenditures [20]

#### 3.5.2. OPEX

Operating expenditure is all the costs related to operational activities, such as maintenance and crew costs. Wang et al. stated that, unlike onshore LNG plants, the FGSS has varying fuel consumption during the voyage [20].

The total expenditure in the lifetime system for LCC includes CAPEX and OPEX costs [18]. Furthermore, when the CRF has been determined, it is multiplied by the CAPEX using the formula:

$$\text{LCC} = \text{CAPEX} \times \text{CRF} \times \text{n} + \text{OPEX} \tag{3}$$

The use of LNG fuel after conversion is perceived as an annual saving despite the different fuel price increments per remaining vessel life cycle [42]. It includes emission reduction with respect to the alternative fuels used in the vessel.

## 4. Case Study: Economic Analysis 600 TEU Fuel Transition from MFO to LNG

Based on an economic perspective, this study analyzed the existing 600 TEU container vessel transition from MFO to LNG to determine the life cycle cost (LCC) of a 10-year investment retrofit scheme. The container vessel plies from Tanjung Priok, Indonesia, to Singapore, at a distance of 591 nautical miles. Therefore, the existing vessel will have to be retrofitted to the LNG system, and the major information extracted from the technical analysis of the owners' plans to change fuel, specifically the data concerning the Specific Fuel Oil Consumption from both main and auxiliary engines. The type of fuel used is MFO and Table 1 shows the estimated price of the new build 600 TEU container vessel in the market. PT Samudera Indonesia purchased this vessel for 8.5 million USD in 2018 from Jingjiang Nanyang Shipbuilding China [43]. This container vessel was an MFO-fueled vessel. Table 2 shows 2% inflation per year for new build vessels manufactured with FGSS installed on board culminating in two million USD. The estimated cost is based on 600 TEU newly build MFO fuel container vessel prices from 2018 from PT Samudera Indonesia and calculates 2% inflation each year.

**Table 2.** Estimated new build 600 TEU container vessel.

| New Build with MFO Fuel (MUSD) | New Build with LNG Fuel (MUSD) |
|---|---|
| 9,180,000 | 11,180.000 |

Furthermore, Figure 9 adapted from the Moore Index 2021 illustrates a container vessel OPEX with various sizes. There are no data for those below 1000 TEU as per the subject size vessel used in this study. For the new build, the analysis was carried out using vessel between 1000 TEU to 1999 TEU. The Moore Index was used to determine each sub-category independently. This study calculated the total OPEX expenditure, which includes maintenance and crew costs using the Moore Index as a reference.

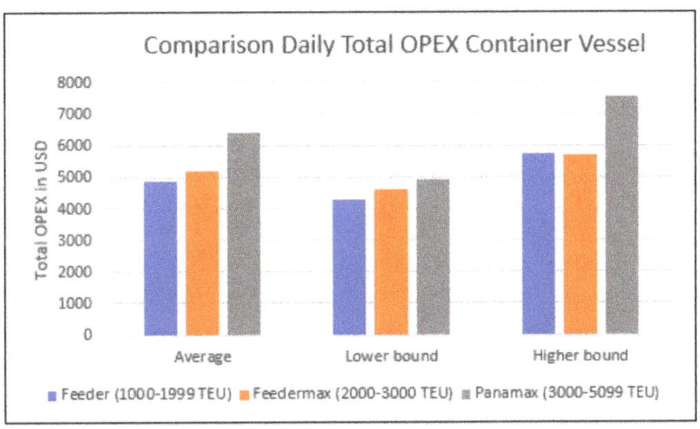


**Figure 9.** OPEX cost for various container vessels.

### 4.1. Cost Assessment for Retrofitting 600 TEU Container Vessel

This analysis was centered on the assumption of retrofit cost in 2022, which encompasses main, and auxiliary engines, fuel gas supply system and installation. The conversion cost is USD 200 to USD 340 per HP, based on the upper bound assumption [39] Furthermore, this retrofit has an estimated cost of USD 3,600,000. Figure 10 is adapted from Lagemann et al. [26] and illustrates retrofit costs for various fuel types.

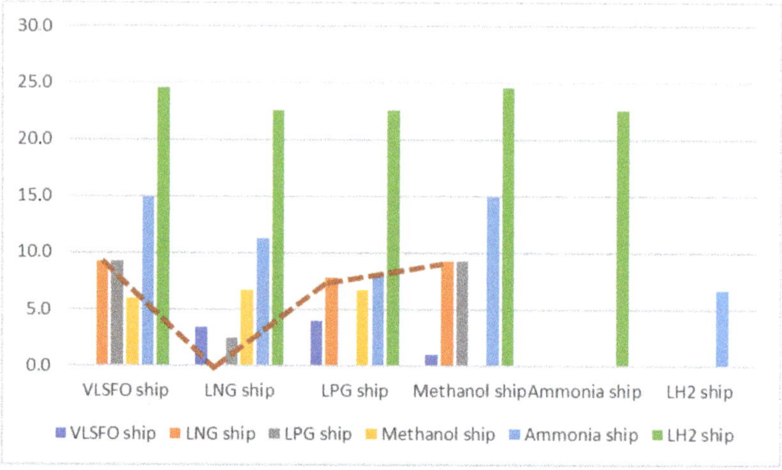


**Figure 10.** Retrofit cost various types of fuel vessel.

Banawan et al. stated that using gas as fuel reduces the deposit of organic material in the combustion chamber [43]. The reduction in hydrocarbons and other particles from the fuel affects its mass deposit. Some studies carried out on LNG as a marine fuel stated that the maintenance cost is reduced because a small amount of lubricating oils is applied on the spare part compared to the engine system using MFO or HFO. By reducing emissions, the annual expenditure determines the entire cost of natural gas applied on the main fuel onboard, including the capital expenses due to conversion [25].

Based on OPEX per year, the equivalent loading and offloading per year is 312 days. This is because the sailing duration from Jakarta to Singapore lasts for approximately three days. Based on an interview session held with one of the owners, the loading and offloading duration is usually two days for one trip, with the assumption that the in-container at the terminal is 60 containers/day.

Table 3 shows the engine information that it used for investment calculation which consist of data regarding power, number of unit, specific fuel oil consumption and type of fuel.

**Table 3.** Engine information.

| | Main Engine | Auxiliary Engine |
|---|---|---|
| Power | 2500 kW | 450 kW |
| Unit | 1 | 1 |
| SFOC | 183 | 213 g/kWh |
| Fuel | MFO | MFO |

Table 4 shows that the total OPEX/year after retrofit is usually within the range of minimum, average and maximum variables. All tend to be reduced according to the acquired data. This Table 4 ilustrate operation cost of spares, repair, maintenance, and lubricant based on the Moore Maritime Index [34]. The total OPEX per year was calculated using three variables, namely minimum, average and maximum.

**Table 4.** OPEX Calculation per year.

| | Minimum USD/Day | Average USD/Day | Maximum USD/Day |
|---|---|---|---|
| Spares | 344 | 407 | 407 |
| Repair and maintenance | 154 | 309 | 442 |
| Lubricant | 2301 | 2488 | 2738 |
| Total OPEX/year after retrofit | Million USD/day 0.3146 | Million USD/day 0.3679 | Million USD/day 0.4143 |

The fuel consumption of a 600 TEU container vessel that uses MFO is 6296.66 tons per year based on daily SFOC multiplied by 312 days of operation. For the calculation, the yearly consumption of LNG, based on the heating value of diesel oil, is 4958.2 tons per year or 228,871.27 MMBtu. Table 5 shows yearly fuel consumption LNG and MGO.

**Table 5.** Yearly fuel consumption LNG and MGO.

| Yearly Fuel Consumption LNG and MGO | |
|---|---|
| MGO (Ton/Year) | LNG (MMBtu/Year) |
| 6296.66 | 228,871.27 |

The annual pilot diesel fuel at the terminal is approximately 10% of the total MFO consumption per year or 629.67 tons [25]. The annual fuel price for LNG is 6.27 million USD compared to MFO, which is 8.13 million USD. Therefore, there is a difference of 1.86 million USD between LNG and MFO. It simply implies that the use of LNG is more economical compared to MFO. Supposing the annual OPEX is calculated yearly based on Moore Maritime Index, 2.228 million USD will be realized, meaning LNG is more economical.

Based on an interview with one of the vessel owners, the economic analysis for a 10-year scheme is shown in Table 6.

**Table 6.** Period 10-Year Scheme for Economic Analysis.

| | | Data |
|---|---|---|
| Cost Component | Unit | Total Cost |
| CAPEX | USD | 3,617,624 |
| OPEX Change | USD/Year | 367,860 |
| Project Duration | Years | 10 |
| Annual Depreciation | USD/Year | 180,881 |
| Disposal/Salvage Value | USD | 1,808,812 |
| Tax Cost | %/Year | 22% |
| Inflation Rate | %/Year | 4.5% |

CAPEX was calculated using an approach based on a literature review, and USD 3,617,624 was realized. OPEX data were not given, and the interviewee only mentioned the profit per container, which is 10 USD. It simply implies that only 80% of the container vessel space is occupied by a total of 600 TEU. Assuming the vessel uses LNG as fuel, only 480 TEU container is conveyed on every single trip from Jakarta to Singapore. The total number of trips from Jakarta to Singapore is 48 trips per year. Target BEP (Break Even Point) for this analysis is 10 years with equity from the company of approximately 40% and 60% loan. Table 6 shows that the CAPEX obtained is USD 3,617,624, while the OPEX realized for 3 years is USD 183,900. Meanwhile, 60% of CAPEX and OPEX amounted to USD 2,280,933. This study used a bank interest of −8%, and in accordance with further calculations, the loan principal is USD 228,093. This loan repayment needs to be taken into consideration during LCC calculation under different vessel financing model scenarios created by each vessel owner.

Figure 11 illustrates a loan payment scheme for 10 years, as follows.

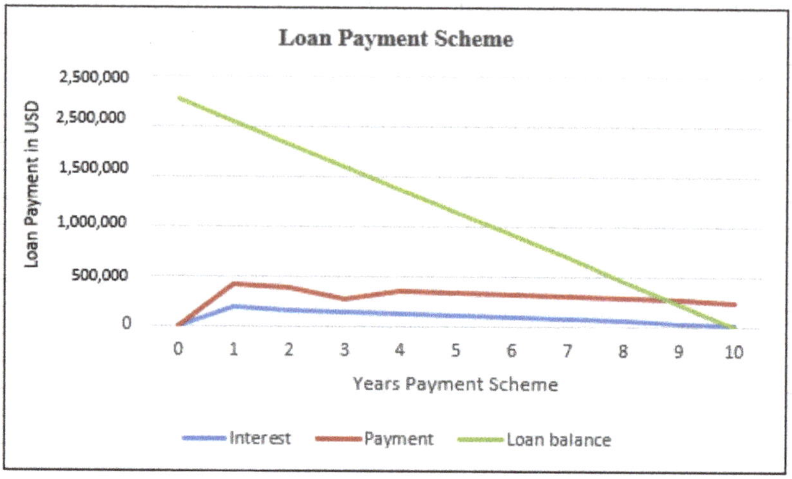


**Figure 11.** Loan payment illustration.

Cash flow is an important indicator of economic feasibility concerning the investment in the retrofit 600 TEU containership. Figure 12 shows an illustration of cash flow for 10 years. Vessel owners will encounter challenges in cash flow until the second year, and from the third, there is bound to be positive cash flow.

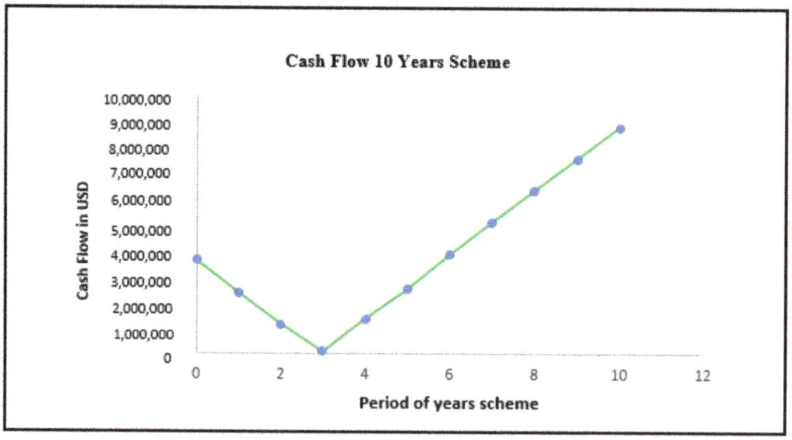


**Figure 12.** Cash flow for 10 years scheme.

For 10 years, the NPV with four Discount Factor (DF) variables tends to provide initial payback, which is used to calculate the Capital Recovery Factor on the investment scenario. Assuming the initial investment has a negative value, it is considered a capital expenditure. However, this scenario's initial cost (-) is USD 3,617,624.

Figure 13 shows the calculated NPV using various discount factors 25%, 30%, 35% and 40% and a positive NPV was realized during the 10 year scheme on the investment. The same interest rate for 10 years of investment was used for the calculation.

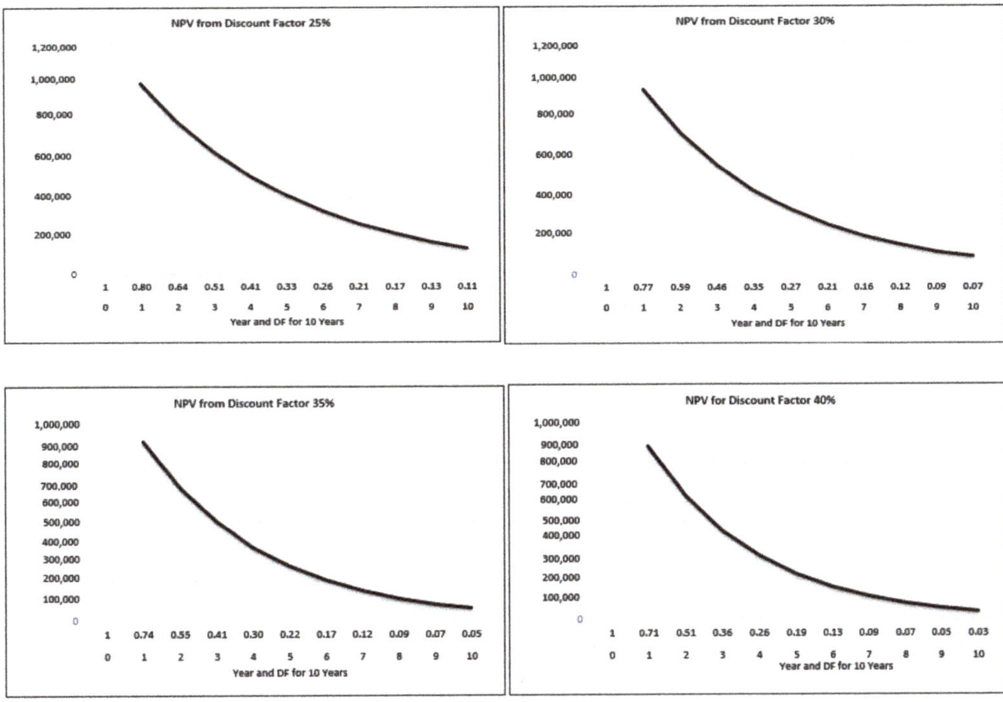


**Figure 13.** Net Present Value (NPV) for 10 Years Investment with various discount factor.

Furthermore, Figure 14 shows the Capital Recovery with three interest types such as 5%, 10%, and 15%. The CRF factor for an interest rate of 0.1 is 0.16, which was realized using the formula [2].

### 4.2. Economic Analysis

In accordance with the data acquired from the retrofit vessel, the new build 600 TEU that uses MFO and LNG fuels are shown in Figure 15. It is evident that the retrofit vessel tends to have a good competitive value compared to the MFO and LNG fuel used in the new build vessel. For the new build LNG fueled vessel, assume the OPEX is similar to the retrofit; the design will use an FGSS gas combustion unit system. The LCC of three 600 TEU container types is shown in Figure 15.

Figure 15 shows that the retrofit vessel with CAPEX on the FGSS only provides low cost with respect to the economic analysis. New build container vessels with LNG fuel consider the initial capital cost compared to the one that uses MFO. However, the OPEX cost on LNG fuel vessels continues to decrease while the vessel experiences low cost compared to the one that uses MFO. The future trend is cost-efficient for LNG fuel vessels.

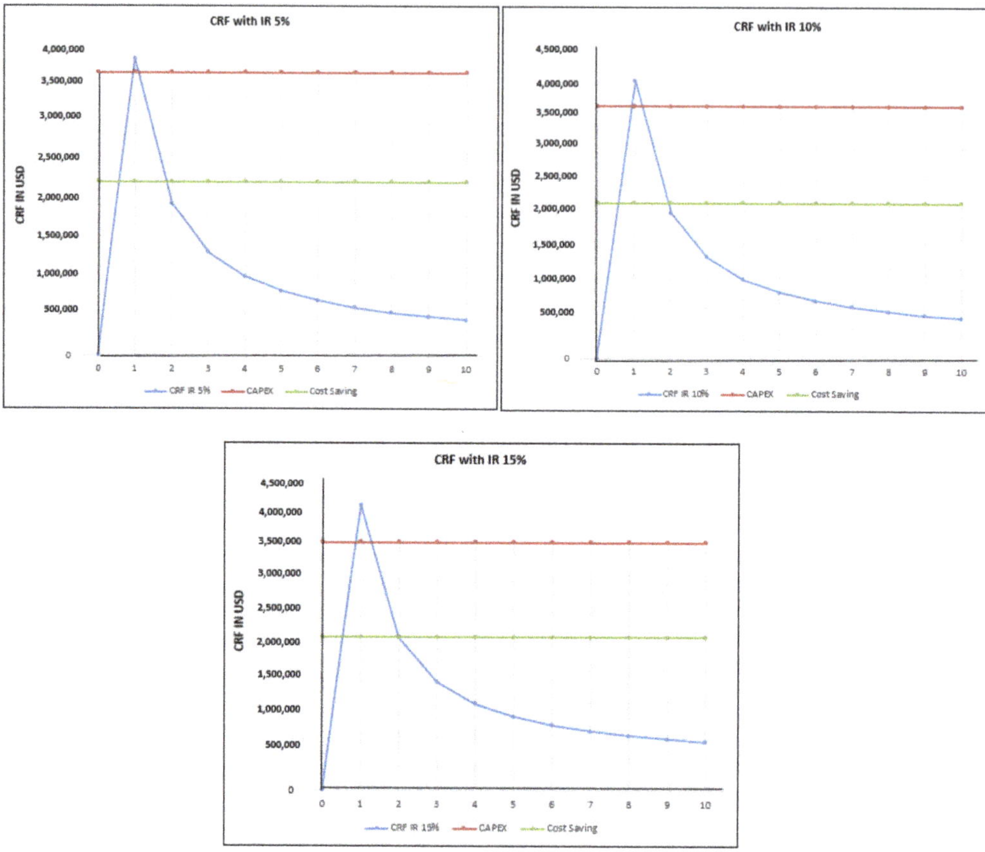


**Figure 14.** Capital Recovery Factor for 600 TEU with various interest.

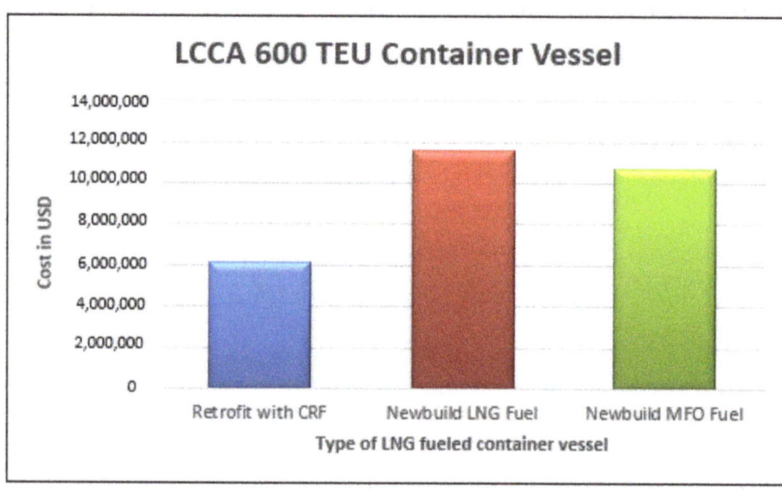


**Figure 15.** LCCA 600 TEU container.

### *4.3. Sensitivity Analysis*

The sensitivity analysis shows the life cycle investment using the retrofit method for the transition process to LNG fuel, alongside some factors that influence the evaluation.

#### 4.3.1. Selection of Technology

The selection of technology has an important impact on retrofit. FGSS with GCU provides low-cost investment while the implemented advanced system depends on the sailing time of the 600 TEU container. Boil-off gas is one of the factors irrespective of whether or not a longer sailing time would have an impact on its loss.

#### 4.3.2. LNG Prices

LNG prices are also a critical analytical factor. The increasing LNG prices also have an impact on the overall LCC analysis. However, its uncertainty is one factor that needs to be considered in the present analysis.

## 5. Conclusions and Recommendations

The prospect of LNG as a marine fuel in Indonesia is growing because LNG is the most advanced technology for alternative marine fuel compared to other alternatives such as hydrogen, methanol and LPG. Furthermore, in terms of investment, LNG has shown good cost efficiency in long-run operations. In vessel design, the ship owner has the option to choose to retrofit technology for their current fleet instead of purchasing new vessels. The life cycle analysis of the retrofit 600 TEU container showed that the retrofit will bring low operational costs for vessel owners. This is aside from the investment, which is mostly for the FGSS on board the vessel. It helps owners to know when to use a retrofit to purchase a new build during the acquisition of a vessel. The FGSS with gas combustion unit is the first option to consider during the selection of technology. However, three comparisons made between retrofit and the other two new build shows that retrofit was the recommended option; the cost of retrofit of USD 6,156,058 is lower than the other two options for LNG fueled 600 TEUs container ship's new build (USD 11,547,860) and MFO fueled 600 TEUs container ship's new build (USD 10,702,872). Other savings that shipowners can obtain from retrofit is less time in dry dock for conversion from current MFO fueled to LNG fueled. The more time that the shipowner can save will provide an opportunity cost for the vessel to return to operation and generate income. LCCA is a tool used for life cycle and low-cost analysis with respect to retrofit investment. This analysis will be affected by LNG prices, especially when the uncertain price of LNG will bring a change in analysis.

LNG is one of the advanced technologies of alternative fuel and several studies proved that it is the most reliable energy source. From the economic analysis, it was discerned that LNG as a marine fuel reduces maintenance and spare part costs. With variable interest rates, the capital recovery factor shows a decrease in payment. The maintenance cost takes significant consideration due to the usage of LNG as fuel. The 600 TEU container vessel capital recovery result served as a reference or guide to vessel owners to be committed to using green fuels such as LNG.

This study already provided information about the challenges that Indonesian vessel owners face when they want to implement green alternative fuels. Some of these challenges are centered on technology, investment and potential profit. However, this is an opportunity for vessel owners to consider the use of LNG as marine fuel due to its long-term impact on cost efficiency and operating activities.

**Author Contributions:** Conceptualization, R.B. and R.O.S.G.; methodology, R.B.; software, R.B.; validation, R.B., R.O.S.G. and S.; formal analysis, R.B.; investigation, R.B.; resources, R.B. and R.O.S.G.; data curation, R.B.; writing—original draft preparation, R.B.; writing—review and editing, R.B. and R.O.S.G.; visualization, R.B.; supervision, R.O.S.G.; project administration, R.B.; funding acquisition, R.B. All authors have read and agreed to the published version of the manuscript.

**Funding:** This research received no external funding.

**Institutional Review Board Statement:** Not applicable.

**Informed Consent Statement:** Not applicable.

**Data Availability Statement:** Not applicable.

**Conflicts of Interest:** The authors declare no conflict of interest.

## References

1. Arefin, M.A.; Nabi, M.N.; Akram, M.W.; Islam, M.T.; Chowdhury, M.W. A Review on Liquefied Natural Gas as Fuels for Dual Fuel Engines: Opportunities, Challenges and Responses. *Energies* **2020**, *13*, 6127. [CrossRef]
2. Smyshlyaeva, K.I.; Rudko, V.A.; Povarov, V.G.; Shaidulina, A.A.; Efimov, I.; Gabdulkhakov, R.R.; Pyagay, I.N.; Speight, J.G. Influence of Asphaltenes on the Low-Sulphur Residual Marine Fuels' Stability. *J. Mar. Sci. Eng.* **2021**, *9*, 1235. [CrossRef]
3. Povarov, V.G.; Efimov, I.; Smyshlyaeva, K.I.; Rudko, V.A. Application of the UNIFAC Model for the Low-Sulfur Residue Marine Fuel Asphaltenes Solubility Calculation. *J. Mar. Sci. Eng.* **2022**, *10*, 1017. [CrossRef]
4. Gl, D.N.V. Maritime Forecast to 2050, Energy Transition Outlook 2019. 2019. Available online: www.dnvgl.com/maritime (accessed on 8 January 2023).
5. Sharples, J. *LNG Supply Chains and the Development of LNG as a Shipping Fuel in Northern Europe*; The Oxford Institute for Energy Studies: Oxford, UK, 2019.
6. Hwang, S.; Jeong, B.; Jung, K.; Kim, M.; Zhou, P. Life Cycle Assessment of LNG Fueled Vessel in Domestic Services. *J. Mar. Sci. Eng.* **2019**, *7*, 359. [CrossRef]
7. Jafarzadeh, S.; Paltrinieri, N.; Utne, I.B.; Ellingsen, H. LNG fuelled fishing vessels; a system engineering approach. *Transp. Res. Part D Transp. Environ.* **2017**, *50*, 202–222. [CrossRef]
8. Dimitrellou, S.C.; Strantzali, E.; Pagonis, D.N.; Livanos, G.A. *Retrofit of Ro-Ro Passenger Ferry to Operate on LNG Fuel: A Greener and Safe Solution for Short Sea Transportation*; Research Gate: Berlin, Germany, 2020.
9. Berkowitz, N. *Fossil Hydrocarbon*; Academic Press: Cambridge, MA, USA, 1997.
10. Oakey, J. *Fuel Flexible Energy Generation: Solid, Liquid and Gaseous Fuels*; Woodhead Publishing: Soston, UK, 2016.
11. Yun, S.; Lee, J. Advancing Societal Readiness toward Renewable Energy Adoption with a Socio-Technical Perspective. *Technol. Forecast. Soc. Chang.* **2015**, *95*, 170–181. [CrossRef]
12. Cheenkachorn, K.; Poompipatpong, C.; Ho, C.G. Performance and emissions of heavy-duty diesel engine fuelled with diesel and LNG. *Energy* **2013**, *53*, 52–57. [CrossRef]
13. Buades, L.S. Implementation of LNG as Marine Fuel in Current Vessels. Perspectives and Improvements on Their Environmental Efficiency. Master's Thesis, Universitat Politecnica De Catalunya Barcelona, Barcelona, Spain, 2017.
14. Chen, S.; Zheng, S.; Zhang, Q. Investment decisions under certainty on LNG-powered vessel for environmental compliance. *J. Shipp. Trade* **2018**, *3*, 5. [CrossRef]
15. Du, Z.; Chen, Q.; Guan, C.; Chen, H. Improvement and Optimization Configuration of Inland Ship Power and Propulsion System. *J. Mar. Sci. Eng.* **2023**, *11*, 135. [CrossRef]
16. Elkafas, A.; Khalil, M.; Elgohary, M.; Shouman, M. Environmental protection and energy efficiency improvement by using natural gas fuel in maritime transportation. *Environ. Sci. Pollut. Res.* **2021**, *28*, 60585–60596. [CrossRef] [PubMed]
17. Banaszklewicz, T.; Chorowski, M.; Gizicki, W.; Jedrusyna, A.; Kielar, J.; Malecha, Z.; Piotrowska, A.; Polinski, J.; Rogala, Z.; Sierpowski, K.; et al. Liquefied Natural Gas in Mobile Application. *Energies* **2020**, *13*, 5673. [CrossRef]
18. Fathallah, A.Z.M.; Ariana, I.M.; Putra, G.H.; Ma'ruf, B. Conceptual design of the LNG dual fuel system for Harbour Tug towards Indonesia Greenport. In *IOP Conference Series: Earth and Environmental Science*; IOP Publishing: Bristol, UK, 2021.
19. Kim, H. A Case Study: An Economic Evaluation of Liquefied Natural Gas (LNG) Fuel for New Ships of Korean Ship Owners. Master's Thesis, World Maritime University, Malmö, Sweden, 2017.
20. Wang, C.; Ju, Y.; Fu, Y. Comparative Life cycle cost analysis of low-pressure fuel gas supply system for LNG fueled ship. *Energy* **2020**, *218*, 119541. [CrossRef]
21. Burel, F.; Taccani, R.; Zuliani, N. Improving sustainability of maritime transport through utilization of liquefied natural gas (LNG) for propulsion. *Energy* **2013**, *57*, 412–420. [CrossRef]
22. Executive, T.M. Retrofit Concepts to Reduce Cost of LNG Conversion and Operations. 2022. Available online: https://www.maritime-executive.com/article (accessed on 8 January 2023).
23. Kou, Y.; Luo, M. Market driven ship investment decision using the real option approach. *Transp. Res. Part A Policy Pract.* **2018**, *118*, 714–729. [CrossRef]
24. Kontan. Samudera Indonesia Mengoperasikan Satu Kapal Baru Berkapasitas 600 TEUs. 2018. Available online: Investasi.kontan.co.id/news (accessed on 10 January 2022).
25. Percic, M.; Frovic, L.; Puksec, T.; Cosic, B.; Li, O.L.; Vladimir, N. Life-cycle assessment and life-cycle cost assessment of power batteries for all electric vessel for short sea navigation. *Energy* **2022**, *251*, 123895. [CrossRef]

26. Favi, C.; Raffaell, R.; Germani, M.; Gregori, F.; Manieri, S.; Vita, A. A lifecycle model to assess costs and environmental impacts of different maritime vessel typologies. In Proceedings of the ASME 2017 International Design Engineering Technical Conferences and Computers and Information in Engineering Conference, Cleveland, OH, USA, 6–9 August 2017.
27. Hsu, P.C. A Lifecycle Cost Analysis of Using Alternative Technologies on Short Sea Shipping Vessels in ECAs. Master's Thesis, University of Rostock, Rostock, Germany, 2014.
28. Park, K.S.; Seo, Y.J.; Kim, A.R.; Ha, M.H. Ship Acquisition of Shipping Companies by Sale & Purchase Activities for Sustainable Growth: Exploratory Fuzzy-AHP Application. *Sustainability* **2018**, *10*, 1763.
29. Snyder, J. Matson Places US$1Bn Order for Three LNG Fuelled Box Ships. 2022. Available online: Rivieramm.com/news-content-hub (accessed on 10 January 2023).
30. Yoo, B.-Y. Economic assessment of liquefied natural gas (LNG) as a marine fuel for $CO_2$ carriers compared to marine gas oil (MGO). *Energy* **2017**, *121*, 772–780. [CrossRef]
31. Parikh, A.; Nikolopoulos, L.; Robertsson, H.; Frorup, U. Alternative fuels: Retrofitting ship engine. *Insight Brief. Ser.* **2021**.
32. Jakobsen, O.; Manager, S.P. The Retrofit Project: Retrofitting to Reduce $CO_2$ Emission—A Case Study of Three Different Vessels. 2020. Available online: www.greenship.org (accessed on 10 January 2023).
33. Tam, I.C.; Dev, A.; Ng, C.; Deltin, L. Concept design of a bulk carrier retrofit with LNG fuel. In Proceedings of the International Conference on Marine Engineering and Technology Oman, Muscat, Oman, 5–7 November 2019.
34. Lagemann, B.; Lindstad, E.; Fagerholt, K.; Rialland, A.; Erikstad, S.O. Optimal ship lifetime fuel and power system selection. *Transp. Res. Part D* **2022**, *102*, 103145. [CrossRef]
35. Zhang, B.; Jiang, Y.; Chen, Y. Research on calibration economy and PM emission of a marine LNG-diesel dual fuel engine. *J. Mar. Sci. Eng.* **2022**, *10*, 239. [CrossRef]
36. Al-Breiki, M.; Bicer, Y. Comparative cost assessment of sustainable energy carriers produced from natural gas accounting for boil-off gas and social cost of carbon. *Energy Rep.* **2020**, *6*, 1897–1909. [CrossRef]
37. Index, M.M. 2021. Available online: www.moore-index.com (accessed on 9 January 2023).
38. Le Fevre, C. *A Review of Demand Prospects for LNG as a Marine Transport Fuel*; The Oxford Institute for Energy Studies: Oxford, UK, 2018.
39. DNV. Current Price Development (Last Prices as of 28 February 2021). 2021. Available online: https://www.dnv.com/maritime/insights/topics/lng-as-marine-fuel/current-price-development-oil-and-gas.html (accessed on 15 January 2023).
40. Watanabe, K. High operation capable marine dual fuel engine with LNG. *J. Jpn. Inst. Mar. Eng.* **2014**, *50*, 738–743. [CrossRef]
41. Wartsila. LNG as a Marine Fuel Boot Profitable While Ensuring Compliance, Business White Paper. 2017. Available online: http://www.wartsila.com/services (accessed on 15 January 2023).
42. Elkafas, A.; Elgohary, M.; Shouman, M. Numerical analysis of economic and environmental benefits of marine fuel conversion from diesel oil to natural gas for container ship. *Environ. Sci. Pollut. Res.* **2020**, *28*, 15210–15222. [CrossRef] [PubMed]
43. Banawan, A.A.; El Gohary, M.M.; Sadek, I.S. Environmental and economical benefits of changing from marine diesel oil to natural gas fuel for short voyage high power passenger ship. *J. Eng. Marit. Environ.* **2010**, *224*, 103–113. [CrossRef]

applied sciences

MDPI

Article

# Determination of Benefits of the Application of CMMS Database Improvement Proposals

Ladislav Stazić *, Nikola Račić, Tatjana Stanivuk and Đorđe Dobrota

Faculty of Maritime Studies, University of Split, 21000 Split, Croatia
* Correspondence: lstazic@pfst.hr; Tel.: +385-(0)21-619-467

**Featured Application: This article concludes the study of CMMS databases and the measures the authors developed to improve data quality in these databases. It includes a calculation of the benefits that the proposed measures can have for improving data quality in CMMS databases. The proposed measures have already been published in several articles.**

**Abstract:** Computerized maintenance management systems (CMMSs) are software packages that support or organize the maintenance tasks of assets or equipment. They are found in the background of any ship maintenance operation and are an important part of maintenance planning, spare parts supply, record keeping, etc. In the marine market, there are a number of CMMSs that are competing fiercely to program a better and more modern program that will capture the market, which has been accompanied by published analyses and scientific papers. At the same time, the quality of the data entered into CMMS databases is questionable, a fact that has been ignored in practice and scientific circles; until recently, there were no published analyses and there was no way to measure the quality of the data entered. This article presents two proposals for improving the quality of CMMS databases and calculates their potential benefits. By implementing the first proposal, the evaluation methodology for the ship's Planned Maintenance System database, between 10% and 15% of databases will have significant financial or safety benefits. This measure will also have an impact on more than 40% of the other databases that can also be improved. The second proposal will have a smaller impact of only 4%. The overall benefit of these proposals is to improve more than 60% of the databases and will result in a significant increase in safety or financial savings.

**Keywords:** computerized maintenance management systems; planned maintenance; database; benefits; quality

**Citation:** Stazić, L.; Račić, N.; Stanivuk, T.; Dobrota, Đ. Determination of Benefits of the Application of CMMS Database Improvement Proposals. *Appl. Sci.* **2023**, *13*, 2731. https://doi.org/10.3390/app13042731

Academic Editor: José A. Orosa

Received: 31 January 2023
Revised: 18 February 2023
Accepted: 19 February 2023
Published: 20 February 2023



## 1. Introduction

Ship maintenance is one of the most researched topics in the industry, and numerous articles have been published on its various aspects [1–3]. An important part of the organization of successful maintenance is performed with the help of CMMS (Computerized Maintenance Management System). The term started long ago as a simple Planned Maintenance System (PMS) and gradually evolved into computerized systems with many modules and multiple functions. Today, there are many different computer programs for CMMS in the maritime industry, the total number of which is estimated to be more than 70. These systems differ in design, quality, and functionality.

PMS and CMMS as tools to reduce downtime and maintenance costs have been widely researched [4–6]. Research has shown that the adoption of PMS brought tremendous financial and safety benefits, and the adoption of CMMS continued this process [7]. At the same time, it is very difficult to find data to measure the benefits that have resulted from the introduction of both systems. The rare values published in scientific articles vary considerably, explaining improvements in maintenance from 30 to 50% (variations of more than 50%) depending on the example (case studied) [8,9].

PMS in paper form was a significant step in improving maintenance and enhancing ship safety. The introduction of CMMSs in shipping brought improvements in terms of ease and speed of communication with the office, easier monitoring of maintenance and procurement, or simplicity of data exchange. Since the communication is mostly done via a satellite link, the size of the exchanged data packets must be very small, usually less than 200 kb [10]. The size of the data packet rarely exceeds the specified values even in the case of major changes to the database. This small size of data packets allowed the introduction of CMMS applications running in the cloud and becoming more and more popular in this market [10].

CMMSs or, as they are also known, computerized Planned Maintenance Systems (PMSs) are in daily use on a wide variety of ships. Although they are widely used, there is neither adequate scientific follow up of these systems nor systematic analysis of the systems and their data. Planned maintenance in shipping was addressed in scientific articles in the late 20th century, mainly in Europe and North America [11–13]. The research topics at that time focused mainly focused on the application of CMMS and aspects of the system used. Today, authors still analyze and research similar topics [14,15]. Another frequently researched topic is the performance of different CMMSs and their comparison [16,17].

Alan Mortimer, a former UK Chief Engineer, echoing various opinions on the quality of CMMS, wrote: "Commercial Planned Maintenance (PM) systems are a collection of very variable beasts, some good, some bad, and some indifferent" [18].

Although this opinion is widely held in the maritime industry, it is hard to believe that there are products (in this case, computer programs) on the commercial market (i.e., they have survived competition) that are poor and do not meet the needs of users. According to this statement, the research team assumed that the cause of the problem can be found in different places and consists of two known facts. In their research, the researchers came across two claims that together describe the problem much better. The first possible cause is declared by Davies, who states that computerization of poor management systems only leads to poor results more quickly [19]. The second possible cause is the well-known fact of the GI–GO (Garbage In–Garbage Out) effect, which is well described in the article by Kilkenny and Kerin [20].

This assumption that poor databases are the root cause of all CMMS problems formed the basis for the research conducted by the CMMS research team at the Maritime Faculty in Split. A large number of CMMS databases had to be examined and analyzed to verify this assumption. The quality of databases and their impact have been studied by many authors [21–24], but only for the land industry. This research topic is very limited or non-existent in the maritime industry.

The first discovery at the beginning of the research was that a large number of ship databases have very poor data and numerous problems. At this point, the team faced a major problem, a major challenge to solve. Although it was clear that the databases were in poor condition, their conclusion was based only on subjective opinion and personal experience. There was no tool or method in the industry to evaluate CMMS databases, measure their quality, identify areas for improvement, and determine the steps needed to improve database quality.

To solve the problem, the team's first task was to develop a universal tool to assess the quality of the CMMS database. The main method used to create the new tool was DQA (Data Quality Assessment), shown in Figure 1, which is based on the idea designed by Pipino et al. [25].

DQA is a methodology developed to provide the general principles for the definition of data quality metrics [26] and the method; according to the authors of the cited text (Batini et al. and Ballou et al.) [27], the main characteristic of the methodology is that it is tailor-made, created specifically for each task. The solution where "one size fits all" in different circumstances cannot be a solution [26]. There are many examples of the DQA methodology in practice and the use of the methodology for different aspects and different types of research [28,29].

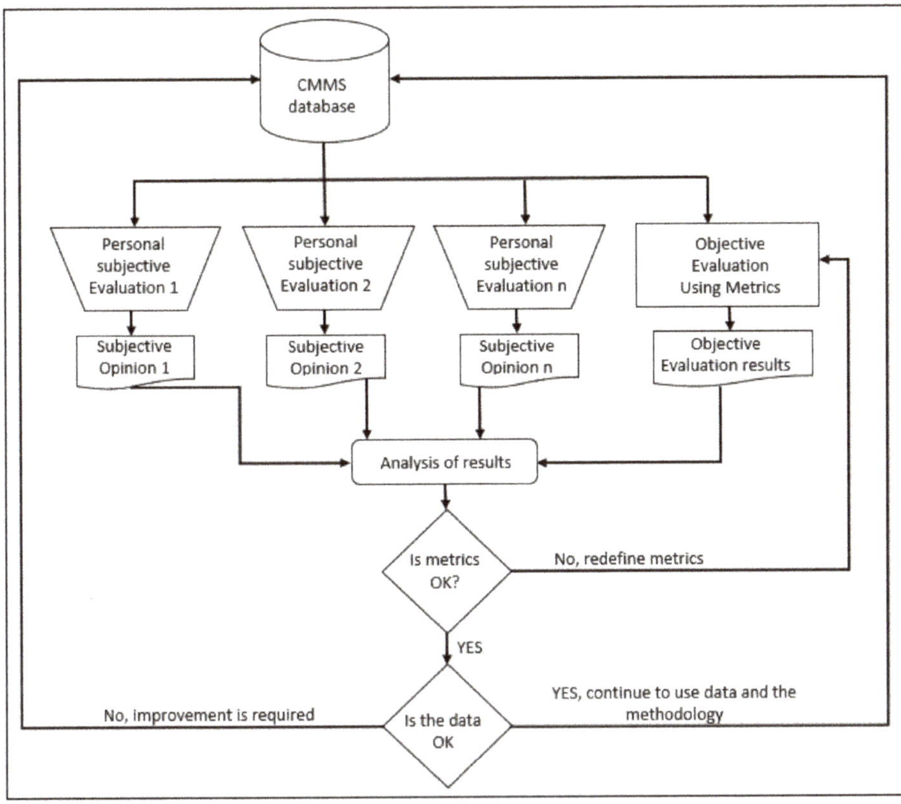


**Figure 1.** DQA in practice, based on [25].

The research team encountered an interesting problem in studying databases to determine how the database improvement proposal program works. In examining 17 vessels from two companies, seven similar improvement requests were found on three vessels, each claiming that there were no manufacturer's maintenance schedules on board and requesting that the company provide them. The number of improvement requests for this type of deficiency is relatively low, mainly due to the fact that both companies only purchase new vessels. This type of issue often occurs when a company buys a used vessel and the previous crew takes all the operating manuals with them, along with the maintenance logs, data, etc., so the new crew starts from scratch, often without the manufacturer's operating manuals. These seven deficiencies were identified during the CMMS system implementation phase and then reported to the company, which worked to correct them. Five of these deficiencies were successfully corrected, while two were not. The reason for the failure to correct this issue was not identified, although the company's SMS was reviewed to determine whether it contained instructions or recommendations for correcting this deficiency.

Consequently, in five out of seven cases, the maintenance plan and spare parts were added to the CMMS by copying the data from the manual received, while in two cases, the items were still missing. Reviewing various articles and books, the research team found that no one has yet answered the question of how to create the equipment maintenance plan without using the manufacturer's manual.

From the above, it can be concluded that a significant improvement in database quality (read: maintenance and safety) can be achieved if these two database problems are solved. These tasks are the focus of the research team, and this paper presents the potential benefits of these two solutions. The design and methods used to create the Evaluation Methodology for the Ship PMS are described in Section 2, while the methods used to solve the second problem are explained in Section 3. The results and discussions are presented in Section 4, followed by the Conclusion, which summarizes the overall benefits of applying these proposals and highlights the importance of the research.

## 2. The Evaluation Methodology

Another example of the application of the DQA methodology is the evaluation methodology for the ship's Planned Maintenance System database. The methodology was developed at the beginning of the research to establish firm rules for CMMS database assessment. All DQA assessment strategies [26] were considered when creating the methodology [30]:

- The acquisition of new data;
- The standardization (or normalization);
- The acquisition of links;
- The integration of data and schemas;
- The trustworthiness of the source;
- The localization and correction of errors;
- The cost optimization.

A tool called the Evaluation Methodology for Ship PMS [30] was developed and field tested to verify its functionality. It consists of the questionnaire with thirty questions divided into six groups: Machinery and Equipment, Jobs inside DB, Special jobs and Rules, DB Jobs general, Spare Parts, and Miscellaneous (Table 1). In front of each question, there is a field (mark) indicating the importance of the question for the quality of the database. The "traffic light" principle (R, Y, G) is used to determine the colors in the field and to describe the importance of the question; red mark has the highest importance, and the deficiencies revealed by these questions have a significant impact on the quality of maintenance. Any deficiencies uncovered by these questions should be taken seriously and corrected to improve the database and the quality of maintenance. The questions with the yellow mark are of medium importance. This group of questions has a lower impact on database quality, and the deficiencies revealed by these questions mainly impact user workload, while the impact on maintenance quality and reliability is negligible. The deficiencies revealed by these questions should be corrected due to unnecessary work of staff [31], which may cause aversion to the system. The questions should be answered with a mark from one to five.

**Table 1.** The questionnaire.

| Group | Color | No. | Question |
|---|---|---|---|
| Machinery and equipment | R | 1 | Are all machines and equipment recorded in the database? |
| | R | 2 | Are all items of equipment properly recorded and clearly identified according to their location on board and their marking? |
| | R | 3 | Are all required machines divided into subcomponents (smaller subsystems) in a logical manner? |
| | Y | 4 | Does any machine or equipment have a greater number of subcomponents than necessary? |
| | Y | 5 | Are machines or equipment listed more than once in the database or do they have the same markings or names? |
| | Y | 6 | Are the manufacturer, type, and serial number data entered for all relevant items? |
| | G | 7 | Do all entries for equipment and machinery have the same style, abbreviations, and identifiers? |

**Table 1.** *Cont.*

| Group | Color | No. | Question |
|---|---|---|---|
| Jobs inside DB | | 8 | Is there a linked maintenance schedule for all equipment on DB according to the manufacturer's recommendations? |
| | R | 9 | Are the manufacturer's recommendations organized by equipment, time periods, and company maintenance requirements? |
| | R | 10 | Is all work required by company policy included in DB (e.g., SSM—Safety Management System)? |
| | Y | 11 | Has all work based on manufacturer's recommendations been modified based on company policy (if applicable)? |
| | R | 12 | Is all work required by flag state regulations included in DB? |
| | Y | 13 | Is all work required by the classification society included in DB? |
| | R | 14 | Are there a number of smaller tasks that can be grouped together? |
| Special jobs and rules | R | 15 | Are fire alarm sensors included in DB along with the test plan? |
| | Y | 16 | Is the alarm system and its test program entered in DB? |
| | R | 17 | Is the PMS self-improvement program entered into DB, and is there a control mechanism for the PMS DB self-improvement program? |
| | R | 18 | Is the critical equipment labeled in accordance with the company's SMS? |
| DB jobs general | R | 19 | Are job descriptions clearly and unambiguously stated? |
| | R | 20 | Are jobs created and grouped according to the multiplier principle? |
| | G | 21 | Are all like jobs that originate from different sources synchronized? |
| | Y | 22 | Are all similar jobs originating from different requirements (sources) merged? |
| Spare parts | Y | 23 | Are all required spare parts included in the database? |
| | Y | 24 | Are the spare parts distributed to the correct equipment and machines? |
| | R | 25 | Are all spare parts correctly identified, do they have sufficient data for ordering? |
| | R | 26 | Is the company critical spare parts list inserted in the DB? |
| | R | 27 | Do all spare parts have the same style, abbreviations, markings, etc.? |
| | R | 28 | Are there spare parts that are entered more than once? |
| Miscellaneous | G | 29 | Are all users entered in the DB, are all access rights correctly defined? |
| | Y | 30 | Are there any other deficiencies in the computerized PMS database? |

The marks should have the following meaning:

Mark 1—Completely negative evaluation result;
Mark 2—Predominantly negative evaluation;
Mark 3—Predominantly positive evaluation with a significant number of irregularities;
Mark 4—Predominantly positive evaluation with a small number of irregularities;
Mark 5—Completely positive evaluation.

Questions rated five and four are considered satisfactory and require no changes to the database. Questions rated four have room for improvement, but DB changes are not recommended (there will be no significant quality improvement). Questions rated three, two, or one are considered unsatisfactory and data improvement should be made here. The schedule for data changes in the database should correspond to the color schedule (R, Y, G).

After the development of the methodology, serious efforts were made to test it in practice and to study various aspects of its application. The methodology was used (from 2017 to 2019) to analyze the state and quality of forty-four CMMS databases in five different shipping companies operating different types of vessels (one company operates passenger vessels, two companies operate a mix of bulk carriers and tankers, one company operates bulk carriers, and one company operates VLCCs). Testing of the methodology in different companies, with different working practices and methods, and on different types of vessels has shown that it can be used as a universal tool for evaluating CMMS databases and paper-based PMSs.

After testing, the following claims about the methodology were made and verified:

- The methodology is a useful tool for evaluating CMMS data and databases [30];
- The methodology is easy to use [30];
- The results obtained are reliable [30];
- The application of the methodology reduces the subjectivity of the evaluator [32];
- The application of the methodology facilitates the evaluation of databases [32];
- The application of the methodology makes the evaluation much more detailed [32];
- The application of the methodology facilitates the identification of deficiencies [32].

*Evaluation Results*

The results of the evaluation of forty-four CMMS databases were published in the article [33]. The testing of the functionality of the methodology is described in the same article and an analysis of the related results is presented. Further analysis of the obtained results was not performed, nor was an analysis of the identified deficiencies. Therefore, the necessary conclusions for maintenance planning that could affect the quality of maintenance were not derived from the evaluation. The deficiencies identified during this evaluation are listed in Tables 2 and 3.

**Table 2.** Deficiencies discovered in companies A, B, and C.

| | Minor Def. | | | Major Deficiencies | | | | | | | | |
|---|---|---|---|---|---|---|---|---|---|---|---|---|
| **Grade** | **4** | | | **3** | | | **2** | | | **1** | | |
| **Color** | **R** | **Y** | **G** | **R** | **Y** | **G** | **R** | **Y** | **G** | **R** | **Y** | **G** |
| A1 | 4 | 4 | - | - | - | - | - | - | - | - | - | - |
| A2 | 3 | 3 | - | 3 | 1 | - | 2 | - | 1 | 2 | - | - |
| A3 | 7 | 4 | 1 | 1 | - | - | 1 | - | - | - | - | - |
| B1 | 1 | 2 | - | 1 | 1 | - | - | - | - | - | - | - |
| B2 | 2 | 2 | - | 1 | 1 | - | - | - | - | - | - | - |
| B3 | - | 2 | - | 1 | 1 | - | - | - | - | - | - | - |
| B4 | 1 | 2 | - | 1 | 1 | - | - | - | - | - | - | - |
| B5 | 1 | 2 | - | 1 | 1 | - | - | - | - | - | - | - |
| C1 | 6 | 6 | 2 | 1 | - | - | - | 1 | - | - | - | - |
| C2 | 7 | 6 | 2 | 1 | 1 | - | - | - | - | - | - | - |
| C3 | 5 | 5 | 2 | 1 | 1 | - | - | 1 | - | - | - | - |
| C4 | 3 | 5 | 1 | 4 | 1 | - | - | 1 | - | 1 | - | - |
| C5 | 6 | 6 | - | 1 | 1 | - | - | - | - | - | - | - |
| C6 | 6 | 6 | 2 | 1 | - | - | - | 1 | - | - | - | - |
| C7 | 5 | 5 | 2 | 3 | 3 | - | - | - | - | 2 | 1 | - |
| C8 | 5 | 6 | 1 | 1 | 1 | - | - | - | - | - | - | - |

**Table 3.** Deficiencies discovered in companies D and E.

| | Minor Def. | | | Major Deficiencies | | | | | | | | |
|---|---|---|---|---|---|---|---|---|---|---|---|---|
| **Grade** | **4** | | | **3** | | | **2** | | | **1** | | |
| **Color** | **R** | **Y** | **G** | **R** | **Y** | **G** | **R** | **Y** | **G** | **R** | **Y** | **G** |
| D1 | - | - | 1 | 3 | 1 | - | 2 | 3 | - | 11 | 2 | - |
| D2 | 5 | 4 | 1 | - | - | - | 1 | - | - | 7 | 2 | - |
| D3 | 5 | 6 | - | 1 | - | - | 1 | - | - | 5 | 1 | - |
| D4 | 3 | 2 | 1 | 2 | 3 | - | 2 | - | - | 7 | 2 | - |
| D5 | 5 | 4 | 1 | 2 | 1 | - | 2 | 1 | - | 7 | 2 | - |
| D6 | 3 | 1 | - | - | - | - | 1 | - | - | 7 | 2 | - |
| D7 | 4 | 3 | - | 1 | - | - | 2 | - | - | 4 | 1 | - |
| D8 | 3 | 5 | 1 | 1 | - | - | 1 | 1 | - | 6 | 1 | - |
| D9 | - | 1 | 1 | 3 | 2 | - | 2 | 1 | - | 10 | 1 | - |
| D10 | 5 | 5 | - | 1 | - | - | 1 | - | - | 4 | 1 | - |
| D11 | 3 | 4 | - | - | - | - | 1 | - | - | 8 | 2 | - |
| D12 | 4 | 2 | - | 1 | - | - | 1 | - | - | 7 | 2 | - |
| D13 | 1 | 4 | - | - | - | - | 1 | - | - | 7 | 2 | - |
| D14 | 5 | 4 | - | 1 | 1 | - | 1 | - | - | 5 | 1 | - |
| D15 | 4 | 3 | 1 | - | - | - | 1 | - | - | 7 | 2 | - |
| D16 | 1 | - | 1 | 2 | 3 | - | 3 | 1 | - | 10 | 2 | - |
| D17 | 7 | 3 | 1 | - | 1 | - | 1 | 1 | - | 7 | 2 | - |
| D18 | 6 | 4 | - | 3 | - | - | 1 | - | - | 4 | 1 | - |
| D19 | 3 | 5 | 1 | 4 | - | - | 2 | 1 | - | 6 | 1 | - |
| E1 | 3 | 5 | 1 | - | - | - | - | - | - | - | 1 | - |
| E2 | 2 | 5 | 1 | - | - | - | - | - | - | - | 1 | - |
| E3 | 2 | 5 | 1 | - | - | - | - | - | - | - | 1 | - |
| E4 | 3 | 5 | 1 | - | - | - | - | - | - | - | 1 | - |
| E5 | 4 | 5 | 1 | - | - | - | - | - | - | - | 1 | - |
| E6 | 2 | 5 | 1 | - | - | - | - | - | - | - | 1 | - |
| E7 | 2 | 5 | 1 | - | - | - | - | - | - | - | 1 | - |
| E8 | 4 | 5 | 1 | - | - | - | - | - | - | - | 1 | - |
| E9 | 4 | 5 | 1 | - | - | - | - | - | - | - | 1 | - |

Tables 2 and 3 reflect this breakdown and represent a cumulative analysis of the identified deficiencies. Each row represents a database, while the columns reflect the total number of deficiencies identified, sorted by the scores obtained and indicated by the color of the group.

In accordance with the recommendations for the application of the methodology described above, all deficiencies rated as four are considered minor and no action is required to correct them. Notwithstanding the fact that no action is required, the CMMS can still be improved in these areas. Tables 2 and 3 show that there is not a single area where no deficiencies were identified, i.e., areas can be improved. At the same time, the lowest number of deficiencies was found to be four, and this was in only one database.

Since the scoring methodology recommends ignoring all items rated four (i.e., there is no need for improvement actions in these areas), new tables have been created (Tables 4 and 5) that include only deficiencies rated three or worse, and all green and yellow boxes have been removed. These tables still contain a very large number of databases and a large number of deficiencies.

**Table 4.** Serious deficiencies in databases A, B and C.

| | Grades | | | |
|---|---|---|---|---|
| **Database** | **1** | **2** | **3** | **Total** |
| A2 | 3 | 2 | 2 | 7 |
| A3 | 1 | 1 | - | 2 |
| B1 | 1 | - | - | 1 |
| B2 | 1 | - | - | 1 |
| B3 | 1 | - | - | 1 |
| B4 | 1 | - | - | 1 |
| B5 | 1 | - | - | 1 |
| C1 | 1 | - | - | 1 |
| C2 | 1 | - | - | 1 |
| C3 | 1 | - | - | 1 |
| C4 | 4 | - | 1 | 5 |
| C5 | 1 | - | - | 1 |
| C6 | 1 | - | - | 1 |
| C7 | 3 | - | 2 | 5 |
| C8 | 1 | - | - | 1 |

**Table 5.** Serious deficiencies in databases D and E.

| | Grades | | | |
|---|---|---|---|---|
| **Database** | **1** | **2** | **3** | **Total** |
| D1 | 3 | 2 | 11 | 16 |
| D2 | - | 1 | 7 | 8 |
| D3 | 1 | 1 | 5 | 7 |
| D4 | 2 | 2 | 7 | 11 |
| D5 | 2 | 2 | 7 | 11 |
| D6 | - | 1 | 7 | 8 |
| D7 | 1 | 2 | 4 | 7 |
| D8 | 1 | 1 | 6 | 8 |
| D9 | 3 | 2 | 10 | 15 |
| D10 | 1 | 1 | 4 | 6 |
| D11 | - | 1 | 8 | 9 |
| D12 | 1 | 1 | 7 | 9 |
| D13 | - | 1 | 7 | 8 |
| D14 | 1 | 1 | 5 | 7 |
| D15 | - | 1 | 7 | 8 |
| D16 | 2 | 3 | 10 | 15 |
| D17 | - | 1 | 7 | 8 |
| D18 | 3 | 1 | 4 | 8 |
| D19 | 4 | 2 | 6 | 12 |
| E1 | - | - | 1 | 1 |
| E2 | - | - | 1 | 1 |
| E3 | - | - | 1 | 1 |
| E4 | - | - | 1 | 1 |
| E5 | - | - | 1 | 1 |
| E6 | - | - | 1 | 1 |
| E7 | - | - | 1 | 1 |
| E8 | - | - | 1 | 1 |
| E9 | - | - | 1 | 1 |

The analysis of Tables 2–5 shows that the analyzed databases have a very large number of deficiencies; in total, there were 220 major deficiencies in the analyzed databases, of which 47 were rated one, 30 were rated two, and 143 were rated three.

Further reflection on the results presented in Tables 1–4 leads to the following findings:

- When evaluating the databases based on methodology, deficiencies were found in all of the databases examined;

- The identified deficiencies varied, some were minor and insignificant, others were very serious;
- Only one of the investigated companies had no red deficiency, and only one deficiency in the yellow group showed that the system in this company was seriously monitored;
- There was a large number of databases that require immediate repair actions (more than 77% of the examined databases);
- On average, there were more than six serious deficiencies per database (to be exact, there were 6.2!!!).

Further review of the assessment results showed that Company D was not paying enough attention to the CMMS, i.e., it had not recognized the benefits that the system can provide.

These poor assessment results show that the CMMS in Company D was neglected both in the offices and on the ships.

Since it is the largest of the companies studied with a large number of vessels, these results could affect the objectivity of the entire research. In order to obtain the most objective picture, the results of the evaluation of Company D's vessels were excluded from the final consideration.

After excluding Company D's vessels from the analysis, the following picture emerges:

- Minor or major deficiencies were found in all the databases examined;
- There was a large number of databases where immediate repair actions were required (in 60% of the analyzed databases);
- A percentage of 63% of all serious deficiencies concerned only four vessels;
- The average number of serious deficiencies was only two;
- Only one database had missing components (4%);
- Only one database was found to lack an adequate maintenance plan (4%).

It can be concluded that more than 60% of all databases could be improved, 16% of them in more than one area. The results of this analysis show that only 1/3 of CMMS databases were in good condition. These poor results were not unexpected, because the only other information found about the condition of CMMS databases of ships declared 1/4 of the databases to be good [34].

## 3. CMMS Development Problem

A possible solution to the missing books problem was published in two articles [1,35], the first [1] showing the preparation of the methodology and the second [35] showing the creation of the maintenance plan.

Fault Tree Analysis (FTA) [36,37] is a widely used method for evaluating the reliability of systems [38], which is used either as static or dynamic. The method is also used to analyze fault causes, improve early fault detection, and improve fault diagnosis during engine operation by reducing false conclusions and inappropriate corrective actions [39]. In this part of the study, the method is used to analyze the turbocharger system of marine diesel engines to identify possible faults in the turbocharger system and determine areas (components) that should be serviced. In this study, the faults identified with the FTA analysis are simulated using the Wartsila-Transas 5000 engine room simulator on the propulsion system of the tanker LCC (Aframax) with the main engine MAN B&W 6560 MC-C [40]. The use of the E/R simulator together with the FTA simplifies the preparation of the fault list and allows its verification from different working aspects.

By combining these two tools, a comprehensive fault list of the turbocharger system of marine diesel engines is created and analyzed in detail. The article [1] once again shows that FTA is extremely useful and practical in analyzing system reliability, energy efficiency, and maintenance costs.

After making a comprehensive list of the faults of the turbocharger system of a marine diesel engine and analyzing what maintenance work needs to be done to avoid these faults, it was necessary to derive the maintenance schedule for the system from the fault list. Each fault from the FTA list is analyzed, and then appropriate preventive maintenance activities are assigned to prevent the occurrence of each fault, resulting in a detailed maintenance plan for the turbocharger system of the marine diesel engine. Several maintenance plans were prepared by the experiment participants (authors of the articles), and each author used his or her own (personal) experience in marine engineering to prepare the maintenance plan.

These plans were compared and a slight variation was found in the maintenance plans for different tasks. These differences are attributed to the different experiences and practices of the authors [41]. To verify the obtained results, the maintenance plans prepared by the authors using the FTA list were compared with the maintenance instructions for the turbocharger system of the marine diesel engine [42,43]. The comparison showed that these schedules differ slightly from the manufacturer's maintenance recommendations, but the overall verdict is that they are very similar and the end goal is achieved in both cases.

The conclusion of this part and the contribution to the overall objective is to show that FTA combined with engineering experience can be a substitute for missing manufacturer's maintenance recommendations when creating the CMMS database. Although the newly created maintenance plan is not the same as the manufacturer's recommended plan, it is very close to it and is a good substitute for it.

## 4. Benefits of These Proposals

The first step in improving the entire CMMS system is a detailed review of the database and the data it contains using the Evaluation Methodology for Ship PMS (Figure 2).

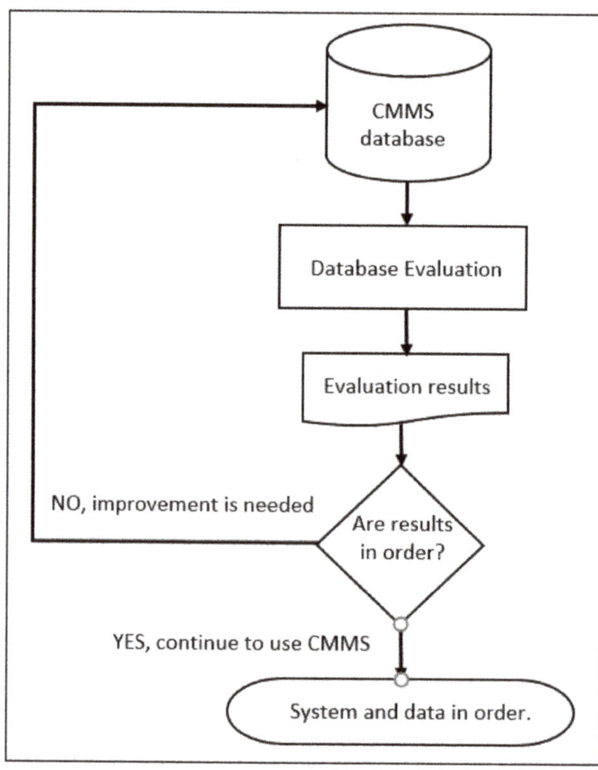


**Figure 2.** CMMS database evaluation process.

This will uncover all the data needed for the improvement effort. This requires expertise, i.e., a good knowledge of the computer programs used and a good knowledge of seamanship, more specifically, marine engineering.

The evaluation methodology for a ship's PMS [30] should be applied during the development of the CMMS database and during the use of the system to avoid deficiencies of the database and to allow proper use with all its benefits. They are relatively easy to calculate using the basic equation:

$$B = \frac{Nsd}{Nq \times Nv} \tag{1}$$

where:
*Nsd*—The number of discovered deficiencies;
*Nq*—Total number of questions;
*Nv*—The number of analyzed vessels.

The application of the methodology will result in the following:

- There will be 4% fewer databases with missing equipment, which will increase maintenance reliability and reduce corrective maintenance;
- There will be 7% fewer databases with missing work orders, which will increase maintenance reliability and decrease corrective maintenance actions;
- There will be 13% fewer databases of missing spare parts, which will increase inventory accuracy, resulting in financial savings and increased vessel safety;
- More than 60% of the databases will have improved data and fewer discrepancies, giving the crew better insight into the system;
- Overall costs will be significantly reduced as fewer repairs will need to be made and/or fewer emergency spare parts will need to be ordered.

In order to calculate the benefit of the second part of the research (missing books problem), it is necessary to determine how many books are still missing when the database is created. According to two database factories (companies that specialize in creating databases), this number varies. It depends on whether the ship is new or used, whether the data is in electronic or paper form, and where the ship was built, etc.

By studying all available databases according to [33] and calculating the number of these cases compared to the number of ship equipment, the estimated benefit of this part of the research will be the potential improvement of 4% of the databases (4% of the equipment will have a maintenance plan that will allow better maintenance of these systems).

The given value was calculated for newbuildings (all analyzed ships were taken as newbuildings), and the value of solving the problem of missing books for second hand ships remains open as a task for future analysis.

## 5. Conclusions

This paper presents two solutions to improve data quality in the CMMS database. The first and far more significant improvement proposal is the evaluation methodology of the ship's PMS, which allows a clear evaluation of the data quality and the identification of areas in the database that can be improved in order to improve the overall maintenance process. The significance of this proposal is that, for the first time, a tool has been created to clearly assess whether the CMMS data is valid and whether the assessment results are the same or similar, even if different people perform the assessment. By incorporating the vessel PMS evaluation methodology into the design and daily use of the CMMS database, the potential benefits described in Section 3 can result in thousands of dollars in maintenance savings if maintenance is not properly adjusted. At the same time, the impact on the safety of the vessel, crew, cargo, and environment can be measured in extremely large amounts (millions or more) if maintenance of certain equipment is properly adjusted and/or performed. The side effect of applying the methodology and improving the quality of the data in the database is to demonstrate to the crew that the CMMS is an important system on board and that it receives the attention it deserves, which further motivates the crew to work with the system on a daily basis. An accurate calculation of the value of this proposed improvement is reflected in the expected improvement of up to 60% of all CMMS databases, including up to 16% in more than one area.

The second proposed improvement is seemingly insignificant, but it is very useful in the case of second-hand vessels, especially those built in failed shipyards or equipped with equipment manufactured by failed companies. The actual financial impact of this proposal is very difficult to calculate after the fact, since each of the possible events can be expressed differently. The benefits of solving the missing books problem calculated in this paper are small, but not insignificant. According to the calculations in this paper, this benefit amounts to 4% of the equipment that will benefit from this proposal, i.e., 4% fewer potential failures and 4% less probability of severe damage.

The calculation of the benefits from the application of these proposals has been made very conservatively, assuming lower values for improvements and for vessels that are purchased as newbuildings. Regardless of how the benefits of these two proposals are calculated, it is clear that both proposals will reduce deficiencies in more than 60% of the databases, improve vessel maintenance, and increase vessel safety.

The main problem with the proposed methods is their current status. Despite the great potential for improvement and the fact that they are publicly available, they are not widely used in practice. The only demonstrated use in practice are the companies that the team contacted personally and the companies that acted as test companies. The next steps the team should take are to analyze why the measures have not been expanded and what should be done to expand their use.

**Author Contributions:** Conceptualization, L.S. and N.R.; methodology, L.S., N.R. and T.S.; validation, Đ.D.; formal analysis, L.S. and T.S.; investigation, L.S.; writing—original draft preparation, L.S., writing—review and editing, L.S., N.R., T.S. and Đ.D.; supervision, N.R., T.S. and Đ.D. All authors have read and agreed to the published version of the manuscript.

**Funding:** This research received no external funding.

**Institutional Review Board Statement:** Not applicable.

**Informed Consent Statement:** Not applicable.

**Data Availability Statement:** No new data were created or analyzed in this study. Data sharing is not applicable to this article.

**Conflicts of Interest:** The authors declare no conflict of interest.

## References

1. Knežević, V.; Orović, J.; Stazić, L.; Čulin, J. Fault tree analysis and failure diagnosis of marine diesel engine turbocharger system. *J. Mar. Sci. Eng.* **2020**, *8*, 1004. [CrossRef]
2. Liu, S.; Chen, H.; Shang, B.; Papanikolaou, A. Supporting Predictive Maintenance of a Ship by Analysis of Onboard Measurements. *J. Mar. Sci. Eng.* **2022**, *10*, 215. [CrossRef]
3. Frangopoulos, C.A. Developments, Trends, and Challenges in Optimization of Ship Energy Systems. *Appl. Sci.* **2020**, *10*, 4639. [CrossRef]
4. Munyensanga, P.; Widyanto, S.A.; Aziz, M.N.A.; Rusnaldy, P. Information management to improve the effectiveness of preventive maintenance activities with computerized maintenance management system at the intake system of circulating water pump. *Procedia CIRP* **2018**, *78*, 289–294. [CrossRef]
5. Parida, A.; Kumar, U. Maintenance performance measurement (MPM): Issues and challenges. *J. Qual. Maint. Eng.* **2006**, *12*, 239–251. [CrossRef]
6. Poór, P.; Šimon, M.; Karková, M. CMMS as an effective solution for company maintenance costs reduction. In *Production Management and Engineering Sciences*; Taylor & Francis Group: London, UK, 2015; pp. 241–246. ISBN 978-1-138-02856-2.
7. Wienker, M.; Henderson, K.; Volkerts, J. The Computerized Maintenance Management System an Essential Tool for World Class Maintenance. *Procedia Eng.* **2016**, *138*, 413–420. [CrossRef]
8. Eti, M.C.; Ogaji, S.O.T.; Probert, S.D. Reducing the cost of preventive maintenance (PM) through adopting a proactive reliability-focused culture. *Appl. Energy* **2006**, *83*, 1235–1248. [CrossRef]
9. Hamilton, J. Early-Stage Transition to Predictive Maintenance: Using CMMS, IR Scans, and Vibration Analysis to Improve Uptime and Lower Maintenance Costs. Bachelor's Thesis, Portland State University, Portland, OR, USA, 2015. [CrossRef]
10. SpecTec. CMMS Presentation. 2020. Available online: https://www.spectec.net (accessed on 11 February 2023).

11. Alleyne, P.; Rhoden, D.; Williams, D. Expert scheduling and planned maintenance systems. *Trans. Inst. Mar. Eng.* **1991**, *103*, 365.
12. Cieri, A.M.; Elfont, M.M. Engineered approach to effective maintenance management. *Nav. Eng. J.* **1991**, *103*, 253–261. [CrossRef]
13. Wireman, T. *Computerized Maintenance Management Systems*; Industrial Press Inc.: New York, NY, USA, 1994; ISBN 978-0831130541.
14. Simion, D.; Purcărea, A.; Cotorcea, A.; Nicolae, F. Maintenance onboard ships using computer maintenance management system. *Sci. Bull. "Mircea Cel Batran" Nav. Acad.* **2020**, *23*, 134A–141A. [CrossRef]
15. Cang, T.; Dung, V.A.; Thien, D.M.; Bich, V.N. Implementation of the Computerized Maintenance Management Systems (CMMS) for the Maritime Industry. In Proceedings of the World Congress on Engineering 2012, London, UK, 4–6 July 2012; International Association of Engineers: London, UK, 2010; Volume 2189, pp. 1103–1106.
16. Gašpar, G.; Poljak, I.; Orović, J. Computerized planned maintenance system software models. *Pomorstvo* **2018**, *32*, 141–145. [CrossRef]
17. Lazakis, I.; Turan, O.; Aksu, S. Increasing ship operational reliability through the implementation of a holistic maintenance management strategy. *Ships Offshore Struct.* **2010**, *5*, 337–357. [CrossRef]
18. Mortimer, A. Planned maintenance, systems and usage. In *Motor Ship*; Mercator Media Ltd.: Fareham, UK, 2014.
19. Davies, C.R. Computer-based planned maintenance programmes. *Prop. Manag.* **1990**, *8*, 40–60. [CrossRef]
20. Kilkenny, M.F.; Kerin, M.R. Data quality: "Garbage in–garbage out". *Health Inf. Manag. J.* **2018**, *47*, 103–105. [CrossRef] [PubMed]
21. Tayi, G.K.; Ballou, D.P. Examining data quality. *Commun. ACM* **1998**, *41*, 54–57. [CrossRef]
22. Batini, C.; Scannapieco, M. *Data and Information Quality*; Springer International Publishing: Cham, Switzerland, 2016.
23. Lee, Y.W.; Pipino, L.L.; Funk, J.D.; Wang, R.Y. *Journey to Data Quality*; The MIT Press: Cambridge, MA, USA, 2006.
24. Marmo, R.; Nicolella, M.; Polverino, F.; Tibaut, A. A methodology for a performance information model to support facility management. *Sustainability* **2019**, *11*, 7007. [CrossRef]
25. Pipino, L.L.; Lee, Y.W.; Wang, R.Y. Data Quality Assessment. *Commun. ACM* **2002**, *45*, 211–218. [CrossRef]
26. Batini, C.; Cappiello, C.; Francalanci, C.; Maurino, A. Methodologies for data quality assessment and improvement. *ACM Comput. Surv.* **2009**, *41*, 1–52. [CrossRef]
27. Ballou, D.; Wang, R.; Pazer, H.; Tayi, G.K. Modeling information manufacturing systems to determine information product quality. *Manag. Sci.* **1998**, *44*, 462–484. [CrossRef]
28. Strigaro, D.; Cannata, M.; Antonovic, M. Boosting a Weather Monitoring System in Low Income Economies Using Open and Non-Conventional Systems: Data Quality Analysis. *Sensors* **2019**, *19*, 1185. [CrossRef]
29. Jan, S.-S.; Tao, A.-L. Comprehensive Comparisons of Satellite Data, Signals, and Measurements between the BeiDou Navigation Satellite System and the Global Positioning System. *Sensors* **2016**, *16*, 689. [CrossRef]
30. Stazić, L.; Komar, I.; Račić, N. Evaluation Methodology for Ship's Planned Maintenance System Database. *Trans. Marit. Sci.* **2017**, *6*, 109–116. [CrossRef]
31. Tipgos, M.A.; Trebby, J.P. Job-Related Stresses and Strains in Management Accounting. *J. Appl. Bus. Res.* **1987**, *3*, 8–14. [CrossRef]
32. Stazić, L.; Stanivuk, T.; Mihanović, V. Testing of the evaluation methodology for Ship's Planned Maintenance System Database. *J. Appl. Eng. Sci.* **2019**, *17*, 273–279. [CrossRef]
33. Stazić, L.; Komar, I.; Mihanović, L.; Mišura, A. Shipowner's impact on planned maintenance system database quality grades resemblance equalization. *Trans. Marit. Sci.* **2018**, *7*, 5–22. [CrossRef]
34. Vučinić, B. MA–CAD, Maintenance Concept Adjustment and Design. Ph.D. Thesis, Faculty of Mechanical Engineering and Marine Technology, Delft, The Netherlands, 1994.
35. Stazić, L.; Knežević, V.; Račić, N.; Orović, J. *Fault Tree Analysis as A Replacement for Manufacturers' Maintenance Instructions, Proceedings of the 2nd International Conference of Maritime Science & Technology Naše More, Dubrovnik, Croatia, 17–18 September 2021*; University of Dubrovnik, Maritime Department: Dubrovnik, Croatia, 2021; pp. 325–331. ISBN 978-953-7153-60-1.
36. Jishkariani, M. Fault Tree Analysis (FTA) For Energy Enterprises. 2020. Available online: https://www.researchgate.net/publication/341494947_Fault_Tree_Analysis_FTA_For_Energy_Enterprises (accessed on 18 July 2022).
37. Boryczko, K.; Szpak, D.; Żywiec, J.; Tchórzewska-Cieślak, B. The Use of a Fault Tree Analysis (FTA) in the Operator Reliability Assessment of the Critical Infrastructure on the Example of Water Supply System. *Energies* **2022**, *15*, 4416. [CrossRef]
38. Čepin, M.; Mavko, B. A dynamic fault tree. *Reliab. Eng. Syst. Saf.* **2002**, *75*, 83–91. [CrossRef]
39. Sánchez-Beaskoetxea, J.; Basterretxea-Iribar, I.; Sotés, I.; de las Mercedes Machado, M.M. Human error in marine accidents: Is the crew normally to blame? *Marit. Transp. Res.* **2021**, *2*, 100016. [CrossRef]
40. Wärtsilä Corporation. *MAN B&W 6S60MC-C Diesel Engine—Tanker LCC (Aframax), Trainee Manual*; Wärtsilä Corporation: Portsmouth, UK, 2018.
41. Trevelyan, J. Reconstructing engineering from practice. *Eng. Stud.* **2010**, *2*, 175–195. [CrossRef]
42. MAN Diesel SE. *TCR 22-2-C1Operating Instructions*; MAN Diesel SE: Augsburg, Germany, 2007.
43. MAN Diesel & Turbo. *TCR Turbocharger, Project Guide Book*; MAN Diesel & Turbo SE: Augsburg, Germany, 2014.

applied sciences

MDPI

*Article*

# Maximum Safe Parameters of Ships in Complex Systems of Port Waterways

**Stanisław Gucma, Maciej Gucma, Rafał Gralak and Marcin Przywarty ***

Faculty of Navigation, Maritime University of Szczecin, 70-500 Szczecin, Poland; s.gucma@am.szczecin.pl (S.G.); m.gucma@am.szczecin.pl (M.G.); r.gralak@am.szczecin.pl (R.G.)
* Correspondence: m.przywarty@am.szczecin.pl

**Abstract:** Context: From the perspective of marine traffic engineering, a system of port waterways is composed of a set of waterways (port areas), such as approach channels, port entrance, inner fairways (port channels, rivers, lakes), turning basins and port basins of various terminals. The sea waterway must be adjusted to the navigation of specific types of ships, characterized by length, breadth, draft and airdraft. The primary requirement for shipping in sea waterways is the safety of navigation. Each sea waterway has traffic restrictions for the ships using it. These restrictions are called conditions of sea waterway operation or conditions of ship operation in the sea waterway. Problem: There are a number of empirical, deterministic or probabilistic methods to determine the safe width of maneuvering areas on port waterways. The direct application of empirical methods to determine the conditions for the safe operation of ships on the complex waterway, such as the Świnoujście–Szczecin fairway, was impossible due to the complexity of the waterway and various restrictions on its individual parts. Method: The paper presents the assumptions and calculation procedure of a method allowing for the determination of maximum safe parameters of ships in existing complex waterways. Results: The proposed method was used in the preparation of port regulations for the dredged and widened Świnoujście–Szczecin waterway. The results of these calculations are presented as a practical application of the method. Conclusions: This article defines conditions for the safe operation of ships in complex port waterways systems and presents the methodology for determining maximum safe parameters of ships in existing complex port waterways systems.

**Keywords:** maritime transport routes; safety of navigation; maritime transport; design of waterways; marine simulation; full-mission ship simulator; maritime traffic engineering; safe maneuvering area; safe operation of the ship; navigational risk

**Citation:** Gucma, S.; Gucma, M.; Gralak, R.; Przywarty, M. Maximum Safe Parameters of Ships in Complex Systems of Port Waterways. *Appl. Sci.* **2022**, *12*, 7692. https://doi.org/10.3390/app12157692

Academic Editor: José A. Orosa

Received: 27 June 2022
Accepted: 28 July 2022
Published: 30 July 2022



## 1. Introduction

### *1.1. Waterway System in Marine Traffic Engineering*

The sea waterway must be adjusted to the navigation of specific type of ships, characterized by length, breadth, draft and airdraft [1]. The primary requirement for shipping in sea waterways is the safety of navigation [2]. Navigational safety comprises all of the issues related to smooth ship conduct from point A to point B of the sea route.

The sea waterway system in marine traffic engineering is composed of a number of separate sections ($n$). A complex port waterways system is usually composed of the following sections:

- Approach channel;
- Port entrance;
- A number of straight sections and bends in the fairway;
- Port basins of specific terminals and related turning basins.

Each of the waterway sections consist of three basic elements [3]:

- Waterway subsystem;

- Subsystem of ship position determination (navigational subsystem);
- Subsystem of traffic control.

These elements affect each other and have an important impact on the system characteristics.

Each sea waterway has traffic restrictions for the ships using it [4,5]. These restrictions are called conditions of sea waterway operation or conditions of ship operation in the sea waterway, and refer to:

- Parameters of ships using the waterway;
- Hydrometeorological conditions, in which, the traffic includes specific types of ships;
- Parameters of vessel traffic intensity in the waterway;
- Conditions of ship maneuvering in the waterway (tug assistance, allowable speeds).

The conditions for the safe operation of ships in a port regarded as a system composed of various types of waterways are dependent on the conditions of the safe operation of ships in each specific waterway section within the port. The parameters of each waterway within the port determine the conditions of the safe operation of ships maneuvering in that area [6].

There are a number of empirical, deterministic or probabilistic methods to determine the safe width of maneuvering areas on port waterways [3]. The most important methods include:

- PIANC;
- ROM (Spanish);
- Japanese;
- CIRM (Centrum Inżynierii Ruchu Morskiego).

They are widely used, in particular, at the preliminary stage of waterway design. For example, the most widely known PIANC method was used during the initial localization of the LNG terminal in India [7] to determine the approach fairway in the Chinese port of Panjin [8] and to determine the parameters of fairways in Korean waters [9]. However, the use of these methods was limited to determining the parameters of one or a maximum of several types (sections) of waterways (e.g., straight sections, bends, turning basins).

The direct application of the above methods to determine the conditions for the safe operation of ships on a complex waterway, such as the Świnoujście–Szczecin fairway, was impossible due to the complexity of the waterway and various restrictions on its individual parts.

This article:

- Defines conditions for the safe operation of ships in complex port waterways systems;
- Presents the methodology for determining maximum safe parameters of ships in existing complex port waterways systems.

### 1.2. Conditions of Safe Operation of Ships in Port Waterways Systems

From the perspective of marine traffic engineering, a system of port waterways is composed of a set of waterways (port areas), such as approach channels, port entrance, inner fairways (port channels, rivers, lakes), turning basins and port basins of various terminals.

In a port regarded as a system of sea waterways, different areas make up separate waterway sections that can be grouped into several criteria [3]:

- Technical parameters of the port area;
- Technical parameters of navigation systems used;
- Prevailing hydrometeorological conditions;
- Conditions of safe operation.

The conditions for the safe operation of ships in the $i$-th section of the waterway can be written as a set:

$$\mathbf{W}_i = \left[t_{yp}, L_{ci}, B_i, T_i, H_{sti}, V_i, C_i, \mathbf{H}_i\right] \tag{1}$$

where:

| | | |
|---|---|---|
| $t_{yp}$ | — | Type of 'maximum ship'; |
| $L_{ci}$ | — | Overall length of 'maximum ship'; |
| $B_i$ | — | Breadth of 'maximum ship'; |
| $T_i$ | — | Draft of 'maximum ship' in i-th section of waterway; |
| $H_{sti}$ | — | Airdraft in i-th section of waterway (useful while passing under a bridge or overhead power line); |
| $V_i$ | — | Allowable speed of 'maximum ship' in i-th fairway section; |
| $C_i$ | — | Tug assistance in i-th section of fairway, if it is required (required number and bollard pull of tugs); |
| $\boldsymbol{H}_i$ | — | Set of hydrometeorological conditions acceptable for 'maximum ship' in i-th fairway section. |

$$\boldsymbol{H}_i = \left[d/n_i, s_i,\ \Delta h_i,\ V_{wi}, V_{pi},\ h_{fi}\right] \tag{2}$$

where:

| | | |
|---|---|---|
| $d/n_i$ | — | Allowable time of day in *i*-th section of waterway; |
| $s_i$ | — | Allowable visibility in *i*-th section of waterway; |
| $\Delta h_i$ | — | Allowable drop of water level in *i*-th section of fairway; |
| $V_{wi}$ | — | Allowable wind speed in *i*-th section of fairway; |
| $V_{pi}$ | — | Allowable speed of current in *i*-th section of fairway; |
| $h_{fi}$ | — | Allowable wave height in *i*-th section of fairway. |

The conditions for the safe operation of ships passing through the waterways system consisting of a set of *n* sections can be written in this form:

$$\boldsymbol{W} = \left[t_{yp}, L_c, B, T, H_{st}, V_i, C_i, \boldsymbol{H}\right] \tag{3}$$

where:

| | | |
|---|---|---|
| $t_{yp}$ | — | Type of 'maximum ship'; |
| $L_c$ | — | Maximum overall length of ships that can safely pass through the waterway system (port entrances); |
| $B$ | — | Maximum breadth of ships that can safely pass through the waterway system; |
| $T$ | — | Maximum draft of ships that can safely pass through the waterway system; |
| $H_{st}$ | — | Maximum airdraft of ships that can safely pass through the waterway system. |

$$\boldsymbol{H} = [d/n,\ s,\ V_w] \tag{4}$$

where:

| | | |
|---|---|---|
| $d/n$ | — | Allowable time of day in the waterway system; |
| $s$ | — | Allowable visibility in the waterway system; |
| $V_w$ | — | Allowable wind speed in the waterway system. |

Additional restrictions may occur, causing temporary changes in vessel traffic in the waterway system (or restrictions of maximum ship parameters). These are:

- Drop in water level in specific sections ($\Delta h_i$);
- Exceeded speed of the sea current in specific sections ($V_{pi}$);
- Exceeded wave height in specific sections ($h_{fi}$).

The conditions for the safe operation of ships, defining the parameters of waterway system components, are defined separately for one-way and two-way traffic. In two-way traffic sections, the conditions for the safe operation of fairways can be written as follows:

$$\boldsymbol{W}_i = f_2\begin{bmatrix} t_{yp}^{in}, L_c^{in}, B^{in}, T^{in}, V_i^{in}, C_i^{in}, \boldsymbol{H} \\ t_{yp}^{out}, L_c^{out}, B^{out}, T^{out}, V_i^{out}, C_i^{out}, \boldsymbol{H} \end{bmatrix} \tag{5}$$

where *in* means a ship entering the port and *out* refers to a departing ship.

The state vector of safe ship operation conditions in waterway systems is a function of the parameters of this system [10,11]:

$$\mathbf{W} = F \begin{bmatrix} \boldsymbol{A}_i \\ \mathbf{N}_i \\ \boldsymbol{Z}_i \end{bmatrix} \tag{6}$$

where:

| | | |
|---|---|---|
| $W$ | — | Conditions of safe ship operation (state vector); |
| $A_i$ | — | Subsystem of *i*-th section of the waterway, determining the area parameters and the type of maneuver performed in that area (area subsystem); |
| $N_i$ | — | Subsystem of ship position determination, characterizing parameters of navigational systems in use (navigational system); |
| $Z_i$ | — | Subsystem of traffic control, characterizing its parameters and waterway capacity. |

In cases where port waterways are covered by the identical system of regulations, each of the waterway sections consist of two basic components [12,13]:

1. Waterway subsystem;
2. Navigational subsystem (ship position determination subsystem).

These elements affect each other and have a vital impact on the system characteristics.

The parameters of subsystems of individual port waterway sections determine the conditions for the safe operation of ships in the waterway system:

$$\boldsymbol{W} = F \begin{bmatrix} \boldsymbol{A}_i \\ \boldsymbol{N}_i \end{bmatrix} \tag{7}$$

Conditions for the safe operation of ships in seaports are subject to two restrictions [11,12]:

1. The basic maximum parameters of ships that can safely pass through the waterway system cannot be greater than maximum parameters of ships safely passing through all of the sections of the system. Therefore:

$$\begin{aligned} L_c &= \min_i L_{ci} \\ B &= \min_i B_i \\ T &= \min_i T_i \end{aligned} \tag{8}$$

2. The hydrometeorological conditions that allow for maneuvering in the given waterway system of ships with maximum parameters are identical. This applies to the time of day (*d/n*), visibility (s) and wind speed ($v_w$).

## 2. Methods

The problem of maximum safe parameters of ships arises in the case of the construction, conversion or modernization of a port waterways system. This particularly applies to complex port waterways systems, composed of several port basins (cargo handling terminals), fairways (straight sections and bends) and turning basins.

Individual sections of waterways (system components) differ in technical and operational parameters:

- Cargo-handling terminals (port basins): type and maximum parameters of ships handled ($t_{yp}$, $L_C$, $B$, $T$);
- Fairways (bends and straight sections): depth (*h*), width at bottom (*D*) and slope angle;
- Turning basins: depth (h), length and width ($l_{obr}$, $b_{obr}$) or diameter.

In addition, mooring ships are taken into account in port basins and quays or piers located along the fairways of the system, anchorages and lay-by berths.

A complex port waterway system is usually composed of the following sections:

- Approach channel (from an anchorage);

- Port entrance;
- A number of inner fairway sections leading to various terminals;
- Port basins of the terminals;
- Turning basins for ships handled at a given terminal.

In order to take into account all constraints on individual sections of the fairway, the following procedure is proposed. It allows us to determine the maximum safe parameters of ships that may use the system between specific turning basins:

1. Preliminary determination of ships' maximum drafts in specific fairway sections:

$$T_{wi} = h_i - \Delta_{wi} \tag{9}$$

The underkeel clearance $\Delta_{wi}$ was determined for the pre-defined ship speed $V_{wi}$.

2. Determination of maximum overall lengths $L_{co}$ of ships in turning basins, taking into account the possibility of turning in port basins of the terminals; location of turning basins in the fairway accounting for the length overall of ships safely turned:

$$L_{co}max, \ldots, L_{co}min \tag{10}$$

It was assumed that maximum lengths of turning ships decrease along with the increase in turning basin distance from the fairway entrance.

3. The division of the fairway into sections between the turning basins ($j$ turning basins):
    - First section: from port entrance (into fairway) to the turning basin where the maximum length ship can be turned ($L_{co}max$);
    - Next sections were determined between subsequent turning basins in the fairway.

   Further calculations were made separately for each section between the turning basins.

4. Determination of length overall $L_{cz}$ and breadth of $B_z$ of ships safely maneuvering in fairway bends ($z$). Making calculations by starting from the turning basin where a vessel of greater length can turn, we should assume the safe overall length and breadth of the ship on the fairway section running from the considered turning basin as:

$$L_c^{zak} = \min_z L_{cz} \tag{11}$$

$$B^{zak} = \min_z B_z \tag{12}$$

5. Determination of maximum safe breadths of ships in straight one-way fairway sections ($B_p$); here, we should assume the safe ship breadth for all straight fairway sections ($p$) from the considered turning basin as:

$$B^{pr} = \min_p B_p \tag{13}$$

6. Determination of maximum safe lengths and breadths of ships in particular sections between the considered $j$-th turning basin:

$$L_{cj} = L_c^{zak} \text{ for } L_{cj} < L_{coj} \tag{14}$$

$$B_j = \min\left(B^{zak}, B^{pr}\right) \tag{15}$$

7. Determination of ship's safe draft and allowable speeds in specific fairway sections of the considered waterway, taking into account ship and fairway section parameters. The ship speed in individual waterway sections is calculated from this formula:

$$T_i = h_i - \Delta_i \tag{16}$$

The underkeel clearance is a function of ship speed and other parameters, fairway parameters and the performed maneuver:

$$\Delta_i = f(V_i) \tag{17}$$

The maximum safe draft of the ship in the considered $j$-th section of the waterway to be assumed:

$$T_i = \min_i T_i \tag{18}$$

8. Determination of maximum parameters of two-way traffic in each fairway section.
9. Determination of the conditions of safe operation of ships passing through the waterway system between the examined turning basins:

$$\boldsymbol{W}_j = \left[t_{ypj}, L_{cj}, B_j, T_j, H_{sr}, V_i, C_i, \boldsymbol{H}\right] \tag{19}$$

The algorithm of the parameters determination is shown in Figure 1.

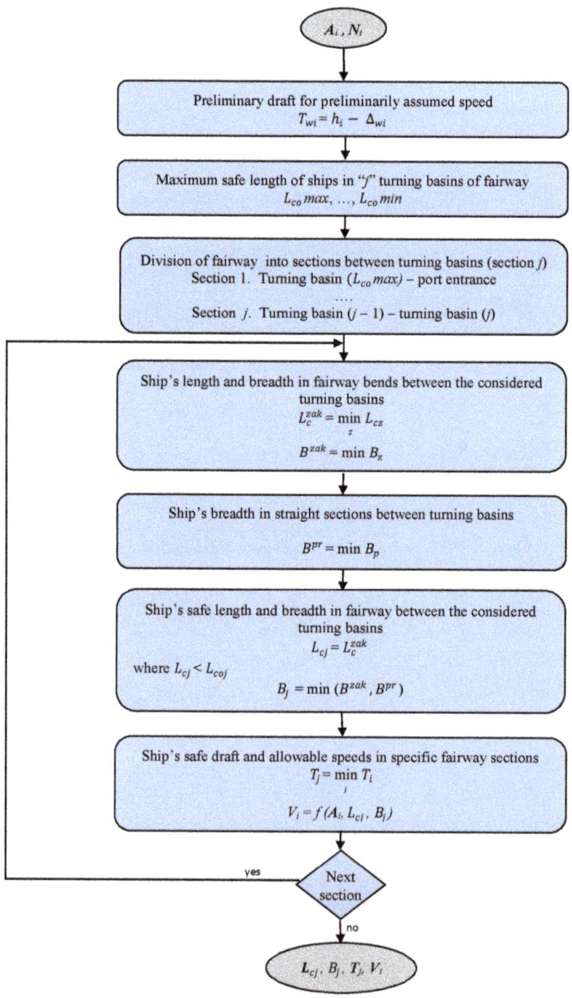


**Figure 1.** The algorithm of the process of determining maximum parameters of the ship in complex port waterways systems.

Notably, safe parameters of ships ($L_c$, $B$) maneuvering in turning basins and fairway bends can be determined by simulation or empirical methods [3,14,15]. Simulation methods using full mission bridge simulators are more accurate than empirical methods. In straight sections of the fairway, empirical methods are sufficiently accurate.

## 3. Results

The developed method for determining the maximum safe parameters of ships in complex port waterways systems was used for the calculation of the conditions for safe ship operations in the modernized Świnoujście–Szczecin fairway, dredged from 10.5 m to 12.5 m and widened. The requirements for determining the conditions for the safe operation of ships included parameters of 'maximum ships' passing through this fairway.

The modernized Świnoujście–Szczecin fairway is 68 km in length, and 12.5 m deep. It includes the Świnoujście entrance channel and a number of straight sections and bends with various technical parameters. The main sections of the fairway, together with their parameters, are shown in Table 1. The basic sections are connected with transition sections, on which, fairway parameters change. The whole waterway system comprises four turning basins. In addition, turning is possible in the Świnoujście–Szczecin fairway for ships entering and departing from two other ports (Figure 2):

- Police (49.4 km of the fairway), handling bulk carriers, chemical tankers and LPG tankers;
- Newly designed Skolwin port (55.7 km) to serve product, chemical and LPG tankers.

**Table 1.** Sections and the parameters of the modernized Świnoujście–Szczecin fairway.

| Kilometer of Fairway from | to | Type of Section | Radius [m] | Width Max [m] | Width Min [m] | Depth Max [m] | Depth Min [m] |
|---|---|---|---|---|---|---|---|
| 0 | 5.3 | Port Świnoujście | | | | | |
| 5.3 | 7.6 | Straight | | 130 | 130 | 14.0 | 12.5 |
| 8.1 | 9.1 | Straight | | 110 | 110 | 12.5 | 12.5 |
| 9.8 | 10.9 | Bend | 2300 | 120 | 120 | 14.0 | 12.5 |
| 11.4 | 17.0 | Straight | | 110 | 110 | 18.0 | 12.5 |
| 17.4 | 23.0 | Straight | | 100 | 100 | 12.5 | 12.5 |
| 23.8 | 28.8 | Straight | | 250 | 250 | 12.5 | 12.5 |
| 29.6 | 40.9 | Straight | | 100 | 100 | 12.5 | 12.5 |
| 41.2 | 43.0 | Bend | 2200 | 155 | 110 | 12.5 | 12.5 |
| 43.0 | 48.6 | Straight | | 110 | 110 | 12.5 | 12.5 |
| 48.8 | 49.5 | Bend + Turning basin | | 350 | 150 | 12.5 | 12.5 |
| 49.5 | 51.5 | Straight | | 350 | 220 | 14.0 | 12.5 |
| 52.0 | 53.0 | Bend | 1680 | 150 | 150 | 13.0 | 12.5 |
| 53.2 | 54.4 | Straight | | 130 | 130 | 12.5 | 12.5 |
| 54.7 | 55.4 | Bend | 1730 | 150 | 150 | 12.5 | 12.5 |
| 55.6 | 59.0 | Straight | | 100 | 100 | 12.5 | 12.5 |
| 59.4 | 60.5 | Bend | 1600 | 150 | 150 | 12.5 | 12.5 |
| 60.8 | 62.9 | Straight | | 100 | 100 | 12.5 | 12.5 |
| 63.4 | 64.0 | Turning basin | | 360 | 360 | 12.5 | 12.5 |
| 64.3 | 66.6 | Straight | | 100 | 90 | 12.5 | 12.5 |
| 66.7 | 67.0 | Straight | | 130 | 130 | 12.5 | 12.5 |
| 67.1 | 67.4 | Turning basin | | 330 | 330 | 12.5 | 12.5 |

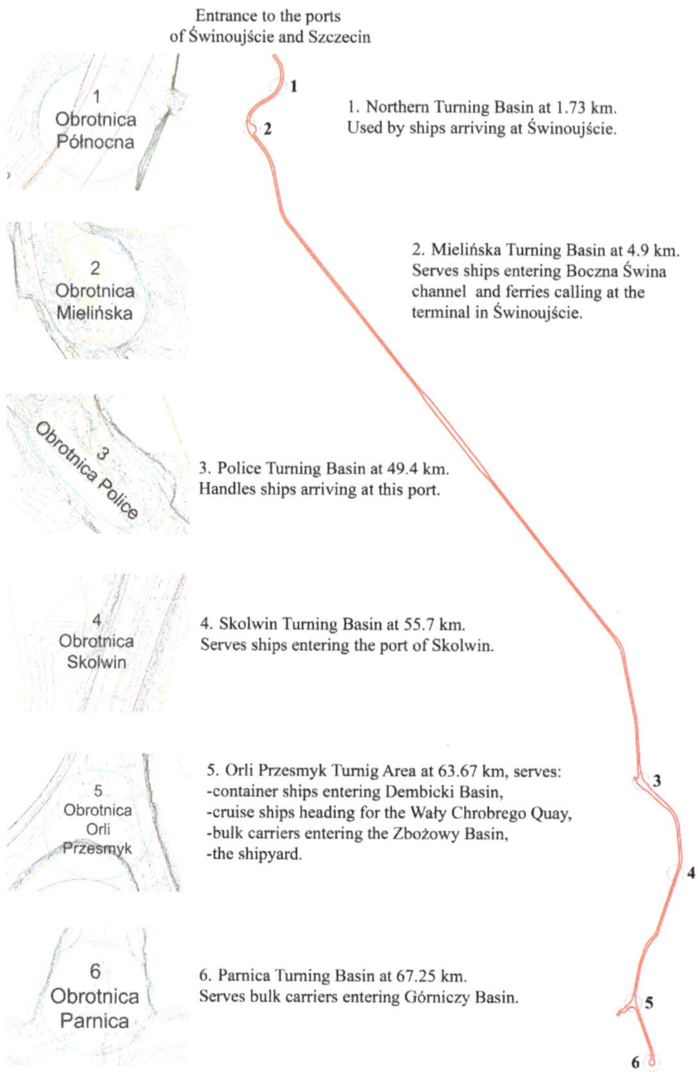


**Figure 2.** A simplified model of the complex port waterways system (four terminals).

Maximum safe parameters of ships passing through the Świnoujście–Szczecin fairway were determined in a procedure composed of the following steps:

1. Preliminary maximum draft of ships is:

$$T_w = 11 \text{ m}$$

   where the underkeel clearance ($\Delta_{wi}$ = 1.5 m) was calculated using the method of components and adopting:

   - Preliminary ship speed in individual sections $V_i$ = 8 knots;
   - Water level drop $\Delta h_i = -0.5$ m [16,17].

2. The maximum length of ships that can safely turn in four Świnoujście–Szczecin fairway turning basins is, respectively:

- Northern Turning Basin (1.7 km of the fairway), $L_{co}$ = 300 m, $B_o$ = 50 m—bulk carriers under ballast;
- Mielińska Turning Basin (4.9 km of fairway), $L_{co}$ = 270 m, $B_o$ = 40 m, $T_o$ = 11.0 m;
- Przesmyk Orli Turning Basin (63.7 km), $L_{co}$ = 260 m, $B_o$ = 33 m, $T_o$ = 9.0 m—cruise ships and $L_{co}$ = 250 m, $B_o$ = 33 m, $T_o$ = 11.0 m—container ships
- Parnica Turning Basin (67.3 km of the fairway), $L_{co}$ = 230 m, $B_o$ = 33 m, $T_o$ = 11.0 m—bulk carriers;
- Port of Police (49.4 km of the fairway), $L_{co}$ = 230 m, $B_o$ = 33 m, $T_o$ = 11.0 m—bulk carriers and chemical tankers;
- Designed Skolwin Port (55.7 km of the fairway), $L_{co}$ = 230 m, $B_o$ = 35, $T_o$ = 11.2 m—product tankers and LPG tankers.

The maximum ship lengths in the turning basins of Northern, Mielińska and Przesmyk Orli were determined using simulation methods, whereas the maximum ship length in the Parnica Turning Basin was estimated using the empirical method.

Example results of simulation tests are presented for the Northern Turning Basin. The tests were conducted on the FMBS simulator from Kongsberg. The preliminary assumed conditions for the safe operation of bulk carriers entering the Port of Świnoujście were as follows:

- Ingoing vessel entering and berthing port side alongside—loaded bulk carrier: $L_c$ = 300 m, T = 13.2 m;
- Outgoing—bulk carrier under ballast: $L_c$ = 300 m, $T_D$ = 7.4 m, $T_R$ = 9.0 m; the ship turns around the starboard side in the Northern Turning Basin.

Two simulation models of this bulk carrier were built to conduct three test series of ship arrivals and departures. The simulation experiment consisting of 12 passages in one series was performed by port pilots. Each test series was conducted in different least favorable hydrometeorological conditions. Figure 3 presents statistically developed test results of the port entry by a loaded bulk carrier, and Figure 4 depicts the turning of the bulk carrier under ballast, at wind N 10 m/s and ingoing current 0.8 knots. The test results are shown as safe maneuvering areas determined at three levels of confidence: maximum (red line), mean (green line) and 95% (magenta line).

3. The maximum overall length and breadth of ships safely maneuvering in fairway bends were determined for fairway sections between the turning basins. The calculations were made in this order:

   - Northern Turning Basin (1.7 km)—port entrance (0.0 km of the fairway). The maximum overall length and breadth of ships entering the port and approaching the Northern Turning Basin were determined by the simulation method (Figure 3), $L_c$ = 300 m, B = 50 m, T = 13.2 m;
   - The Mielińska Turning Basin (4.9 km of the fairway)—Northern Turning Basin. The maximum overall length and breadth of ships safely passing from the Northern to Mielińska Turning Basins were determined by the simulation method $L_c$ = 270 m, B = 40 m, T = 11.0 m [16]. In addition, the navigational risk of maximum ship passage was examined in connection with the planned arrivals of a ferry $L_c$ = 220 m at berth No 2 of the ferry terminal in Świnoujście. The risk that the maneuvering ship moves out of the available navigable area and passenger fatalities occur $R = 2.8 \times 10^{-7}$ [year$^{-1}$] is lower than the acceptable risk [18].
   - Przesmyk Orli Turning Basin (63.7 km of the fairway)—Mielińska Turning Basin. The maximum overall length $L_c^{zak}$ and breadth $B^{zak}$ of ships safely maneuvering in the fairway bends (turns) between Mielińska and Przesmyk Orli Turning Basins were determined by the simulation method. The tests were conducted using a Kongsberg-made FMBS simulator in three fairway bends: Mańków, Mańskie-Babina and Święta. A simulation experiment was conducted for a cruise ship $L_c$ = 260 m, B = 33.0 m, T = 9.0 m, and container ship $L_c$ = 250 m, B = 33.0 m, T = 11.0 m [16]. Two series of tests were conducted in least favorable hydrom-

eteorological conditions for each bend. The simulation experiment consisting of 12 passages in one series was performed by port pilots. Figure 5 shows statistically processed test results for the Święta bend. The results refer to the safe maneuvering areas for a cruise ship $L_c^{zak}$ sailing through the bend in the least favorable hydrometeorological conditions;

- There are no bends between Parnica (57.3 km) and Przesmyk Orli Turning Basins.

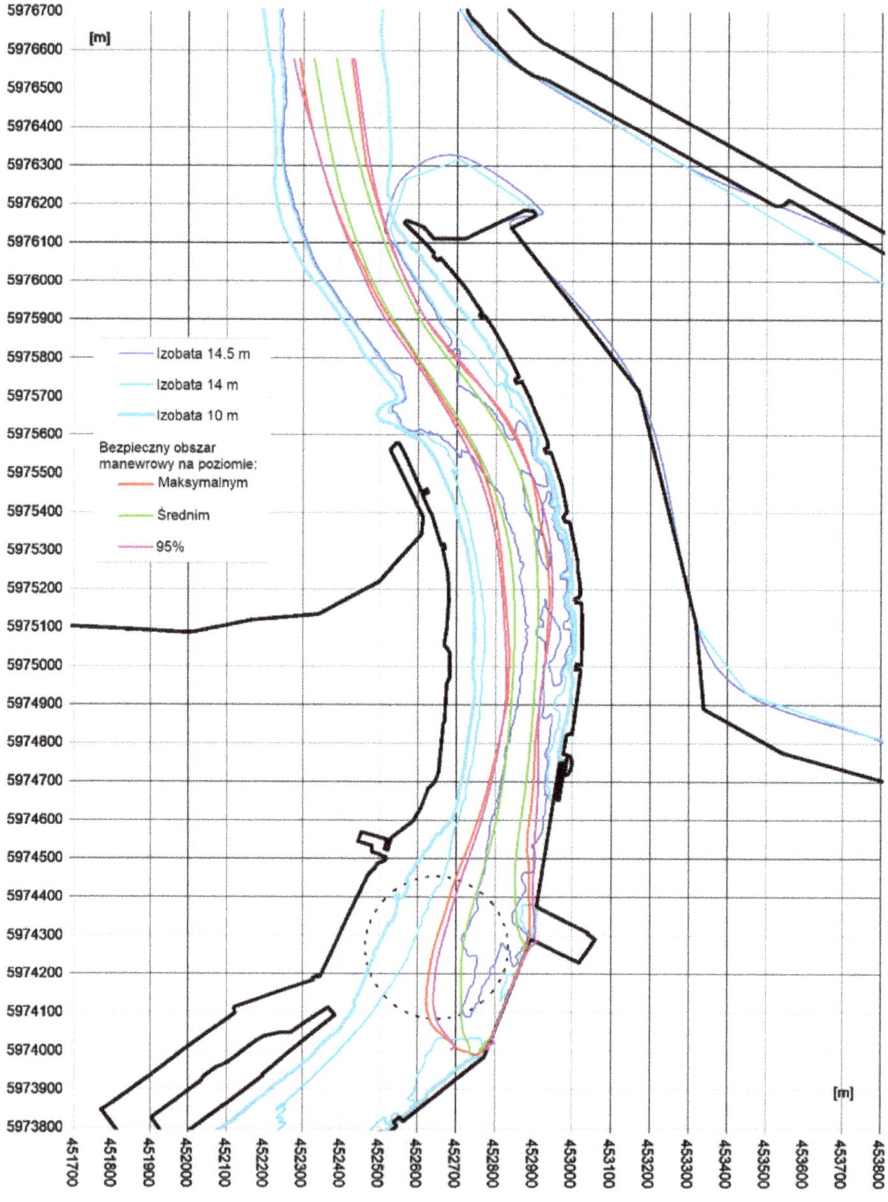


**Figure 3.** Safe maneuvering areas of the loaded carrier $L_c$ = 300 m, T = 13.5 m at the entrance to Port Świnoujście. Wind N 10 m/s, ingoing current 0.8 knots, UTM coordinates (zone 33U).

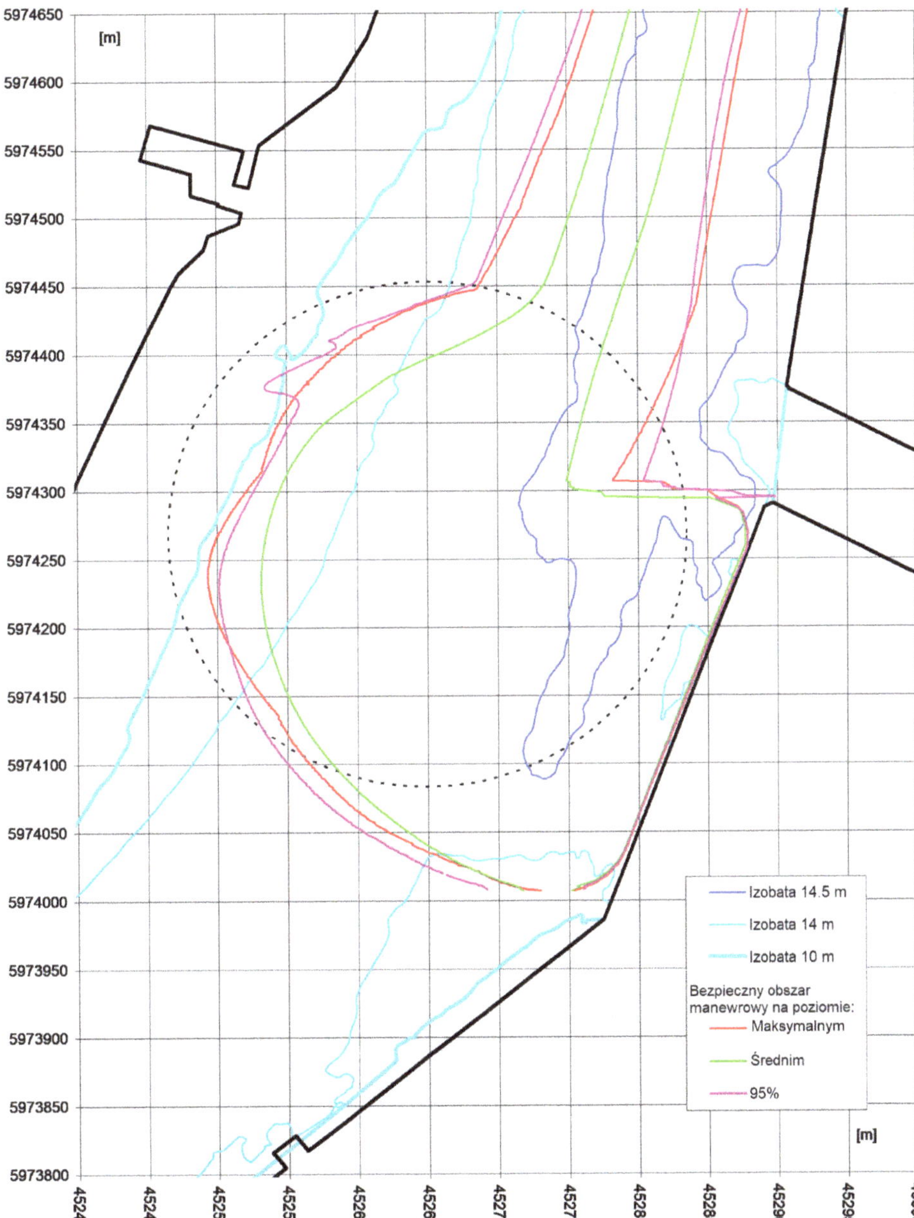


**Figure 4.** Safe maneuvering areas of turning in the Northern Turning Basin in Świnoujście, bulk carrier under ballast $L_c = 300$ m, $T_D = 7.4$ m, $T_R = 9.0$ m/s, ingoing current 0.8 knots, UTM coordinates (zone 33U).

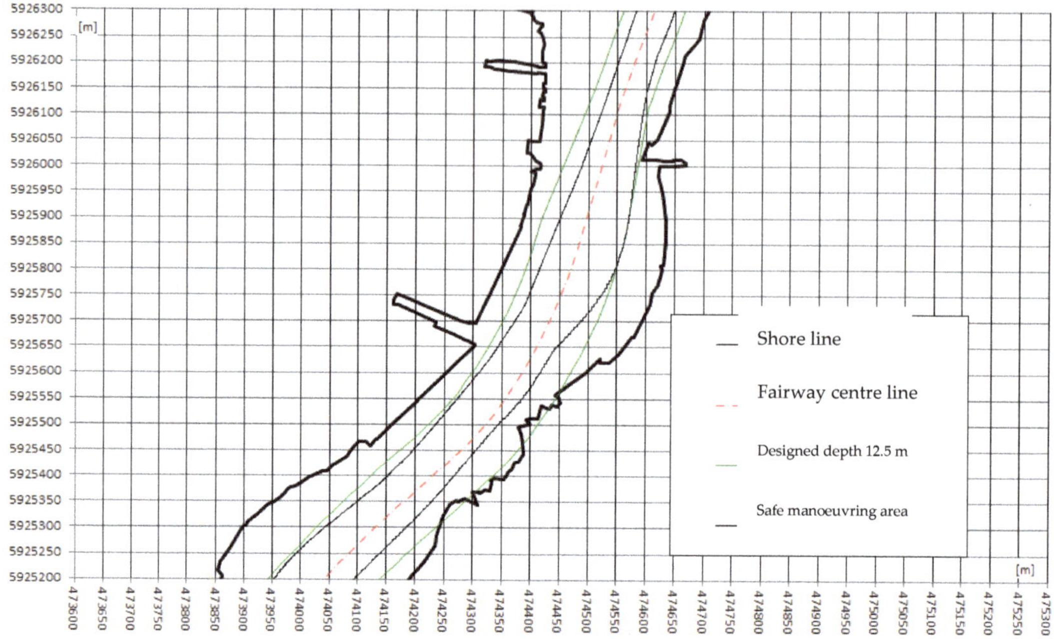


**Figure 5.** Święta fairway bend. Safe maneuvering area of cruise ship passage $L_c$ = 260 m in least favorable hydrometeorological conditions. Wind S and W, 10 m/s, outgoing current 0.7 knots, UTM coordinates (zone 33U).

4. The maximum breadth of ships safely maneuvering in straight fairway sections was determined after transforming the condition of navigational safety, defined in the empirical CIRM method [3,15]. This condition can be written as:

$$D_j \geq d_m + 2d_{n(1-\alpha)} + d_r^p + d_r^l \tag{20}$$

where:

| | | |
|---|---|---|
| $D_j$ | — | Width at bottom of j-th point of the fairway center line for safe depth (available width of the navigable area) [m]; |
| $d_m$ | — | Deterministic maneuvering component of the swept path width [m]; |
| $d_{n(1-\alpha)}$ | — | Probabilistic navigational component of the swept path width [m] of the ship at a given confidence level $(1-\alpha)$ [m]; |
| $d_r^p$ | — | Reserve of maneuvering area width on the right-hand side [m]; |
| $d_r^l$ | — | Reserve of maneuvering area width on the left-hand side [m]; |

The maneuvering component is the sum of the basic width of the swept path $d_{mp}$ and additional corrections of the swept path width $d_{md}$

$$d_m = d_{mp} + d_{md} \tag{21}$$

while:

$$d_{mp} = k \cdot B \tag{22}$$

where:

| | | |
|---|---|---|
| $k$ | — | Coefficient determining the ship's maneuverability, dependent on ship type $k = f\,(t_{yp})$. |

Additional corrections of the swept path width depend on the ship speed, technical parameters of the fairway and hydrometeorological conditions, i.e.:

$$d_{md} = f(V_i, A_i,\ H_i) \tag{23}$$

The width allowances depend on the ship speed and technical parameters of the fairway:

$$d_r = f(V_i,\ A) \tag{24}$$

The navigational component is determined by the subsystem of ship position determination in the given fairway section.

$$d_n = f(N_i) \tag{25}$$

Transforming this relationship (20) leads to the determination of the maximum safe width of the ship in straight fairway sections between the turning areas: (*p* sections):

$$B_p = \left(D_p - d_{md} - 2d_{n(1-\propto)} - d_r^p - d_r^l\right)/k \tag{26}$$

where all data necessary for determining the right-hand side terms of the equation are known:

$$B_p = f\ \left(A_i, N_i,\ H_i, V_{wi}, t_{yp}\right) \tag{27}$$

and these were used to determine the maximum safe breadths Bp of ships in each section and between the examined turning areas $B^{pr}$.

$$B^{pr} = \min_p B_p \tag{28}$$

5. Maximum safe lengths, breadths and drafts of ships between Świnoujście–Szczecin fairway turning areas were calculated as above and are equal to:
    - Northern Turning Area—port entrance channel: $L_c$ = 300 m, B = 50 m, T =13.2 m;
    - Mielińska to Northern Turning Areas: $L_c$ = 270 m, B = 40 m, T = 11.0 m;
    - Przesmyk Orli to Mielińska Turning Areas:
      $L_c$ = 260 m, B = 33 m, T = 9.0 m—cruise carrier;
      $L_c$ = 240 m, B = 33 m, T = 11.0 m—container ship;
      $L_c$ = 230 m, B = 33 m, T = 11.0 m—bulk carrier;
    - Parnica to Przesmyk Orli Turning Areas: $L_c$ = 230 m, B = 33 m, T = 11.0 m—bulk carrier.
6. Safe allowable speeds of ships in specific sections of the Świnoujście–Szczecin fairway were determined by starting from the basic condition of navigational safety

$$T \geq h - \Delta \tag{29}$$

where:

Δ – Underkeel clearance.

There are two methods for determining the underkeel clearance:

- Static method;
- Dynamic method.

The safe allowable speeds of ships in specific fairway sections were determined using the static method of components [15]. In the method, the allowance Δ was divided into two components: static allowance $\Delta_s$ and dynamic allowance $\Delta_d$:

The static allowance does not depend on ship movement and is constant for the given area and the ship. Its components include:

$\Delta_1$ — Water allowance for sounding error, depending on the depth of the area;
$\Delta_2$ — Navigational allowance due to the discontinuity of soundings;
$\Delta_3$ — Allowance for siltation, depending on the area;
$\Delta_4$ — Allowance for the height of tide determination error, depending on the shipping area;
$\Delta_5$ — Allowance for the error of water level determination due to water level fluctuations relative to chart datum;
$\Delta_6$ — Allowance for the draft determination error, depending on the type of ship;
$\Delta_7$ — Allowance for ship's list assessment error, depending on ship parameters.

The components of the dynamic allowance include:

$\Delta_8$ — Allowance for moving ship squat, depending on ship's speed, draft, depth and other area parameters;
$\Delta_9$ — Allowance for waves, depending on wave and ship parameters.

In the Świnoujście–Szczecin fairway, no water allowance for waves ($\Delta_9 = 0$) was made, so the condition of navigational safety can be written as:

$$T \leq h - \sum_{l=1}^{7} \Delta_l - \Delta_8 \quad (30)$$

The underkeel clearance was determined by five empirical methods recommended by PIANC in the given conditions [19]:

- Tuck;
- Huuska/Guliev;
- ICORELS;
- Barras3;
- Eryuzlu2.

It was calculated for container ships (block coefficient $C_b = 0.62$) and bulk carriers ($C_b = 0.8$) by adopting appropriate bottom profiles for individual fairway sections. Given the obtained results, specific safe allowable speeds were identified in all fairway sections. For ships T > 10 m, the following values apply:

- Container ships, general cargo ships, LPG tankers ($C_b \leq 0.65$) V = 8 to 12 knots;
- Bulk carriers, tankers ($Cb \geq 0.75$) V = 6 to 10 knots.

The guidelines were also formulated for the implementation of the dynamic underkeel clearance system, which will increase the allowable maximum draft to T = 11.2 m.

7. The passing of two ships in the Świnoujście–Szczecin fairway require the determination of maximum safe breadths of the meeting ships, depending on their draft and the fairway channel slope (Table 2).

The breadths of ships safely passing each other in straight fairway sections were determined based on the condition of the navigational safety of two-way traffic [13]:

$$D_{it} \geq d_m^{in} + d_m^{out} + 2d_{n(1-\alpha)}^{in} + 2d_{n(1-\alpha)}^{out} + d_r^{in} + d_r^{out} + d_r^{s} \quad (31)$$

where:

$D_{it}$ — Width of available navigable area for ships with draft $T$ in $i$-th section of the fairway;
$d_r^s$ — Separation zone allowing for the suction force effect;
$in, out$ — Indices marking incoming and outgoing ships.

**Table 2.** Determination of safe breadths of passing ships in straight sections of the Świnoujście–Szczecin fairway.

| $T_1 + T_2$ [m] | $B_1 + B_2$ [m] Slope Inclination | | |
|---|---|---|---|
| | 1:2 | 1:3 | 1:4 |
| 8 | 32 | 40 | 47 |
| 9 | 31 | 38 | 45 |
| 10 | 30 | 36 | 43 |
| 11 | 29 | 34 | 41 |
| 12 | 28 | 33 | 39 |
| 13 | 27 | 32 | 37 |
| 14 | 26 | 31 | 35 |
| 15 | 25 | 30 | 34 |
| 16 | 24 | 28 | 32 |
| 17 | 23 | 27 | 30 |
| 18 | 22 | 26 | 28 |
| 19 | 21 | 25 | 27 |
| 20 | 20 | 23 | 25 |

## 4. Discussion

The main objective of the research presented in the article is to develop a general method for determining the maximum parameters of ships in complex port water systems. This goal was achieved by developing a computational procedure (algorithm) that takes into account various constraints specific to individual sections of the fairway. The developed method stands out from the methods known in the literature (PIANC, ROM, Japanese, CIRM), which focus on determining the maximum parameters of a ship on individual types of waterways, but do not include the procedure for their use for complex waterways.

The developed method was used to determine the conditions for the safe operation of ships on the modernized Szczecin–Świnoujście fairway. The obtained results showed that the deepening and widening of the waterway will, obviously, allow for the passage of larger ships. However, the increase in ship dimensions relates more to the draft and length of the ship than to its breadth. For example, for the port of Szczecin, the maximum vessel draft increased from 9.15 m to 11.0 m and the length from 215 m to 260 m, and the width of the maximum vessel only increased from 31 m to 33 m. This is due to the fact that the horizontal dimensions of the safe maneuvering area depend not only on the width of the vessel, but also on its length, the draft ratio and the available depth.

The proposed method does not introduce new dependencies allowing for the determination of safe ship parameters, but, thanks to the systematization of calculations, it allows for the determination of these parameters on complex waterways. The limitations of its use result from the limitations of the detailed methods used; however, due to the possibility of using various methods (both empirical and simulation), it can be treated as a universal method allowing for the determination of safe parameters of ships on complex waterways. In the future, work on the development of this method will focus on the partial automation of calculations, which will allow us to shorten the time necessary to obtain results.

In conclusion, the developed method allows us to determine the maximum safe parameters of ships in complex port waters, which has been practically confirmed by defining the conditions for the safe operation of ships on the modernized Szczecin–Świnoujście fairway.

## 5. Conclusions

The article presents a new method for determining the maximum safe parameters of ships in existing complex port waterway systems and the conditions for safe ship operations.

The developed method can be described with the following general procedure:

1. The determination of maximum safe lengths of ships in specific turning basins and ship drafts;

2. The waterway division between turning basins, depending on the maximum length of ships turned there;
3. The determination of maximum lengths and breadths of ships safely maneuvering in fairway bends (turns) between the turning basins;
4. The determination of maximum safe breadths of ships in straight fairway sections between the turning basins;
5. The determination of the allowable speed of ships in specific fairway sections;
6. The determination of ship parameters in two-way traffic lanes;
7. The determination of conditions for the safe operation of ships in a complex port waterway system.

The prosed method was used for the determination of conditions for the safe operation of ships on the modernized 68-kilometer Świnoujście–Szczecin fairway. The modernization of the fairway resulted in deepening the channel from 10.5 m to 12.5 m and an appropriate widening of the channel. By dredging the fairway to 12.5 m, the maximum permissible draught of vessels calling at Szczecin was increased to approx. 11.0 m (before 9.15 m), and, thus, the availability of the Szczecin port to a certain group of large vessels was ensured. There will be no need for them to be discharged at Świnoujście before continuing on to the Szczecin port.

The results were used to draw up detailed port regulations defining the conditions for safe fairway operations of ships heading for the ports of Świnoujście, Szczecin, Police and Skolwin.

**Author Contributions:** Conceptualization, S.G.; methodology, S.G.; software, M.G., R.G. and M.P.; validation, S.G.; formal analysis, S.G., M.G., R.G. and M.P.; investigation, S.G., M.G., R.G. and M.P.; resources, R.G. and M.P.; data curation, M.G, R.G. and M.P.; writing—original draft preparation, S.G.; writing—review and editing, M.G., R.G. and M.P.; visualization, M.P.; supervision, S.G.; project administration, R.G. and M.P.; funding acquisition, M.G., R.G. and M.P. All authors have read and agreed to the published version of the manuscript.

**Funding:** This research received no external funding.

**Institutional Review Board Statement:** Not applicable.

**Informed Consent Statement:** Not applicable.

**Data Availability Statement:** Not applicable.

**Conflicts of Interest:** The authors declare no conflict of interest.

## References


1. Olba, X.B.; Daamen, W.; Vellinga, T.; Hoogendoorn, S.P. State-of-the-Art of Port Simulation Models for Risk and Capacity Assessment Based on the Vessel Navigational Behaviour through the Nautical Infrastructure. *J. Traffic Transp. Eng. (Engl. Ed.)* **2018**, *5*, 335–347.
2. Formela, K.; Weintrit, A.; Neumann, T. Overview of Definitions of Maritime Safety, Safety at Sea, Navigational Safety and Safety in General. *TransNav Int. J. Mar. Navig. Saf. Sea Transp.* **2019**, *13*, 285–290. [CrossRef]
3. Gucma, S.; Zalewski, P. Optimization of Fairway Design Parameters: Systematic Approach to Manoeuvring Safety. *Int. J. Nav. Archit. Ocean. Eng.* **2020**, *12*, 129–145. [CrossRef]
4. Reshnyak, V.; Sokolov, S.; Nyrkov, A.; Budnik, V. Inland Waterway Environmental Safety. *J. Phys. Conf. Ser.* **2018**, *1015*, 042049. [CrossRef]
5. Froese, J. Safe and Efficient Port Approach by Vessel Traffic Management in Waterways. In *Transport of Water versus Transport over Water: Exploring the Dynamic Interplay of Transport and Water*; Ocampo-Martinez, C., Negenborn, R.R., Eds.; Operations Research/Computer Science Interfaces Series; Springer International Publishing: Cham, Switzerland, 2015; pp. 281–296. ISBN 978-3-319-16133-4.
6. Shang, J.; Wang, W.; Peng, Y.; Tian, Q.; Tang, Y.; Guo, Z. Simulation Research on the Influence of Special Ships on Waterway through Capacity for a Complex Waterway System: A Case Study for the Port of Meizhou Bay. *Simulation* **2020**, *96*, 387–402. [CrossRef]
7. Rushidh, M.D.; Suhuraa, S.; Reddya, L.R.; Dwarakisha, G.S. Planning of Marine Facilities for an LNG Terminal in India. In Proceedings of the GITA-2K15, SMVITM, Bantakal, India, 16–17 October 2015.
8. Zhang, L. Discussion on the Method for Determining the Harbor Channel Design Width-Take the Proposed 250,000 ton Waterway of Panjin Harbor for Instance. *Adv. Mater. Res.* **2015**, *1065*, 480–485. [CrossRef]

9. Kang, W.S.; Park, Y.S. A Study on the Design of Coastal Fairway Width Based on a Risk Assessment Model in Korean Waterways. *Appl. Sci.* **2022**, *12*, 1535. [CrossRef]
10. Gucma, S.; Gralak, R.; Muczyński, B. Areas of Emergency Maneuvers and the Navigational Risk of Accidents in Fairways Due to Ship Technical Failures Determined by the Ship Movement Simulation Method. *Zesz. Nauk. Akad. Mor. Szczec.* **2020**, *61*, 39–47.
11. Gucma, S. Conditions of Safe Ship Operation in Seaports–Optimization of Port Waterway Parameters. *Pol. Marit. Res.* **2019**, 3, 22–29. [CrossRef]
12. Gucma, S.; Gralak, R. Probabilistic-Deterministic Method of Rescaling Ship Manoeuvring Simulation Data Defining Parameters of Fairway Bends. *New Trends Prod. Eng.* **2018**, *1*, 127–133. [CrossRef]
13. Akram, M.A.; Liu, P.; Wang, Y.; Qian, J. GNSS Positioning Accuracy Enhancement Based on Robust Statistical MM Estimation Theory for Ground Vehicles in Challenging Environments. *Appl. Sci.* **2018**, *8*, 876. [CrossRef]
14. Thoresen, C.A. *Port Designer's Handbook*; Thomas Telford: London, UK, 2010.
15. Gucma, S. Studying Sea Waterway System with the Aid of Computer Simulation Methods. *Pol. Marit. Res.* **2015**, *22*, 10–15. [CrossRef]
16. Maritime University of Szczecin (Szczecin, Poland). Navigational Analysis of Modernization of the Waterway Swinoujscie-Szczecin (Deepening to 12.5 m). 2015; not intended for publication.
17. Maritime University of Szczecin (Szczecin, Poland). Analysis of Ship Operation Safety in the Świnoujście—Szczecin Fairway. 2021; not intended for publication.
18. Bąk, A.; Gucma, S.; Przywarty, M. The Safety of Navigation in Restricted Areas—Method of Risk Estimation and Analysis. Maritime University of Szczecin: Szczecin, Poland, 2022; *Unpublished*.
19. McBride, M.; Boll, M.; Briggs, M. *Harbour Approach Channels—Design Guidelines*; PIANC: Brussels, Belgium, 2014.

applied sciences

MDPI

*Article*

# Human-Related Hazardous Events Assessment for Suffocation on Ships by Integrating Bayesian Network and Complex Network

**Weiliang Qiao [1,*], Hongtongyang Guo [1], Enze Huang [1], Wanyi Deng [2], Chuanping Lian [3] and Haiquan Chen [1]**

1 Marine Engineering College, Dalian Maritime University, Dalian 116026, China; ghty@dlmu.edu.cn (H.G.); huangenze@dlmu.edu.cn (E.H.); chenapec@dlmu.edu.cn (H.C.)
2 School of Maritime Economics and Management, Dalian Maritime University, Dalian 116026, China; dwy@dlmu.edu.cn
3 COSCO Shipping Seafarer Management Co., Ltd., Dalian Branch, Dalian 116026, China; linlili@sino-ocean.com
* Correspondence: xiaoqiao_fang@dlmu.edu.cn

**Abstract:** To investigate the human-related factors associated with suffocation on ships during docking repair, a comprehensive analysis model composed of a Bayesian network (BN) and a complex network (CN) is proposed in the present study. The principle of event tree analysis (ETA) is firstly applied to identify the hazardous events involved in the accident according to the accident report, based on which the CN would then be developed with the logic relationships among the hazardous events. The improved K-shell decomposition algorithm is utilized to determine the criticality of nodes in the CN, the results of which are then used to develop the BN model within the framework of a human factor analysis classification system (HFACS). Then, the developed BN model can be simulated with the probability distribution of all the nodes within the BN, which are obtained on the basis of node criticality. Finally, the results of the BN simulation are interpreted from the perspectives of a brief analysis, backward analysis and sensitivity analysis. The results are verified with existing studies and the accident investigation report issued by authority, which are presented as evidence to verify the effectiveness of the proposed methodology to evaluate the human-related risk involved in the suffocation on ships. The methodology proposed in this study integrates the advantages of BN and CN to investigate the human-related hazardous events involved in maritime accidents, which can be seen as the main innovation of this work.

**Keywords:** suffocation on ships; human-related hazardous events; Bayesian network; complex network; HFACS

**Citation:** Qiao, W.; Guo, H.; Huang, E.; Deng, W.; Lian, C.; Chen, H. Human-Related Hazardous Events Assessment for Suffocation on Ships by Integrating Bayesian Network and Complex Network. *Appl. Sci.* **2022**, *12*, 6905. https://doi.org/10.3390/app12146905

Academic Editor: José A. Orosa

Received: 17 June 2022
Accepted: 5 July 2022
Published: 7 July 2022



## 1. Introduction

The operational scenarios in shipyards are characterized by high risk level due to the frequent interaction among various stakeholders and the different work types involved during docking repair [1–3]. During docking repair, it is common that a lot of professional and dangerous operations are carried out simultaneously in the same space, meaning that risk factors are highly interconnected and harmful, especially human-related factors [4]. In Singapore, accidents and incidents which occur during docking repair are receiving attention, and a legislative amendment was passed by the attorney general's chambers [5] to prevent or minimize the occurrence of such accidents. Later, in 2019, the International Labor Organization (ILO) adopted a revised code of practice for safety and health in shipbuilding and ship repair at the 329th session [6], which aimed to improve safety and health practices by providing good practice for governments, employers, and workers. However, occupational incidents and accidents still occur in the shipyard during docking repair, which indicates that safety management processes during docking repair need to be further improved [7]. In addition, with the influence of some important regulations and codes adopted by International Maritime Organization (IMO), such as the revised MARPOL Annex VI [8] and the International Convention for the Control and Management of Ships Ballast Water and Sediments (BWM) [9], a large number of operating ships have to be refitted during docking repair to satisfy the new regulations. It can be reasonably predicted that the ship repair industry will continue to run with relatively high demand. For instance, according to the statistics of the China Association of the National Shipbuilding Industry (CANSI), in 2019, the shipbuilding industry and ship repair industry accounted for 65.4% and 4.2% of the overall business income in China, respectively. Compared with the figures of 2015, the proportion of these increased significantly [10]. It is important and urgent to reduce the rate of occupational accidents and accurately and efficiently promote industrial safety management.

### 1.1. Related Studies

In the scientific literature, the general characteristics and common causes of accidents can be summarized in terms of the time, place, type of work, age of workers, work experience, etc., through statistics and research on historical accidents [4,11,12]. When the average temperature is above 25 °C from June to September every year, the death toll is the highest. These aspects can be taken as accident-related risk factors for analyzing the cause and effect of occupational accidents from a comprehensive perspective [13,14]. However, such studies often lack pertinence due to the variety of accident types; thus, some studies began to focus on causal analysis based on a typical accident or operation category [15–17]. Still, accidents that may occur in confined spaces such as cargo holds have not received enough attention. In fact, it is estimated that there are 0.05–0.08 deaths per 100,000 workers in such working conditions [18]. In addition, nearly 38 people die of poisoning or suffocation accidents in confined spaces every year [19]. In Virginia, the highest death rate per million employees is occupational confined-space accidents in shipbuilding and repair facilities, with a probability of 23.2% [20]. Therefore, it is especially necessary to analyze and solve the safety dilemma surrounding this kind production activity.

When there is a production accident, people tend to investigate the causes of accidents on the surface, rather than analyze the root causes, which leads to simply attributing the accidents to worker violations or errors. As a result, the risk control strategy is always inadequate, and accidents endangering the safety of employees occur repeatedly. Consistent with these observations, organizational and management factors such as the weak safety awareness of workers, high work pressure, inadequate professional familiarization and training, and lax supervision of processes [1,17,21] are increasingly being recognized as the deep-seated causes of accidents that need to be acknowledged. These factors are closely related to human behavior, not only human errors. Here, the related risks are collectively referred to as human factors. It is generally acknowledged that, with the economic environment and advanced science and technology providing technical guarantees and a material basis for the realization of safe conditions, human factors play an increasingly prominent role in the occurrence and evolution of accidents [22,23].

In addition, the risks that lead to incidents and accidents in shipbuilding and repair, including human factors, are extremely complex and uncertain. Therefore, it is crucial to systematically study the formation mechanism of accidents and the causal relationship between risk factors. Risk assessment is a useful technology in the field [3,13,24]. Some qualitative research methods involving literature reviews, field surveys of practitioners, and expert interviews can assist in the identification of potential risk factors related to accidents [25]. They provide factual information about working conditions, occupational characteristics, and management vulnerabilities that may not be covered in official accident reports, promoting a further understanding of the risks and hazards faced by employees. Meanwhile, studies have shown that the causes of accidents are multifaceted and variable [23,26,27]; hence, there is a need to classify these factors from different research perspectives. The decomposition and classification of human factors have been studied and widely applied in various fields by [28] and other researchers. The HFACS is widely regarded as having great value in the systematic analysis of the causes of accidents, which takes the continuous influence of high-level factors on low-level factors into account [29]. However, the complex logic relationships among factors allocated at different hierarchical layers under the HFACS framework are frequently ignored. Therefore, this study considers the logic relationships among different factors to implement risk evaluation.

The methodologies employed to investigate the causation for shipyard accidents are mainly represented by statistics-based techniques, such as those used in [1,4,12,30,31]. To further analyze shipyard accidents, the hybrid methodologies proposed in recent years for marine accident analysis can be referred to for investigation of the potential risks involved in shipyard accidents. Some of these hybrid methodologies are listed in Table 1.

**Table 1.** Hybrid methodologies used within the scope of marine accident analysis in recent years.

| Source | Hybrid Methodology | Risk Scenario |
|---|---|---|
| [32] | AcciMap-ANP | Ship grounding accident |
| [33] | BN-TRACEr | Ship collision |
| [34] | ANP-HFACS | LPG leakage accident from gas carrier |
| [35] | Fuzzy theory-FTA | Ship mooring operation |
| [36] | FTA-Modified CREAM | Oil tanker collision accident |
| [37] | BN-TOPSIS | Maritime accident prevention |
| [38] | FAHP-spatial fuzzy multi-criteria approach | Maritime accidents in the South China sea |
| [23] | Fuzzy theory-BN | Maritime accidents associated with sand carriers |
| [25] | FT-ANN-HFACS | Maritime accidents associated with sand carriers in coastal waters in China |
| [39] | DT-BN | Ship oil spill accident |
| [40] | FTA-BN | Ship grounding accidents |
| [27] | ETA-CN | Ship grounding accidents |
| [41] | SEM-BN | Human-related factors in ship grounding accidents |
| [42] | FRAM-BN | Maritime liquid cargo leakage accidents |

Lots of complex causal relationships and uncertain risks are involved in shipyard accidents, which can hardly be addressed by statistics-based approaches. Therefore, it is necessary to consider hybrid methodologies to improve the safety management in shipyards by analyzing human-factor-related accidents. With reference to similar studies in the maritime industry, as listed in Table 1, it can be observed that BN is widely accepted as an effective technique to model maritime accidents. To implement the Bayesian inference, the probability distribution for the nodes within the BN is required; for this purpose, fuzzy theory and expert elicitation are frequently applied. However, in these cases, the subjective bias and knowledge limitation may inevitably reduce the accuracy of the Bayesian inference. Therefore, in the present study, the exploratory application of a CN approach is conducted to quantify the probability of all the nodes within BN. The focus of this study is to establish a comprehensive accident analysis and risk assessment model that can reflect the risk propagation path, and then combine with the fuzzy algorithm and information theory to quantify and address the uncertainty of human factors in a ship repair system. Moreover, an accident that occurred in a Chinese shipyard is used as an empirical case to verify the proposed method. A classic method of accident analysis, fault tree analysis (FTA), is used and combined with fuzzy extent AHP to systematically establish the causal relationship of accidents and address the randomness of risk factors in this study. In addition, BN is adopted to predict the probability of risk because of its powerful learning and reasoning capability in many intelligent algorithms. In particular, the introduction of canonical probabilistic models has come to be recognized as an ingenious method to reduce the difficulty of obtaining the conditional probability between nodes in a BN, and it can simplify and deal with the problems of complex systems.

### 1.2. Innovative Contribution

The purpose of this study is to develop a comprehensive, human-related hazardous events assessment methodology with full consideration of the influences from social and technical aspects. Using this methodology, the risks stemming from on-site operations and management level can be quantitatively analyzed for accidents which occur during docking repair. For this purpose, a suffocation accident which occurred in a Chinese shipyard is exampled as a case study to illustrate the application of the proposed methodology. By the application of the methodology, the accident report is firstly qualitatively interpreted according to the basic principle of ETA to identify the human-related hazardous events involved in the suffocation accident. Additionally, the logic relationships among the identified hazardous events are analyzed to establish the CN, which is then evaluated by the improved K-shell decomposition algorithm to obtain the node criticality. Meanwhile, the identified hazardous events are reorganized within the framework of the HFACS with reference to the node criticality to develop the framework of the BN. Subsequently, the above-obtained node criticality is utilized to calculate the probability distribution of the nodes involved in the BN, which is essential for Bayesian inference. The innovative contribution of the proposed methodology is summarized as follows.

1. Theoretical exploration of quantitatively describing the human-related hazardous events involved in the maritime accidents which have occurred during docking repair with a full consideration of factors from social–technological aspects.
2. The application of the CN in this study is able to conquer the shortage of the ETA, in which it fails to consider the logic relationships among hazardous events allocated at different levels.
3. The determination of prior probability and conditional probability of the Bayesian inference is implemented by the improved K-shell decomposition algorithm.
4. An entire solution to assess human-related hazardous events in shipyard accidents is established, and the consistent precision of this is validated with a case study of a suffocation accident.

### 1.3. Organization

The remainder of this paper is organized as follows. Section 2 describes the materials and methods for the evaluation of human-related hazardous events which contribute to suffocation. The results and discussion are presented in Section 3. Finally, the paper is concluded in Section 4.

## 2. Materials and Methods

### 2.1. Risk Scenario and Overview for the Proposed Methodology

#### 2.1.1. Risk Scenario Description

On 5 May 2021, at approximately 14:45, a newly built 180,000-tonnage bulk carrier named "H1502" suffered from a poisoning and suffocation accident caused by high-density nitrogen during docking repair in a shipyard in Shanghai. Unfortunately, this accident caused two deaths. The following details were extracted from the accident report which can be accessed from Office of Emergency Management of Shanghai [43].

In the morning of the 5 May, the operators were assigned to purge the pipes with pure nitrogen in the stem of the ship by the coordinator on site, in the absence of a representative of the ship's owner. After noon, the operators on site implemented the purge in the cabin containing the bathymeter, a space regarded as a typical enclosure space, without taking effective safety measures. At approximately 14:30, two hydraulic pipes which connected together with the "U" pipe, marked as A and B, respectively, were purged fully with nitrogen. Then, the outlet valve of nitrogen was closed. Subsequently, the operators on site dismantled the "U" pipe connecting the hydraulic pipes A and B, and intended to purge hydraulic pipe A. For this purpose, the outlet valve of nitrogen was opened again. At this moment, a large amount of nitrogen leaked into the cabin. Five minutes later, the operators were found syncope in the cabin.

2.1.2. Overview for the Proposed Methodology

In the present study, a novel methodology integrated by CN and BN is proposed to investigate the human-related hazardous events involved in the suffocation on ships from the perspective of the complex social–technological system. The general principle for the proposed methodology is illustrated in Figure 1, and there are four main steps for its implementation.

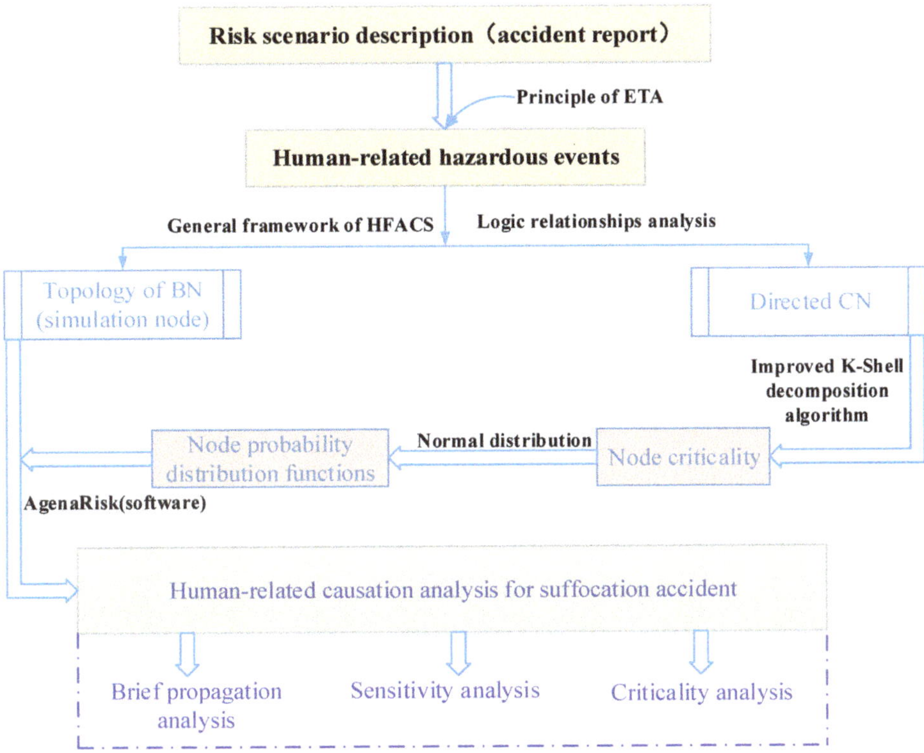


**Figure 1.** Overview for the proposed methodology.

Step 1—Risk scenario description: The human-related hazardous events involved in the accident report are firstly identified based on the principle of ETA. Then, the directed CN is developed on the basis of hazardous events, with the nodes and logic relationships among hazardous events being directed edges. In addition, the topological description for the developed CN is also conducted with the aspects of weights and degrees.

Step 2—Human-related hazardous events: The criticality of the nodes involved in the directed CN is then calculated by means of an improved K-shell decomposition algorithm based on the values of weights and degrees.

Step 3—Development of BN: The framework of the BN is first developed by the integration of HFACS and the results of node criticality. Then, the prior probability distribution for root nodes and the conditional probability tables for other nodes can be determined with application of the probability-related method proposed in this study, on the basis of node criticality.

Step 4—Human-related causation analysis based on BN simulation: The developed BN is finally simulated to investigate the human-related hazardous events involved in the suffocation accident, which is beneficial in improving the safety management for operation on board the ships.

### 2.2. Development of CN

In the present study, the CN is developed by the human-related hazardous events identification and causal logic analysis among these events. The principle of ETA is utilized to identify the human-related hazardous events involved in the accident report. As a result, these events are considered as the nodes within the CN, and the causal logic among the events would be regarded as the directed edges.

#### 2.2.1. Human-Related Hazardous Event Identification

According to the principle of ETA, the suffocation accident report was interpreted in detail to identify the human-related hazardous events contributing to the occurrence of the objective accident. Finally, a total of 40 hazardous events were identified, all of which are presented in Table 2.

**Table 2.** Hazardous events identified from the accident report.

| Item | Description |
| --- | --- |
| N_1 | A large amount of nitrogen remained in hydraulic pipes A and B, left by the operators |
| N_2 | The inappropriate design of hydraulic pipes A and B meant they ran through the ship from ahead to astern |
| N_3 | Hydraulic pipes A and B were connected as a loop by the operator to implement the pipe cleaning |
| N_4 | The safety supervisor on board the ship did not perceive the unsafe actions of the operators and failed to correct the inappropriate operations |
| N_5 | The representative of the ship owner was absent during the operation |
| N_6 | The volume of the fore peak tank was relatively small |
| N_7 | The hull length was approximately 290 m; as a result, the hydraulic pipe to be purged was longer than common pipes |
| N_8 | Nitrogen leakage |
| N_9 | A large amount of nitrogen was stored on board the ship—a total of 25 sets of nitrogen cylinders |
| N_10 | The safety management department of the shipyard failed to strictly implement all safety measures during the holiday season |
| N_11 | The safety management department of the shipyard did not attach great importance to the safety of the operation on site, and the safety issues were not paid much attention |
| N_12 | The quality management system in the safety management department was found be defective in the aspect of the required process guidance documents |
| N_13 | The shipyard failed to effectively supervise the operators on site to strictly implement the safety management system and the operation instruction |
| N_14 | The safety management department of the shipyard did not strictly implement the safety management regulations—there was no confirmation of the key operation |
| N_15 | The safety training and drilling in the safety management department of the shipyard had not been implemented for a long time |
| N_16 | The superintendent of the civil marine project failed to effectively supervise the issues in risk prevention |
| N_17 | The managers and officers in the civil marine project failed to pay much attention to the preventive measures in the field of safety when formulating the operation plan |
| N_18 | The superintendent of the civil marine project did not eliminate the potential dangers for the common operation in time |
| N_19 | Personnel suffocation |
| N_20 | The nitrogen accumulated in the enclosed space on site |
| N_21 | The operators on site did not take any measures to ventilate the enclosed space |
| N_22 | The person in charge of the operation on site did not implement safety-related regulations, such as confirmation, lighting, and supervision |
| N_23 | The person in charge of the operation on site failed to give input on the operation environment and provide caution to the operators |

**Table 2.** *Cont.*

| Item | Description |
|---|---|
| N_24 | The person in charge of the on-site operation did not confirm the ventilation |
| N_25 | The operators on site did not implement the required risk-prevention measures for the operation in the limited space |
| N_26 | The operator on site did not apply for a permit for the operation procedures |
| N_27 | The person in charge of the operation on site failed to check the operation permit in the limited space before the operation |
| N_28 | The person in charge of the operation on site did not confirm the implementation of gas detection |
| N_29 | The person in charge of the operation on site did not effectively perform their designated responsibility during the operation |
| N_30 | The work associated with risk identification before the operation was not performed by the person in charge of the operation |
| N_31 | The operators on site failed to perform gas testing |
| N_32 | The removing of the "U" pipe containing nitrogen in the enclosed space is usually characterized by high risk, which was not did not receive due attention from the operators on site |
| N_33 | The risk-prevention measures applicable for the enclosed space were not in place before the operation, and various potential risks were not effectively identified |
| N_34 | The process guidance documents for the officers in the general assembly department were absent |
| N_35 | The officers in the general assembly department failed to identify all the risks associated with the temporary operation |
| N_36 | The officers in the general assembly department failed to implement the safety-related measures designed for the holiday season |
| N_37 | The person on duty in the general assembly department did not perform their responsibilities effectively |
| N_38 | The officers in the general assembly department failed to implement the safety training for the temporary operators in relation to operative environments and the potential risks |
| N_39 | The officers in the general assembly department did not effectively perform their supervision and risk monitoring responsibilities |
| N_40 | Most of the people involved in the accident were found to have low awareness of the safety-related issues during the "May 1st" Labor Day |

### 2.2.2. Topology for the CN

The logic relationships among various human-related hazardous events listed in Table 2 were used to develop a CN. For this purpose, five experts were employed in this study to determine the logic relationships, whose basic information is summarized in Table 3. To conduct the expert elicitation, a safety meeting was held, during which the accident report was introduced. The logic relationships are summarized in Figure 2.

**Table 3.** Basic information for the experts employed in the present study.

| Expert | Age | Occupation | Educational Level | Certificate Rank | Job Tenure |
|---|---|---|---|---|---|
| Expert 1 | 47 | Shipbuilding engineer | Master of naval engineering | Chief engineer | He has 20 years of experience in shipbuilding and is responsible for the formulation of plans related to safe operation in shipbuilding |
| Expert 2 | 52 | Ship surveyor | Doctor of marine engineering | Ship surveyor | He has been engaged in ship risk assessment for nearly 20 years and has participated in the investigation of many major ship safety accidents |
| Expert 3 | 45 | Security incident investigator | Master of Naval Architecture and Marine Engineering | Senior captain | He has worked in a shipyard for 15 years and is responsible for the safety operation of ships and was recently ranked as senior in ship safety accident analysis |
| Expert 4 | 49 | Security incident investigator | Master of Naval Architecture and Marine Engineering | Captain | He is ranked as the captain because he has 5 years of experience in the investigation and analysis of ship safety incidents |

**Table 3.** *Cont.*

| Expert | Age | Occupation | Educational Level | Certificate Rank | Job Tenure |
|---|---|---|---|---|---|
| Expert 5 | 53 | Professor in university | Doctor of Naval Architecture and Marine Engineering | Captain | He is a professor employed in a maritime university and his research interests involve operational safety aspects in shipbuilding |

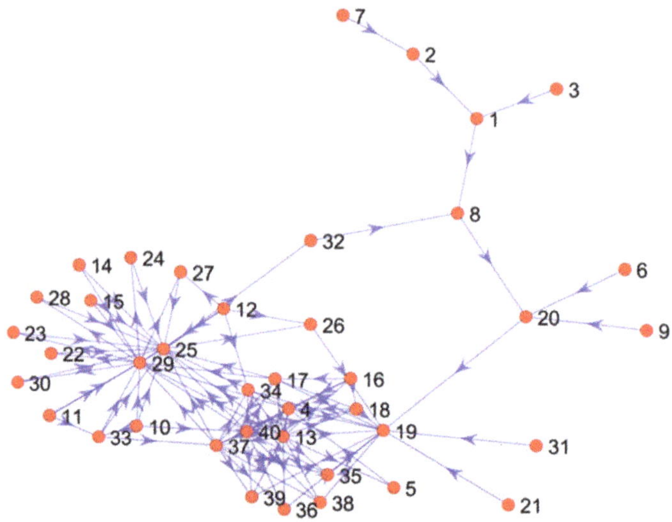


**Figure 2.** CN obtained from the accident report.

### 2.3. *Calculation for the Node Criticality*

#### 2.3.1. Basic Criticality Calculation for Nodes within CN

The topology of the CN can be described by the parameters associated with degree and weight. It is noticeable that the weight-related parameter is equal to 1 in the case of a single accident report. To implement the topological description for the developed directed CN, the network is represented as:

$$G = (V, E) \tag{1}$$

where $V = \{v_1, v_2, \ldots, v_N\}$ and $E = \{e_1, e_2, \ldots, e_M\}$ represent the set of nodes and the set of edges, respectively. In the present study, $v_i$ indicates the identified hazardous event, while $e_i$ denotes the logic relationship. Then, an adjacency matrix A is employed to quantitatively describe $G = (V, E)$, and the elements in the adjacency matrix are determined by:

$$A_{ij} = \begin{cases} a_{ij} \cdot w_{ij} \ i \to j \\ 0 \ \ else \end{cases} \tag{2}$$

where $a_{ij}(i, j \in V)$ is equal to 1 when $v_j$ is triggered by $v_i$, and is zero otherwise. The weight between the node $v_i$ and the node $v_j$ is expressed by $w_{ij}$. The meaning of weight is to make the connection strength between the nodes are quantified. The weight is divided into two kinds: in-weight and out-weight. In general, the characteristics of the CN are obtained by node strength and cross-weight and can be determined by the correlation calculation of the weight.

$$T_w(i) = S_{in}(i) + S_{out}(i) = \sum_{j \in V} w_{ji} + \sum_{j \in V} w_{ij} \tag{3}$$

$$C_w(i) = S_{in}(i) \cdot S_{out}(i) = \sum_{j \in V} w_{ji} \cdot \sum_{j \in V} w_{ij} \tag{4}$$

Another important parameter for a typical, directed CN refers to the degree of node, which is defined as the sum of nodes connecting or being connected to the objective node. In the present study, the degree of the objective node is denoted by $d(i)$. Similarly, the degree is also divided as in-degree and out-degree, which can be determined, respectively, by:

$$d_{\text{in}}(i) = \sum_{j \in V} a_{ji} \tag{5}$$

$$d_{\text{out}}(i) = \sum_{j \in V} a_{ij} \tag{6}$$

According to the equations expressed in (5) and (6), the in-degree of a node is the total number of nodes affecting the objective node, and the out-degree of a node refers to the total number of nodes affected by the objective node. Then, the total degree of the objective node can be obtained by:

$$T_{\text{d}}(i) = d_{\text{in}}(i) + d_{\text{out}}(i) = \sum_{j \in V} a_{ji} + \sum_{j \in V} a_{ij} \tag{7}$$

The total degree is generally used to evaluate the activeness of nodes within a network according to graph theory; however, the difference between nodes characterized by the same total degree cannot be distinguished from the perspective of activeness. Therefore, the concept of the cross-degree is introduced in this study, which is mainly applied to distinguish nodes with the same total degree:

$$C_{\text{d}}(i) = d_{\text{in}}(i) \cdot d_{\text{out}}(i) = \sum_{j \in V} a_{ji} \cdot \sum_{j \in V} a_{ij} \tag{8}$$

In the present study, node criticality was determined by the improved K-shell decomposition algorithm, which is implemented with the precondition of calculating the basic criticality as the input for the improved K-shell decomposition algorithm. To calculate the values of basic criticality for the identified nodes, a group of control equations associated with degree and weight are proposed here, which are expressed as:

$$f_1(w_{ji}, w_{ij}) = \left| \sum_{j \in V} w_{ji} - \sum_{j \in V} w_{ij} \right| \tag{9}$$

$$f_2[d_{\text{in}}(i), d_{\text{out}}(i)] = |d_{\text{in}}(i) - d_{\text{out}}(i)| \tag{10}$$

$$f_3(w_{ji}, w_{ij}) = \sum_{j \in V} w_{ji} + \sum_{j \in V} w_{ij} \tag{11}$$

$$f_4[d_{\text{in}}(i), d_{\text{out}}(i)] = d_{\text{in}}(i) + d_{\text{out}}(i) \tag{12}$$

The nodes are generally characterized as critical in case of being valued by larger $f_3$ and $f_4$, and by smaller $f_1$ and $f_4$. However, the ideal values of $f_1$ and $f_4$ cannot be obtained simultaneously. Therefore, it is essential to search for a balance between $f_1$ and $f_4$. For this purpose, two balance coefficients are defined as [27]:

$$B_{\text{d}}(i) = \left[ 1 - \left( \frac{d_{\text{in}}(i)}{d_{\text{in}}(i) + d_{\text{out}}(i)} - 0.5 \right)^2 \right] \tag{13}$$

$$B_{\text{w}}(i) = \left[ 1 - \left( \frac{\sum_{j \in V} w_{ij}}{\sum_{j \in V} w_{ij} + \sum_{j \in V} w_{ji}} - 0.5 \right)^2 \right] \tag{14}$$

where $B_d(i)$ is the degree balance coefficient, and $B_w(i)$ is the weight balance coefficient. When the control Equation (10) and cross-strength of the node is considered synthetically, the weight balance index is defined as:

$$\Psi_{\mathrm{w}}(i) = B_{\mathrm{w}}(i) \cdot [T_{\mathrm{w}}(i) + C_{\mathrm{w}}(i)] \tag{15}$$

The degree balance index is similar to the weight balancing index, which can be obtained with the cross-degree and control Equation (11):

$$\Psi_{\mathrm{d}}(i) = B_{\mathrm{d}}(i) \cdot [T_{\mathrm{d}}(i) + C_{\mathrm{d}}(i)] \tag{16}$$

As a result, the basic criticality of node can be determined by:

$$\Psi_{\mathrm{B}}(i) = \Psi_{\mathrm{d}}(i) \cdot \Psi_{\mathrm{w}}(i) \tag{17}$$

#### 2.3.2. Node Criticality Determination by Improved K-Shell Decomposition Algorithm

In the present study, the aforementioned basic criticality would be considered as the input of K-shell decomposition algorithm to determine the node criticality. The K-shell decomposition algorithm is widely applied to analyze the node criticality based on the topology of the CN [44]. Generally, the basic K-shell decomposition algorithm is applicable for undirected unweighted networks, and aims to decompose the node set into subsets based on node centrality [45]. Meanwhile, a layer sequence number $\xi_{\mathrm{K}}$ is assigned to each node according to its centrality value in the network and used to assess the criticality of objective nodes [27]. Nodes assigned larger $\xi_{\mathrm{K}}$ values are closer to the center of the network, and nodes assigned smaller $\xi_{\mathrm{K}}$ values tend to be at the periphery of the network. The basic K-shell decomposition algorithm can be improved to be applicable for the directed, weighted CN.

In the present study, the obtained basic criticality of the objective node is considered as the decomposition function, which is then embedded into K-shell decomposition algorithm, as a result, the layer sequence number can be determined as:

$$\Psi_{\mathrm{K}}(i) = \mathrm{K}[\Psi_{\mathrm{B}}(i)] \tag{18}$$

According to the principle of the K-shell decomposition algorithm, $\Psi_{\mathrm{K}}(i)$ is the criticality value of the layer containing the $i$th node, and the layer sequence number is denoted by $\xi_{\mathrm{K}}$. All the nodes at the same layer are regarded as a set of $G_{\xi_{\mathrm{K}}} = (\text{node } i|\xi_{\mathrm{K}}(i) = \xi_{\mathrm{K}})$. To distinguish the nodes sharing the same layer sequence number, in the present study, an iteration factor is introduced into the traditional K-shell decomposition algorithm. As a result, the improved K-shell decomposition algorithm can be obtained and is expressed as:

$$\Psi_{\mathrm{K}i}(i) = \Psi_{\mathrm{K}}(i) + \Gamma_{\xi_{\mathrm{K}}}(i) \tag{19}$$

where $\Psi_{\mathrm{K}i}(i)$ is the criticality of the objective node, and the iteration factor is represented by $\Gamma_{\xi_{\mathrm{K}}}(i)$, which can be obtained by [27]:

$$\Gamma_{\xi_{\mathrm{K}}}(i) = \begin{cases} \frac{(\sigma_i - 1)}{\gamma} \cdot (\Psi_{\mathrm{K}+1} - \Psi_{\mathrm{K}}) \text{ node } i \in G_{\xi_{\mathrm{K}}}, \xi_{\mathrm{K}} \text{ is not the outermost layer} \\ \frac{(\sigma_i - 1)}{\gamma} \cdot (\Psi_{\mathrm{K}} - \Psi_{\mathrm{K}-1}) \text{ node } i \in G_{\xi_{\mathrm{K}}}, \xi_{\mathrm{K}} \text{ is the outmost layer} \end{cases} \tag{20}$$

where $\gamma$ is the number of nodes contained in layer $\xi_{\mathrm{K}}(i)$, and $\sigma_i$ is the order in which node $v_i$ is assigned to layer $\xi_{\mathrm{K}}(i)$.

According to the improved K-shell decomposition algorithm represented by Equations (9)–(20), the algorithm is coded and implemented with the application of MATLAB. Then, the criticality for nodes within the CN can be obtained, and the results of all the nodes in the CN are summarized in Table 4. It is noticeable that the criticality value for the node coded by "N_19" (personnel suffocation) is marked as 1. Moreover, in this study, the "N_19" node is regarded as the top event in the BN in the following section.

**Table 4.** Criticality values for nodes.

| Item | Node Criticality | Item | Node Criticality |
|---|---|---|---|
| N_1 | 0.1245 | N_21 | 0.0461 |
| N_2 | 0.1059 | N_22 | 0.3940 |
| N_3 | 0.0873 | N_23 | 0.3940 |
| N_4 | 0.8433 | N_24 | 0.3940 |
| N_5 | 0.3391 | N_25 | 0.6745 |
| N_6 | 0.0598 | N_26 | 0.5494 |
| N_7 | 0.1010 | N_27 | 0.5220 |
| N_8 | 0.5038 | N_28 | 0.3940 |
| N_9 | 0.0598 | N_29 | 0.6745 |
| N_10 | 0.4764 | N_30 | 0.3940 |
| N_11 | 0.3598 | N_31 | 0.0461 |
| N_12 | 0.3418 | N_32 | 0.4489 |
| N_13 | 0.9474 | N_33 | 0.4297 |
| N_14 | 0.2255 | N_34 | 0.5526 |
| N_15 | 0.2255 | N_35 | 0.5388 |
| N_16 | 0.9337 | N_36 | 0.5388 |
| N_17 | 0.9337 | N_37 | 0.8352 |
| N_18 | 0.9337 | N_38 | 0.5388 |
| N_19 | 1.0000 | N_39 | 0.5388 |
| N_20 | 0.5313 | N_40 | 0.8215 |

### *2.4. Bayesian Inference*

#### 2.4.1. Development of Topology for BN

It is difficult to transfer the CN directly into BN due to the complex logic relationships in the CN. Therefore, in this study, the HFACS was introduced to connect the CN and the BN. In detail, the human-related hazardous events involved in the CN were categorized under the framework of HFACS in the aspects of unsafe acts, precondition for unsafe acts, unsafe supervision and organizational influences. As a result, in the developed BN, the human-related hazardous events are considered as the root nodes, the four aspects of HFACS serve as the middle nodes and the accident is the top node. It is notable that not all the nodes within the CN can be transferred into the BN as the root nodes.

To develop a BN, the human-related hazardous events listed in Table 2 were first categorized according to the principle of HFACS model. Under the framework of HFACS, all the human-related factors contributing to the accident are summarized in the aspects of unsafe acts, precondition for unsafe acts, unsafe supervision and organization influence. In the present study, those nodes characterized by the least 10 importance are deleted for the establishment of the BN, and the results are summarized in Table 5.

**Table 5.** HFACS framework for the identified hazardous events.

| Aspect | Contents |
|---|---|
| Unsafe acts (UA) | N_8, N_20, N_22, N_23, N_24, N_25, N_26, N_27, N_28, N_29, N_30, N_32 |
| Precondition for unsafe acts (UP) | N_5, N_33, N_40 |
| Unsafe supervision (US) | N_4, N_16, N_17, N_18, N_34, N_35, N_36, N_37, N_38, N_39 |
| Organizational influence (OI) | N_10, N_11, N_12, N_13, N_14 |

According to the category illustrated in Table 5, the "N_19" node (personnel suffocation) is regarded as the top node in the BN; the four aspects of HFACS are considered as the intermediate nodes; and the nodes contributing to every aspect of the HFACS are set as the root nodes. As a result, the basic framework of the BN is developed as shown in Figure 3.

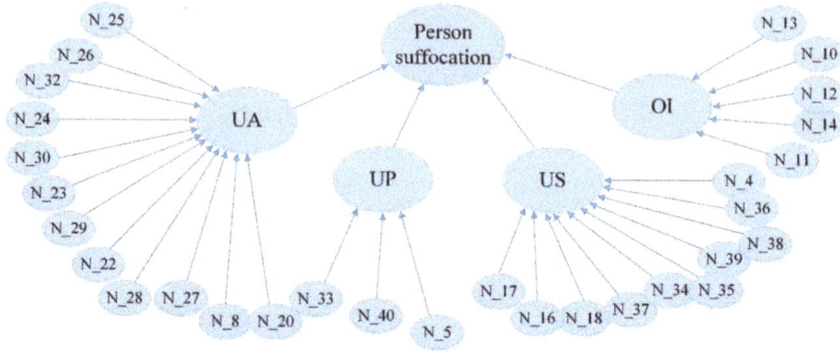


**Figure 3.** BN developed for the human-related factors analysis.

2.4.2. Determination of Probability Distribution for Nodes within BN Based on Criticality

The basic principle for Bayesian inference is presented in [23]. In the present study, all the nodes within the BN are considered as simulation nodes rather than discrete nodes. Therefore, the traditional conditional probability tables for the middle nodes are presented as the continuous probability distribution functions. Additionally, the probability distribution of all the nodes needed for the BN simulation in this study are determined with reference to [42]. It is assumed that the probability distribution of all the nodes, including the root nodes, intermediate nodes and top-level nodes, obeys the normal distribution. For the root nodes, the mean value of probability distribution can be calculated by normalizing the maximum value of the criticality for the root nodes, which is expressed as:

$$P_{\mathrm{Ki}}(i) = \frac{\Psi_{\mathrm{Ki}}(i)}{\mathrm{Max}(\Psi_{\mathrm{Ki}})} \tag{21}$$

where $\Psi_{\mathrm{Ki}}(i)$ denotes the node criticality, and $\Psi_{\mathrm{Ki}}(i)$ can be valued according to the algorithms proposed in Section 2.3. For the intermediate nodes within the BN, the mean values of probability distribution can be obtained on the basis of probability distribution of the parent nodes contributing to the objective intermediate node, which is expressed as:

$$\chi(j) = \sum \omega_{\mathrm{Ki}}(i) \cdot \Lambda(i) \tag{22}$$

where $\Lambda(i)$ denotes the sampling value of parent nodes directing to the $j$th intermediate node. In the present study, $\Lambda(i)$ refers to the mean value of the probability distribution for the $i$th node. The weight of $i$th root node is represented by $\omega_{\mathrm{Ki}}(i)$ that can be calculated by:

$$\omega_{\mathrm{Ki}}(i) = \frac{\Psi_{\mathrm{Ki}}(i)}{\sum \Psi_{\mathrm{Ki}}} \tag{23}$$

Similarly, the mean value of probability distribution for the top nodes can be calculated by the following equation:

$$\Omega(t) = \sum \lambda_{\mathrm{Ki}}(j) \cdot \chi(j) \tag{24}$$

where $\lambda_{\mathrm{Ki}}(j)$ denotes the weight of the intermediate node, and can be calculated by the following equation:

$$\omega_{\mathrm{Ki}}(j) = \sum_{i \in j} \frac{\Psi_{\mathrm{Ki}}(i)}{\sum_{i \in R} \Psi_{\mathrm{Ki}}(i)} \tag{25}$$

(1) Determination of probability distribution for root nodes:

According to the methodology proposed here, the prior probability for root nodes can be determined on the basis of the node criticality. In the present study, the prior probability distribution of the root nodes is assumed to be normal distribution, which can be expressed as:

$$F_i \sim N(P_i, \sigma) \quad (26)$$

where $F_i$ is the prior probability distribution of the $i$th root node, and $P_i$ denotes the mean value of the probability distribution, which is valued as the node criticality presented in Table 4. The variance of normal distribution is donated by $\sigma$, which is determined as one-third of the mean value according to the principle of $3\sigma$ (standard deviation). According to Equation (26), the prior probability distribution for the root nodes in the BN can be obtained, and the results are summarized in Table 6.

**Table 6.** The criticality and probability distribution for root nodes.

| Node | Criticality | Normal Distribution | | Node | Criticality | Normal Distribution | |
|---|---|---|---|---|---|---|---|
| | | Mean | Variance | | | Mean | Variance |
| N_4 | 0.8433 | 0.8901 | 0.2967 | N_25 | 0.6745 | 0.7120 | 0.2373 |
| N_5 | 0.3391 | 0.3579 | 0.1193 | N_26 | 0.5494 | 0.5799 | 0.1933 |
| N_8 | 0.5038 | 0.5318 | 0.1773 | N_27 | 0.5220 | 0.5509 | 0.1836 |
| N_10 | 0.4764 | 0.5028 | 0.1676 | N_28 | 0.3940 | 0.4158 | 0.1386 |
| N_11 | 0.3598 | 0.3797 | 0.1266 | N_29 | 0.6745 | 0.7120 | 0.2373 |
| N_12 | 0.3418 | 0.3608 | 0.1203 | N_30 | 0.3940 | 0.4158 | 0.1386 |
| N_13 | 0.9474 | 1.0000 | 0.3333 | N_32 | 0.4489 | 0.4738 | 0.1579 |
| N_14 | 0.2255 | 0.2380 | 0.0793 | N_33 | 0.4297 | 0.4535 | 0.1512 |
| N_16 | 0.9337 | 0.9855 | 0.3285 | N_34 | 0.5526 | 0.5832 | 0.1944 |
| N_17 | 0.9337 | 0.9855 | 0.3285 | N_35 | 0.5388 | 0.5687 | 0.1896 |
| N_18 | 0.9337 | 0.9855 | 0.3285 | N_36 | 0.5388 | 0.5687 | 0.1896 |
| N_20 | 0.5313 | 0.5608 | 0.1869 | N_37 | 0.8352 | 0.8816 | 0.2939 |
| N_22 | 0.0461 | 0.4158 | 0.1386 | N_38 | 0.5388 | 0.5687 | 0.1896 |
| N_23 | 0.3940 | 0.4158 | 0.1386 | N_39 | 0.5388 | 0.5687 | 0.1896 |
| N_24 | 0.3940 | 0.4158 | 0.1386 | N_40 | 0.8215 | 0.8671 | 0.2890 |

(2) Determination of probability distribution for intermediate nodes and the top node:

The conditional probability distribution for the child nodes (intermediate nodes and the top node) in the developed BN can be determined based on the algorithm proposed in this section. Specifically, the probability distribution of the root nodes listed in Table 6 is set as the input for Equations (21)–(25). Then, the probability distribution for the four intermediate nodes and the top node can be determined, which is the function of the probability distribution of the root nodes, as shown in Table 7. It is notable that the probability distribution expressions presented in Table 7 can be then entered directly into the software of AgenaRisk for the simulation of BN.

**Table 7.** Probability distribution expression for intermediate nodes and the top node.

| Node | Probability Distribution Expression |
|---|---|
| UA | 0.1148 × N_25 + 0.0935 × N_26 + 0.1148 × N_29 + 0.0904 × N_20 + 0.0888 × N_27 + 0.0858 × N_8 + 0.0764 × N_32 + 0.0671 × N_22 + 0.0671 × N_23 + 0.0671 × N_24 + 0.0671 × N_28 + 0.0671 × N_30 |
| UP | 0.5166 × N_40 + 0.2702 × N_33 + 0.2132 × N_5 |
| US | 0.1299 × N_16 + 0.1299 × N_17 + 0.1299 × N_18 + 0.1173 × N_4 + 0.1162 × N_37 + 0.0769 × N_34 + 0.0750 × N_35 + 0.0750 × N_36 + 0.0750 × N_38 + 0.0750 × N_39 |
| OI | 0.4030 × N_13 + 0.2026 × N_10 + 0.1530 × N_11 + 0.1454 × N_12 + 0.0959 × N_14 |
| Personnel suffocation | UA × 0.3455 + UP × 0.0935 + US × 0.4227 + OI × 0.1383 |

## 3. Results and Discussion

### 3.1. Results

#### 3.1.1. Topology Analysis of CN

According to the calculation principle for the degree and weight described in Section 2.3, the degree distribution and weight distribution can be discussed according to [44]. In the present study, the weight is set as 1 because the analysis sample is limited to a single accident report; therefore, the topology analysis for the developed CN is limited to the degree distribution. According to [44], the degree of each node in the developed CN can be determined. Following this, the frequencies for the in-degree and out-degree can be plotted, as illustrated in Figure 4.

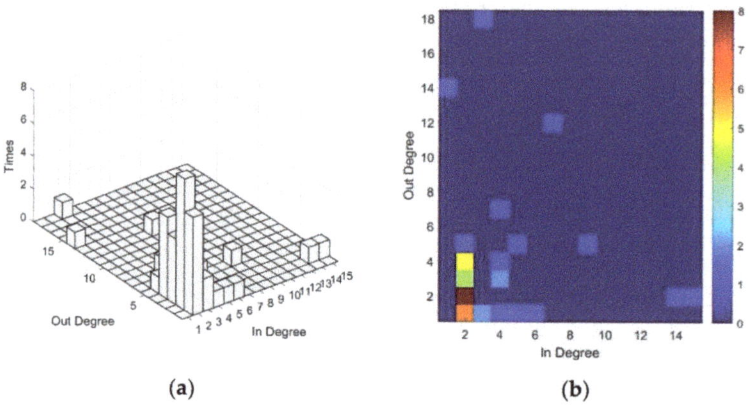


**Figure 4.** Degree distribution for the developed CN. (**a**) Degree distribution of the CN; (**b**) phase diagram of degree distribution.

The phase diagram of Figure 4b is essentially the projection of Figure 4a along the Y–Z axis. Based on the contents of Figure 4, the in-degree and out-degree of most nodes are less than 4, and the number of nodes with $d_{in} = 2$, $d_{out} = 2$ is found to be larger than other nodes. In addition, some nodes illustrated in Figure 4 are supposed to receive significant attention. For instance, the nodes with large in-degree are characterized by being caused or affected easily, which indicates that these nodes are difficult to control; meanwhile, the nodes with large out-degree are capable of affect other nodes easily, which shows that these nodes are important in improving the safety level. To distinguish the nodes with different in-degree and out-degree, the degrees for the nodes involved in the CN are then calculated, and the results are illustrated in Figure 5.

#### 3.1.2. Belief Propagation Analysis

According to [46], the belief propagation analysis of Bayesian inference usually contains forward-propagation and backward-propagation analysis. The forward-propagation analysis is mainly aimed at investigating the influence of parent nodes on their child nodes, while the backward-propagation analysis focuses on the sensitivity of the parent nodes to their child nodes, which is implemented by updating the probability distribution of the parent nodes by setting their child nodes as "evidence". In the present study, the criticality-based values are utilized to assess the human-related factors involved in the accident, and the higher values of a node, the more important the objective factors. The results presented in Figure 6 are set as the base for comparison, which can be found in Table 8. According to the results of criticality evaluation summarized in Table 4, the nodes with largest criticality value within each category of human-related factor under the framework of HFACS are selected as the given evidence for the backward-propagation analysis. As a result, these nodes coded by "N_13", "N_16", "N_17", "N_18", "N_25", "N_29" and "N_40" were selected as the given evidence in the present study. Then, different risk scenarios are developed by decreasing the criticality values of these selected nodes by 50% one by one, and the different scenarios are summarized in Table 8. Subsequently, the various risk scenarios are simulated by the developed BN on the software of AgenaRisk, and the variations in the node of "personnel suffocation" and the four aspects of HFACS are presented in Table 8.

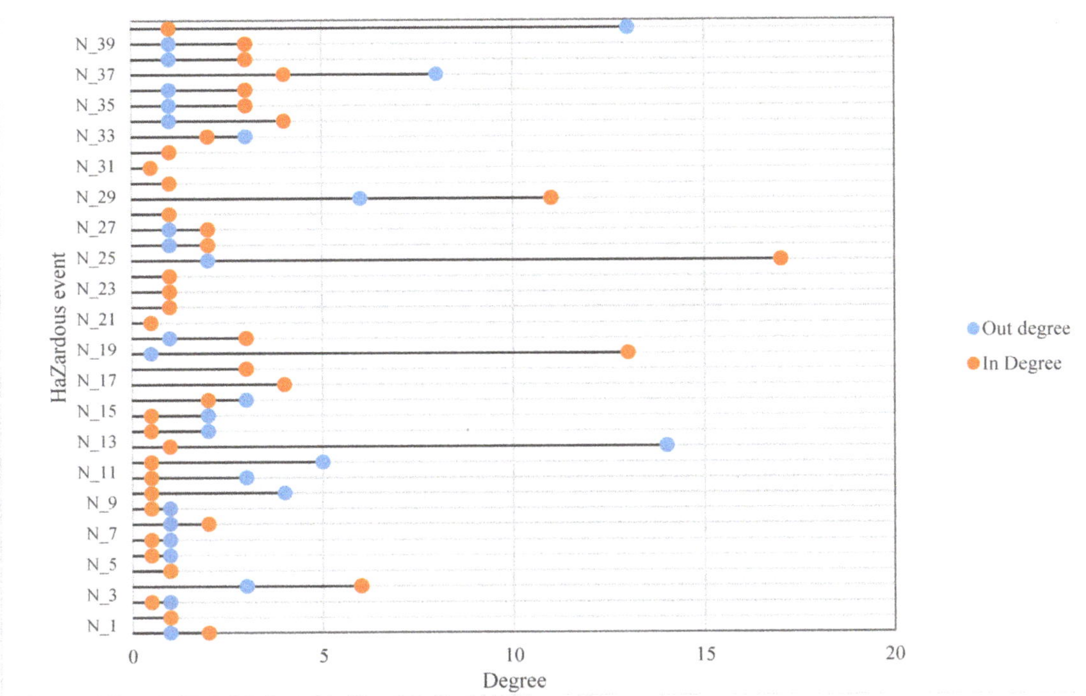


**Figure 5.** Node evaluation from degree.

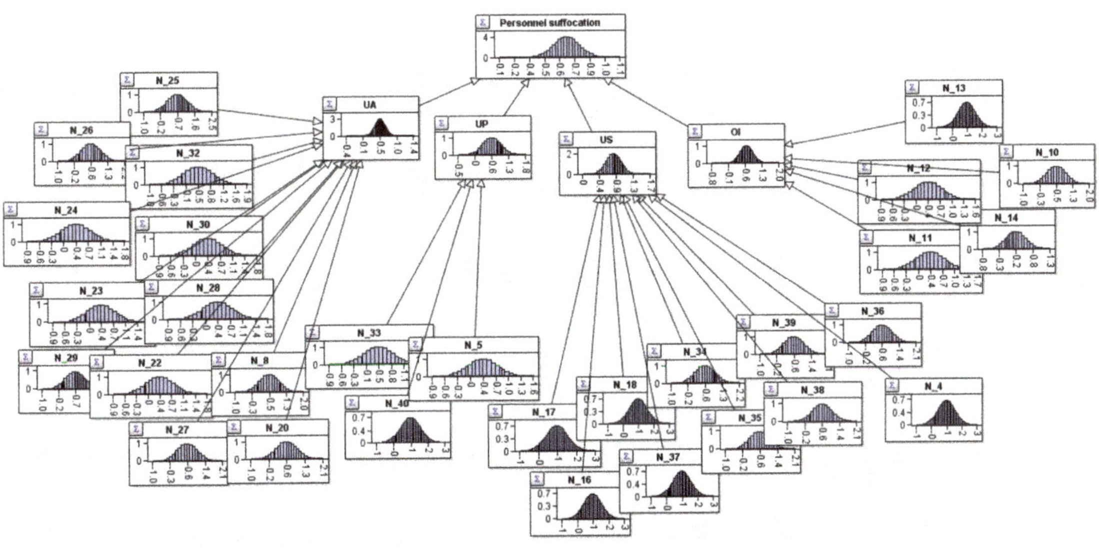


**Figure 6.** The results of running the BN model.

**Table 8.** Different scenarios for propagation analysis.

| Scenario | Criticality (↓ 50%) | UA | UP | US | OI | Personnel Suffocation |
|---|---|---|---|---|---|---|
| Base | — | 0.538 | 0.646 | 0.806 | 0.638 | 0.675 |
| 1 | N_25 | 0.497(↓) | 0.646 | 0.806 | 0.638 | 0.661(↓) |
| 2 | N_29 | 0.497(↓) | 0.646 | 0.806 | 0.638 | 0.661(↓) |
| 3 | N_40 | 0.538 | 0.415(↓) | 0.806 | 0.638 | 0.654(↓) |
| 4 | N_16 | 0.538 | 0.646 | 0.743(↓) | 0.638 | 0.649(↓) |
| 5 | N_17 | 0.538 | 0.646 | 0.743(↓) | 0.638 | 0.649(↓) |
| 6 | N_18 | 0.538 | 0.646 | 0.743(↓) | 0.638 | 0.649(↓) |
| 7 | N_13 | 0.538 | 0.646 | 0.806 | 0.440(↓) | 0.648(↓) |

Note: ↓ indicates the reduction of a certain variable value in the specific scenario compared with the value in the base case.

### 3.1.3. Sensitivity Analysis

The sensitivity analysis was implemented in the present study to quantify the sensitivity of the identified human-related hazardous events to the personnel suffocation accident using the backward-propagation analysis of the developed BN. During the sensitivity analysis, the top event, which refers to the personnel suffocation, was set as the given evidence, and the developed BN was simulated with AgenaRisk. In the present study, sensitivity nodes were selected according to the node criticality, and the top ten root nodes were considered as the sensitivity nodes for the sensitivity analysis. The mean value of the target node (referring to personnel suffocation) was selected to indicate the summary statistics. Finally, the results of the sensitivity analysis are presented by the tornado graph, which is illustrated in Figure 7. The criticality of the selected nodes is larger than 0.5526, and some nodes share the same criticality, such as "N_16", "N_17" and "N_18". According to the basic principle of the sensitivity analysis, the sensitivity level of the nodes is represented by the length of the blue bars illustrated in Figure 7. The longer the blue bar is, the more sensitive the objective node is, and the nodes characterized by high sensitivity would be paid much attention to improve the safety management system to prevent the similar accidents.

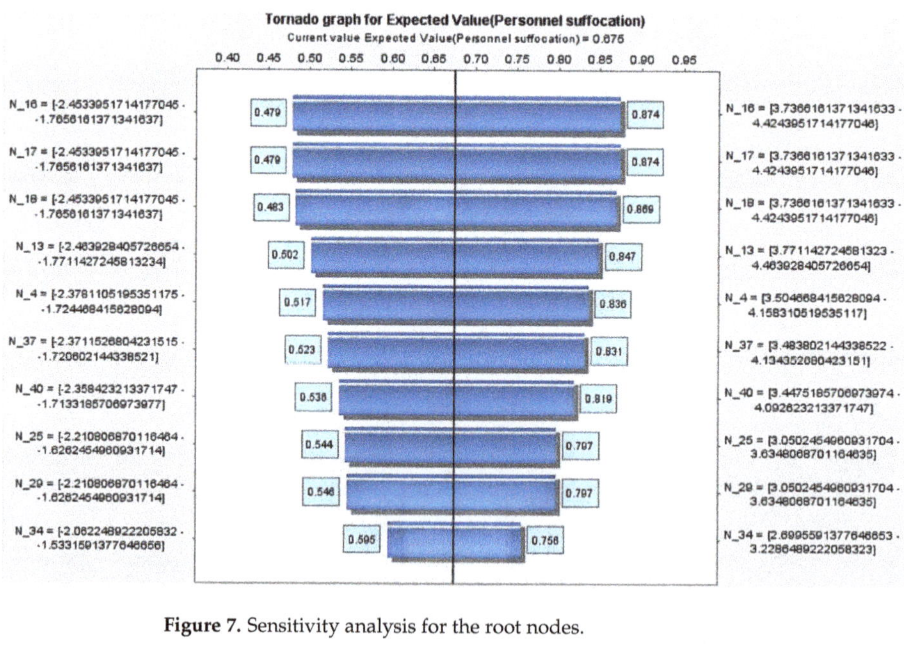


**Figure 7.** Sensitivity analysis for the root nodes.

In the present study, the sensitivity analysis was implemented to investigate the sensitivity of the intermediate nodes which were set as the sensitivity nodes, while the top event was again regarded as the target node. The results are shown in Figure 8 in the form of a tornado graph, which was obtained by simulating the developed BN with the mean value of the top node criticality representing the summary statistics.

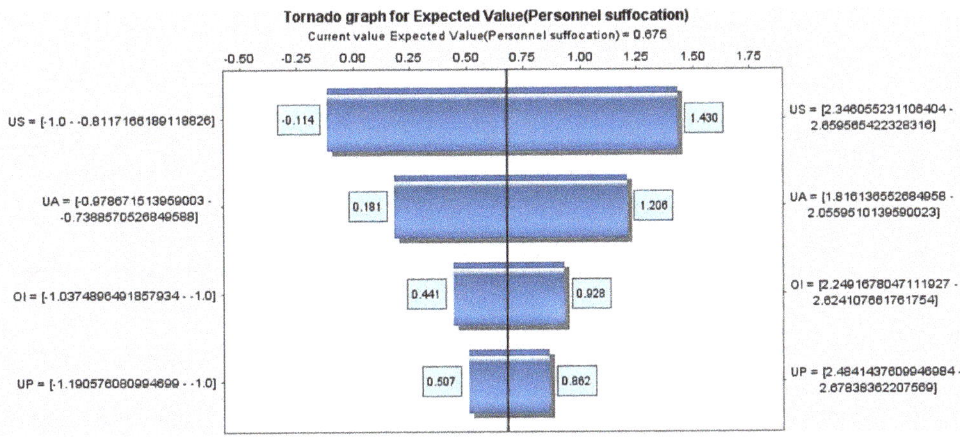


**Figure 8.** Sensitivity analysis of the intermediate nodes.

### 3.2. Discussion

In the aspect of the topology of the CN, according to Figures 4 and 5, it is clear that the node of "N_25" is valued as the maximum in terms of in-degree, which indicates that the risk-prevention measures taken by the operators on site should have been supervised and instructed by many other stakeholders. In addition, the node of "N_13" can be observed as the maximum in terms of out-degree, and it can be reasonably inferred that the supervision of the operators on the site of the shipyard is critical for accident prevention. Moreover, most operators on site in the shipyard are affiliated with subcontractors, which has been identified as a primary risk according to [12]. In addition, it can also be seen that the safety culture, especially during the holiday season, is essential for safety, which is proved by the second largest out-degree of "N-40", which is similar with the study conducted by the authors of [12]. In their study, the lunch effect attenuated the concentration of the operators on-site, and the holiday effect also functioned by a similar principle.

The results of belief propagation analysis summarized in Table 8 can be utilized to assess the sensitivity of the human-related hazardous events for the suffocation accident. Taking the comparison of the base and scenario 1 as an example, the status of UA would decrease from 0.538 to 0.497 in the case of lowering the node criticality of "N_25" by 50%, which would subsequently cause the change of the top event by 2.1% (from 0.675 to 0.661). According to the above-mentioned principle of the improved K-shell decomposition algorithm, the decrease in the node criticality indicates that the corresponding node would be less active in the network. According to the contents in Table 8, the influence of "N_13" on the top event is the most obvious, which would lower the probability of the "personnel suffocation" by 4.0%. Meanwhile, the node of "N_13" is also identified as the most active node in terms of out-degree discussed in Section 3.1.1. Therefore, the performance of the shipyard's supervision of the operators, strictly implementing the safety management system, is essential for safe operation, which was identified as the most important factor in the field of OI (organization influence under the framework of the HFACS). Similar advice was presented in [12] for ensuring a safe work environment. The second most influential nodes coded by "N_16", "N_17" and "N_18" would lower the probability of personnel suffocation by 3.9%, all of which are categorized as US (unsafe supervision). Even though it is difficult to distinguish these nodes in terms of their influence on personnel suffocation, the correlation between these nodes and the personnel suffocation accident can be observed according to the results summarized in Table 8, which indicates that the superintendents and the officers of the operation department are critical to the supervision procedures on site. The last influential nodes presented in Table 8 were found to be "N_25" and "N_29", both of which are categorized as UA (unsafe acts), which implies the concentration of unsafe acts from operators in controlling and managing human-related risk factors is difficult to improve the safety level.

In the aspect of sensitivity analysis, according to the tornado graph illustrated in Figure 7, the top three sensitive nodes coded by "N_16", "N_17" and "N_18" are associated with the superintendents and officers in the civil marine project affiliated with the shipyard, especially the risks which occur in relation to the supervision by superintendents and officers of the operators run by subcontractors. Actually, the ILO identified the contracting service of subcontractors as one of main hazards in shipyard operations, and supervision for risk-prevention measures should be implemented at regular intervals [6]. It is interesting to find that, even though the three most sensitive nodes share the same criticality, the sensitivity level of N_18 is slightly lower than that of "N_16" and "N_17", which indicates that the identification and prevention of the risks associated with the planned operation on-site are more important than the management for the potential risks associated with the similar operation, although the latter is also important in the safety management system. In addition, the sensitivity of the node coded by "N_13" is ranked behind that of the nodes of "N_16", "N_17" and "N_18", even though the hazardous event denoted by "N_13" (shipyard failed to supervise operators on-site to implement strictly the safety management system) is regarded as being active in the developed CN. It is also noticeable that the absence of risk-prevention measures taken by the operators on-site coded by "N_25" is characterized by high in-degree and low sensitivity level, which indicates that the human-related risk management should focus on the causation for "N_25", such as "N16" and "N_17". It can be observed that the unsafe supervision coded by US in Figure 8 is found as the most sensitive node for the personnel suffocation, which implies the well-implemented safety management system is essential for the operation on-site. The similar conclusions were obtained by Fragiadakis et al., who argued that "bad information of subcontractors about safety rules" played a key role in shipyard injury accidents [13]. Therefore, the supervision for the operators from subcontractors is an effective way to maintain safe operations in a shipyard. The sensitivity level of US is followed by UA, which can be interpreted as the initial causation for the personnel suffocation in the unsafe actions of operators on site. However, the unsafe actions of the operators were not identified as the most active or sensitive events in the present study, which indicates that the consequences caused by the unsafe acts of the operators can be eliminated by the supervision or monitoring arrangements which followed. In addition, according to the contents of Figure 8, the organizational influence (coded by OI) is less sensitive than that of unsafe supervision and unsafe acts based on the accident report, and the difference in terms of sensitivity level between US and OI is as high as 3.17 times; therefore, much more attention should be paid to the factors involved in unsafe supervision rather than factors related to organizational influence. Finally, the least sensitive aspect under the HFACS framework is the preconditions for unsafe acts (UP) which is valued as 0.355, which is followed by the OI with the sensitivity value of 0.487. The sensitivity difference between UP and OI is not considerable for the case in this study, which indicates that the external environment has little influence on the occurrence of the personnel suffocation accident.

Overall, the following lessons can be obtained from the case study:

(1) The significance of "N_13" (shipyard failed to supervise the operator on site) was identified by the proposed methodology in relation to risk propagation based on the perspective of the CN analysis. Therefore, the effective supervision of the shipyard operators on site may be able to intercept risk propagation, despite the occurrence of unsafe acts.

(2) The human-related hazardous events coded by "N_16", "N_17" and "N_18" were identified as the most sensitive events for the occurrence of the personnel suffocation accident case study, which indicates that the superintendents or managers of the operators were essential for the safety of the operation. Therefore, an important lesson can be learned: superintendents or managers must pay attention to the behaviors of operators on site.
(3) Another important lesson learned from the personnel suffocation case study is that the focus of safety management or causation investigations should emphasize unsafe supervision instead of the traditional perspective, which focuses on the unsafe acts of operators. Unsafe acts are inevitable without well-implemented supervision. The consequences stemming from unsafe acts can be eliminated by well-designed supervision, which was reflected in the first lesson.

## 4. Conclusions

The present study proposes an innovative methodology to bridge the CN and the BN on the basis of an event tree analysis under the HFACS framework. This methodology quantitatively assesses the human-related hazardous events involved in maritime accidents. For this purpose, a personnel suffocation accident on board a ship was selected as a case study, and the ETA was applied to identify the human-related hazardous events according to the accident report. Then, the logic relationships among the identified hazardous events were mapped into an event tree. As a result, the CN was developed. The improved K-shell decomposition algorithm was subsequently proposed to quantify the criticality of the nodes contained in the developed CN. Meanwhile, the framework of the BN was established under the framework of the HFACS with reference to the results of the criticality evaluation. To implement the Bayesian inference, the criticality of the root nodes was used to determine their prior probability distribution with the assumption of the normal distribution. The conditional probability distribution for other nodes in the BN can be also determined on the basis of the probability distribution of the root nodes. Finally, the developed BN was simulated in AgenaRisk, by which the scenarios were simulated, and the sensitivity analyses were implemented.

The advantages of the proposed methodology are mainly characterized by its elimination of the traditional potential bias. This is due to the expert elicitation and our compensation of the constraints of the ETA, without considering the logic relationships among hazardous events. However, the performance of the proposed methodology can be improved by further study in the near future. For instance, the logic relationships among hazardous events can be described in greater detail by a more effective method to aggregate the expert elicitation rather than the safety meeting which was adapted in the present study. In addition, the performance of the proposed methodology can be verified by the implementation of numerous accident reports associated within the industry in the near future.

**Author Contributions:** Methodology and original draft, W.Q.; Formal analysis and modelling, H.G.; Investigation, E.H.; Modelling and calculation, W.D.; Data curation and inveistigation, C.L.; Resource, H.C. All authors have read and agreed to the published version of the manuscript.

**Funding:** This research was funded by National Key Research and Development program of China (grand No. 2019 YFB1600602), the Fundamental Research Funds for the Central Universities (Grand No. 3132022223) and Program of Innovative Talents of Dalian (Grand No. 2021RQ043).

**Institutional Review Board Statement:** Not applicable.

**Informed Consent Statement:** Not applicable.

**Data Availability Statement:** Not applicable.

**Conflicts of Interest:** The authors declare no conflict of interest.

## References


1. Tsoukalas, V.D.; Fragiadakis, N.G. Prediction of occupational risk in the shipbuilding industry using multivariable linear regression and genetic algorithm analysis. *Saf. Sci.* **2016**, *83*, 12–22. [CrossRef]
2. Babur, F.; Cevikcan, E.; Durmusoglu, M.B. Axiomatic Design for Lean-oriented Occupational Health and Safety systems: An application in shipbuilding industry. *Comput. Ind. Eng.* **2016**, *100*, 88–109. [CrossRef]
3. Efe, B. Analysis of operational safety risks in shipbuilding using failure mode and effect analysis approach. *Ocean Eng.* **2019**, *187*, 106214. [CrossRef]
4. Barlas, B. Shipyard fatalities in Turkey. *Saf. Sci.* **2012**, *50*, 1247–1252. [CrossRef]
5. Attorney-General's Chambers (AGC) of Singapore. Workplace Safety and Health (Amendment) Act 2017. 2017. Available online: https://sso.agc.gov.sg/Acts-Supp/44-2017/Published/20171226?DocDate=20171226 (accessed on 20 April 2022).
6. ILO. ILO Code of Practice: Safety and Health in Shipbuilding and Ship Repair. 2019. Available online: https://www.ilo.org/wcmsp5/groups/public/---ed_protect/---protrav/---safework/documents/normativeinstrument/wcms_107897.pdf (accessed on 20 April 2022).
7. Dobbs, S.; Loh, S.K. Unsafety and unions in Singapore's state-led industrialization, 1965–1994. *Labor Hist.* **2020**, *61*, 107–121. [CrossRef]
8. Moncayo, G.A. Testing the boundaries between the Basel and MARPOL regimes: Are they complementary or mutually exclusive? *Transp. Res. Procedia* **2017**, *25*, 233–250. [CrossRef]
9. Kuroshi, L.; Aykut, I.; Òlçer Kitada, M. A tripartite approach to operator-error evaluation in ballast water management system operation. *Int. J. Ind. Ergonom.* **2019**, *69*, 173–183. [CrossRef]
10. Liu, Y.; Ma, X.; Qiao, W.; Luo, H.; He, P. Human Factor Risk Modeling for Shipyard Operation by Mapping Fuzzy Fault Tree into Bayesian Network. *Int. J. Environ. Res. Public Health* **2022**, *19*, 297. [CrossRef]
11. Shin, I.J. Major industrial accidents in Korea: The characteristics and implication of statistics 1996–2011. *Process. Saf. Prog.* **2013**, *32*, 90–95. [CrossRef]
12. Barlas, B.; Izci, B.F. Individual and workplace factors related to fatal occupational accidents among shipyard workers in Turkey. *Saf. Sci.* **2018**, *101*, 173–179. [CrossRef]
13. Fragiadakis, N.G.; Tsoukalas, V.D.; Papazoglou, V.J. An adaptive neuro-fuzzy inference system (anfis) model for assessing occupational risk in the shipbuilding industry. *Saf. Sci.* **2014**, *63*, 226–235. [CrossRef]
14. Zhang, G.; Thai, V.V.; Law, A.W.K.; Yuen, K.F.; Loh, H.S.; Zhou, Q. Quantitative Risk Assessment of Seafarers' Nonfatal Injuries Due to Occupational Accidents Based on Bayesian Network Modeling. *Risk. Anal.* **2019**, *40*, 8–23. [CrossRef] [PubMed]
15. Jacinto, C.; Silva, C. A semi-quantitative assessment of occupational risks using bow-tie representation. *Saf. Sci.* **2009**, *48*, 973–979. [CrossRef]
16. Murat, O. Risk evaluation of pin jig work unit in shipbuilding by using fuzzy AHP method. *J. Brodogr.* **2015**, *66*, 39–53.
17. Kandemira, C.; Celika, M.; Akyuzb, E.; Aydinc, O. Application of human reliability analysis to repair & maintenance operations on-board ships: The case of HFO purifier overhauling. *Appl. Ocean Res.* **2019**, *88*, 317–325. [CrossRef]
18. Selman, J.; Spickett, J.; Jansz, J.; Mullins, B. An investigation into the rate and mechanism of incident of work-related confined space fatalities. *Saf. Sci.* **2018**, *109*, 333–343. [CrossRef]
19. Burlet-Vienney, D.; Chinniah, Y.; Bahloul, A.; Roberge, B. Occupational safety during interventions in confined spaces. *Saf. Sci.* **2015**, *79*, 19–28. [CrossRef]
20. Sahli, B.P.; Armstrong, C.W. Confined space fatalities in Virginia. *J. Saf. Res.* **1993**, *24*, 124–125. [CrossRef]
21. Islam, R.; Abbassi, R.; Garaniya, V.; Khan, F. Development of a human reliability assessment technique for t-he maintenance procedures of marine and offshore operations. *J. Loss Prev. Process. Ind.* **2017**, *50*, 416–428. [CrossRef]
22. Xie, X.; Guo, D. Human factors risk assessment and management: Process safety in engineering. *Process Saf. Environ.* **2018**, *113*, 467–482. [CrossRef]
23. Qiao, W.; Liu, Y.; Ma, X.X.; Liu, Y. Human factors analysis for maritime accidents based on a dynamic fuzzy Bayesian Network. *Risk. Anal.* **2020**, *40*, 957–980. [CrossRef] [PubMed]
24. Moises Fagundes, J.; Figueiredo, M.A.G.; Alberton, A.L.; Jose Antonio, I.; Santos, L. Analysis of accidents t-hrough combination of CAST and TRACEr techniques: A case study. *J. Loss Prevent. Proc. Ind.* **2022**, *74*, 104639. [CrossRef]
25. Qiao, W.; Liu, Y.; Ma, X.X.; Liu, Y. A methodology to evaluate human factors contributed to maritime accident by mapping fuzzy FT into ANN based on HFACS. *Ocean Eng.* **2020**, *197*, 106892. [CrossRef]
26. An, J.; Liu, Y.; Sun, Y.; Liu, C. Impact of work-family conflict, job stress and job satisfaction on seafarer performance. *Int. J. Environ. Res. Public Health* **2020**, *17*, 2191. [CrossRef]
27. Ma, X.X.; Deng, W.Y.; Qiao, W.L.; Lan, H. A methodology to quantify the risk propagation of hazardous events for ship grounding accidents based on directed CN. *Reliab. Eng. Syst. Saf.* **2022**, *221*, 108334. [CrossRef]
28. Shappell, S.A.; Wiegmann, D.A. The Human Factors Analysis and Classification System-HFACS. 2000. Available online: https://commons.erau.edu/cgi/viewcontent.cgi?article=1777&context=publication (accessed on 20 April 2022).
29. Shappell, S.A.; Detwiler, C.; Holcomb, K.; Hackworth, C.; Boquet, A.; Wiegmann, D.A. Human error and commercial aviation accidents: An analysis using the human factors analysis and classification system. *Hum. Factors* **2007**, *49*, 227–242. [CrossRef]
30. Ozkok, M. Risk assessment in ship hull structure production using FMEA. *J. Mar. Sci. Tech-Japan* **2014**, *22*, 173–185. [CrossRef]

31. Romuald Iwańkowicz, R.; Rosochacki, W. Clustering risk assessment method for shipbuilding industry. *Ind. Manag. Data Syst.* **2014**, *114*, 1499–1518. [CrossRef]
32. Akyuz, E. A hybrid accident analysis method to assess potential navigational contingencies: The case of ship grounding. *Saf. Sci.* **2015**, *79*, 268–276. [CrossRef]
33. Sotiralis, P.; Ventikos, N.P.; Hamann, R.; Golyshev, P.; Teixeira, A.P. Incorporation of human factors into ship co-llision risk models focusing on human centred design aspects. *Reliab. Eng. Syst. Saf.* **2016**, *156*, 210–227. [CrossRef]
34. Akyuz, E. A marine accident analysing model to evaluate potential operational causes in cargo ships. *Saf. Sci.* **2017**, *92*, 17–25. [CrossRef]
35. Kuzu, A.C.; Akyuz, E.; Arslan, O. Application of Fuzzy Fault Tree Analysis (FFTA) to maritime industry: A risk analysing of ship mooring operation. *Ocean Eng.* **2019**, *179*, 128–134. [CrossRef]
36. Ung, S.T. Evaluation of human error contribution to oil tanker collision using fault tree analysis and modified fuzzy Bayesian Network based CREAM. *Ocean Eng.* **2019**, *179*, 159–172. [CrossRef]
37. Fan, S.Q.; Zhang, J.F.; Blanco-Davis, E.; Yang, Z.L.; Yan, X.P. Maritime accident prevention strategy formulation from a human factor perspective using Bayesian Networks and TOPSIS. *Ocean Eng.* **2020**, *210*, 107544. [CrossRef]
38. Zhou, X.; Cheng, L.; Li, M. Assessing and mapping maritime transportation risk based on spatial fuzzy multicriteria decision making: A case study in the South China sea. *Ocean Eng.* **2020**, *208*, 107403. [CrossRef]
39. Cakir, E.; Sevgili, C.; Fiskin, R. An analysis of severity of oil spill caused by vessel accidents. *Transp. Res. Part D Transp. Environ.* **2021**, *90*, 102662. [CrossRef]
40. Sakar, C.; Toz, A.C.; Buber, M.; Koseoglu, B. Risk analysis of grounding accidents by mapping a fault tree into a bayesian network. *Appl. Ocean Res.* **2021**, *113*, 102764. [CrossRef]
41. Lan, H.; Ma, X.X.; Qiao, W.L.; Liu, Y. A methodology to assess the causation relationship of seafarers' unsafe acts for ship grounding accidents based on Bayesian SEM. *J. Ocean Coast. Manag.* **2022**, *225*, 106189. [CrossRef]
42. Qiao, W.; Ma, X.X.; Liu, Y.; Deng, W.Y. Resilience evaluation of maritime liquid cargo emergency respons-e by integrating FRAM and a BN: A case study of a propylene leakage emergency scenario. *Ocean Eng.* **2022**, *247*, 110584. [CrossRef]
43. Office of Emergency Management of Shanghai, 2021. Available online: http://yjglj.sh.gov.cn/xxgk/xxgkml/dcpghtj/dcbg/20210929/02d576e5363f4b4e9b26738145401093.html (accessed on 20 April 2022).
44. Boccaletti, S.; Latora, V.; Moreno, Y.; Chavez, M.; Hwang, D.U. Complex networks: Structure and dynamics. *Phys. Rep.* **2006**, *424*, 175–308. [CrossRef]
45. Wan, Z.; Mahajan, Y.; Kang, B.W.; Moore, T.J.; Cho, J.H. A Survey on Centrality Metrics and Their Network Resilience Analysis. *IEEE Access* **2021**, *9*, 104773–104819. [CrossRef]
46. Qiao, W.; Ma, X.X.; Liu, Y.; Lan, H. Resilience assessment for the northern sea route based on a fuzzy Bayesian network. *Appl. Sci.* **2021**, *11*, 3619. [CrossRef]

applied sciences

MDPI

*Article*

# Cruise Industry Trends and Cruise Ships' Navigational Practices in the Central and South Part of the Adriatic East Coast Affecting Navigational Safety and Sustainable Development

**Josip Dorigatti *, Tina Perić and Gorana Jelić Mrčelić**

Faculty of Maritime Studies, University of Split, 21000 Split, Croatia; tperic@pfst.hr (T.P.); gjelic@pfst.hr (G.J.M.)
* Correspondence: josipdorigatti@yahoo.com

**Abstract:** The analysis of cruising trends in the Mediterranean regions shows that the Adriatic is the fastest growing cruise region in terms of the number of passenger movements and cruise ships' port calls among all regions, particularly the central and south part of the east Adriatic coast. The aim of the paper is to analyze leading cruise destination trends in the central and south part of the Adriatic east coast, as well as to identify newly established cruise ships routes, define high-risk navigational and environmental areas and determine cruise traffic density in the vicinity of marine protected areas. The analyses of leading cruise destinations trends are based on four-year (from 2015 to 2019) cruise passenger movement and cruise calls data, whereas analyses of cruise traffic movement are based on one-year cruise ships traffic monitoring (from August 2014 to July 2015). The results of the cruise ship traffic analysis show that cruise ships frequently pass through areas of high navigational and environmental risks that are geographically restricted, navigationally challenging and environmentally sensitive. These routes have become standard navigational practice in newly discovered cruising regions. The obtained results offer a general overview of high-risk cruise ships' navigational practices in coastal navigation that can be associated with any coastal region in the world.

**Keywords:** cruise ships; navigation; safety; sustainability; marine environment; the Adriatic Sea

**Citation:** Dorigatti, J.; Perić, T.; Mrčelić, G.J. Cruise Industry Trends and Cruise Ships' Navigational Practices in the Central and South Part of the Adriatic East Coast Affecting Navigational Safety and Sustainable Development. *Appl. Sci.* **2022**, *12*, 6884. https://doi.org/10.3390/app12146884

Academic Editor: José A. Orosa

Received: 10 June 2022
Accepted: 4 July 2022
Published: 7 July 2022



## 1. Introduction

The popularity of cruise trips has risen since the 1970s [1], and cruise line companies have increased the number of ships as well as the capacities of ships and berths on the market in order to offer broad variety of onboard activities [2]. According to Kovačić and Silveira (2020), the Mediterranean Sea is the most popular cruise destination in Europe (16.7% in 2018) and the second one in the world after the Caribbean (35%) [1]. Italy was the most visited country in the Mediterranean, with 2.4 million cruising passengers, followed by Croatia, with 1.3 million passengers, in 2018 [1]. The analysis of cruising trends in the Mediterranean regions shows that the Adriatic region is the fastest-growing cruise region in the number of passenger movements and cruise ships' port calls [3].

The central and south part of the Adriatic east coast is divided by Croatia, Montenegro and Albania and represents the most valuable natural resource of these countries. Due to its unique beauty and attractiveness, the region is a very popular tourist destination. Natural beauty and cultural–historical diversity are the key factors that attract tourists and represent the main advantage towards competitors [3]. Croatian and Montenegrin economies depend on tourism, as it is their key source of GDP growth. Tourism accounts for some 20% of Croatian GDP [4] and almost 25% of Montenegrin [5].

Globally, the cruising industry experiences higher passenger expectations and higher demands for new bigger cruise ships than ever before. The expansion of the cruising industry is not only related to cruising market expansion but also to the development of new cruising destinations [6]. Due to the strong expansion of cruise traffic and cruise passenger movement in the central and south part of the Adriatic east coast region, it is necessary to analyze established cruise destinations and new ones as well. The development of new cruise destinations and the increment in the number and the capacity of cruise ships puts cruise ships' navigational decisions in newly established cruising regions with high natural and cultural values under question.

Cruising tourism is not only based on natural beauty and cultural–historical diversity but also on ships' experience and navigation [7]. The longitudinal Adriatic corridor in the northwest–southeast direction and the southeast–northwest direction has greater importance than transversal ones [8–10]. According to the International Maritime Organization (IMO) report Routing of ships, ships reporting and related matters—Establishment of new recommended Traffic Separation Schemes and other routing measures in the Adriatic Sea (IMO, 2015), the north Adriatic region has two separation schemes that efficiently direct maritime traffic. The central and south Adriatic have only one separation scheme (the Central Adriatic Separation Scheme), and it is less regulated by navigational aids than the north Adriatic, therefore offering more options to navigation. The IMO report also elaborated upon the environmental consideration of the traffic in the east Adriatic coast with emphasis on the islands of Vis, Jabuka, Svetac, Biševo, Sv. Andrija, Palagruža and Mljet [11]. The main sailing routes in the Adriatic region were analyzed by Lušić and Kos (2006) [9] and by Zec et al. (2016) [12], and the most attention was given to general longitudinal maritime traffic from the Otranto strait to northern Adriatic ports. The authors stated that maritime traffic flow in the central Adriatic was directed mostly through the Central Adriatic Separation Scheme and that maritime accidents were rare, which indicated good maritime traffic coordination. Lušić et al. (2017) [8] analyzed sailing routes and the structure of maritime traffic in the central part of the Adriatic, with focus on maritime traffic inside the Central Adriatic Separation Scheme. The authors concluded that the depth and width is sufficient on a major part of the longitudinal sailing route and that there were no significant navigational risks except the risk of collision and the risk of grounding near the island of Palagruža.

The aim of the paper is to analyze trends of leading cruise destinations in the central and south part of the Adriatic east coast and to identify newly established cruise ship routes as well as their influence on navigational and environmental safety. One of the main goals of the paper is to determine cruise ships' traffic density in the vicinity of marine protected areas and to identify areas of high navigational and environmental risks. The analyses of leading cruise destinations trends are based on four-year (from 2015 to 2019) cruise passenger movement and cruise calls data, whereas analyses of the longitudinal and transversal traffic flow of cruise ships are based on one-year (from August 2014 to July 2015) cruise ships' traffic monitoring in the central and south part of the Adriatic east coast. The data on cruise passenger movement and cruise calls were obtained and analyzed from a MedCruise report (2019) [10]. The data of cruise passenger movement, cruise calls and cruise ships' movement for years 2020 and 2021 are not taken into consideration due to the COVID pandemic.

### 1.1. Global Cruise Trends

A major part of cruising industry is seasonal, which means that most cruise ships follow the season and shift from one region to another in order to achieve optimal passenger occupancy and offer attractive itineraries to passengers. In order to satisfy passenger demand for exciting itineraries and interesting destinations, cruise ships spend the majority of their time in coastal navigation in areas of high natural, environmental and preservation values. They select the most attractive world destinations and navigate along the most attractive coastlines, many of which are environmentally preserved regions with rich biodiversity and high national importance. The usual cruise industry concept is that cruise ships' operation is regional, and itineraries are repetitive, usually in a circular pattern, with limited durations, offering different port experiences every day.

The question of cruise routes' sustainability in costal navigation becomes important due to the fact that cruise ships spend the majority of their time in coastal navigation. In addition to that, globally, the cruising industry records high passenger demand, high demand for new cruise ships and a trend in which cruise ships are becoming bigger. The expansion of the cruising industry is related to cruising market expansion and the development of new cruising destinations.

Figure 1 shows the relation between cruise ships' number and cruise ships' berths. Cruise ship berth in this context represents a bed of any type on a cruise ship [13]. For the period from 2000 to 2020, the relation between the passenger movement trend (+5.5%) and the cruise ships' berths trend (+4.7%) indicates a higher cruise ships' occupancy rate. On the other hand, the relation between cruise ships' number trend (+4.1%) and cruise ships' passenger movement trend (+5.5%) shows that cruise ships are becoming bigger. Predictions from the 2020 to 2027 period estimate that the number of cruise ships will grow at an average rate of 2.48% and the cruise ships' berths will grow at an average rate of 3.67% [10,13]. Growing trends will slightly stabilize, but they still indicate the delivery of larger cruise ships on the market in order to accommodate high passenger demand.

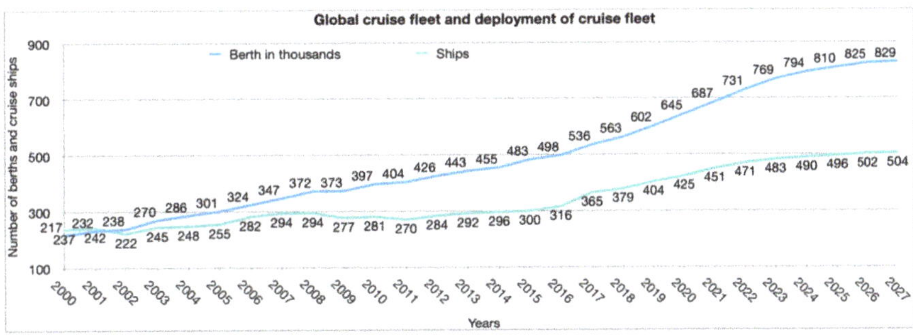


**Figure 1.** Global cruise fleet and deployment of cruise fleet. Adopted from Cruise Industry News 2020–2021.

These factors predict further cruise industry development and cruise traffic expansion. In order to satisfy the demand of the cruise market, it is predicted that popular cruising destinations which are close to reaching their full capacity will maintain the present status or even adjust it, while less prominent cruising regions will continue to grow. It is expected that future cruise industry expansion will focus on new regions that have never been cruising destinations before.

### *1.2. Cruise Trends in the Central and South Part of the Adriatic East Coast*

The Mediterranean region is divided into four sub regions (Figure 2), among which, the Adriatic is the second most visited region after the western Mediterranean region [14]. The Adriatic Sea is located between the West Mediterranean and East Mediterranean zone, in proximity to the central and north European market with high cruising demand and developed infrastructure. Due to the geographical location, infrastructural investment and the further expansion of established destinations, the Adriatic region has promising perspectives in future [1].

Rich cultural heritage, historical value, diversity, natural beauty and prominent traditional touristic locations have established well-known cruising destinations such as Venice (1.56 million passengers) and Dubrovnik (0.732 million passengers) which have been market leaders for a long time [1]. The popularity of Venice and Dubrovnik makes cruise shipping in the Adriatic heavily dependent on these two prominent destinations, in contrast to Rodrigue and Notteboom's (2013) statement that destinations do not sell the cruise industry but itineraries [15]. As a result, these two biggest cruise ports in the Adriatic struggle with city congestion.

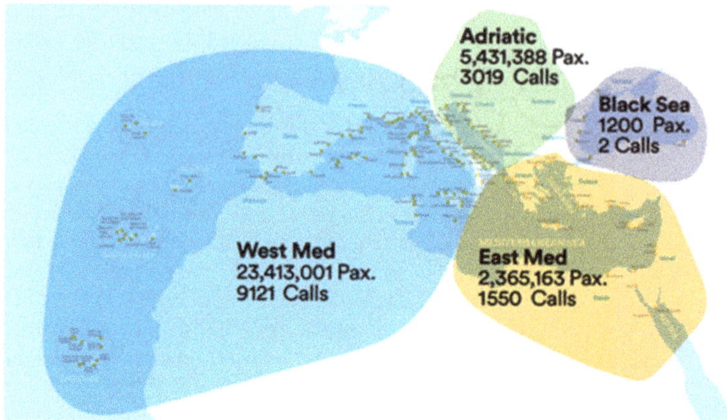


**Figure 2.** Mediterranean cruise market sub regions [11].

The central and south part of the Adriatic east coast extends from the port of Zadar in North Dalmatia to the Otranto strait at the south of the Albanian coast (Figure 3). Croatian and Montenegrin destinations are the most important cruising destinations in this area, while Albanian cruising destinations have not been recognized by the market yet.

**Figure 3.** The central and south part of the east Adriatic coast. Adopted from Google maps.

The area is recognized worldwide for its natural beauty, cultural heritage and authentic, well-preserved ambient. Many marine protected areas have been established here, such as the Kornati National Park, Mljet National Park, Telašćica Nature Park, the Lastovo Islands Nature Park in Croatia, as well as the Tivat Saline Special Nature Reserve in Montenegro. The United Nations Educational, Scientific and Cultural Organization (UNESCO) has recognized the rich history of this area. The UNESCO World Heritage Sites located in the central and south part of the east Adriatic coast are: the Cathedral of St James in Šibenik, the Historic City of Trogir, the Historical Complex of Split with the Palace of Diocletian, the Stari Grad Plain on the island of Hvar and the Old City of Dubrovnik in Croatia, as well as Natural and Cultural–Historical Region of Kotor in Montenegro and Venetian Works of Defence between the 16th and 17th Centuries: Stato da Terra—Western Stato da Mar in Croatia and Montenegro [11].

The number of cruise ships' ports in the Adriatic region is constantly rising; currently, around 30 cruise ports are in use by cruise lines [13]. The Adriatic Sea is the fastest growing region in the Mediterranean, with 18.70% growth in cruise calls (from 2492 cruise calls to 3019 cruise calls) and 20.90% growth in total passenger movements (from 3084 cruise passenger movements to 3748) in the period from 2015 to 2019 [10].

Figure 4 shows total cruise call trends from 2015 to 2019 in five leading ports in the central and south part of the Adriatic east coast. The leading ports are Dubrovnik, Korčula, Kotor, Split, Zadar and Šibenik. The horizontal axis indicates each year of the research, while the vertical axis shows number of cruise calls. Each port is marked with its colored bar, and on the top of each bar is a figure indicating the number of cruise calls of the designated port.

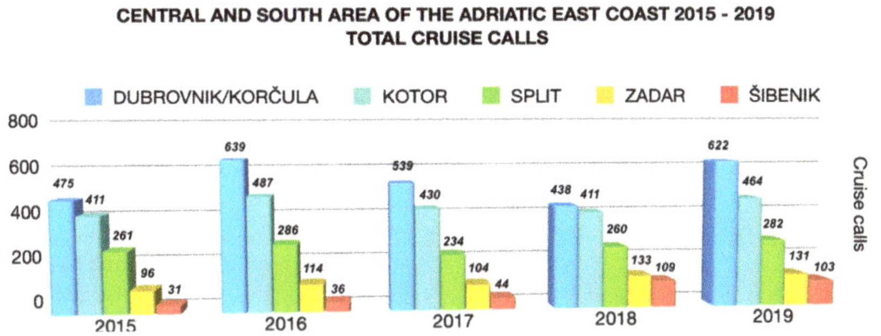


**Figure 4.** Total cruise calls from 2015 to 2019 in five leading ports in the central and south part of the Adriatic east coast.

All destinations in the central and south part of the Adriatic east coast measured growth in cruise calls, with slight oscillation in 2018. The total cruise calls increment in the central and south part of the Adriatic east coast from 2015 to 2019 was 25.75% compared to 18.70% in the Adriatic and only 4.30% in the Mediterranean region (Figure 4) [10].

Figure 5 shows total cruise passenger movement from 2015 to 2019 in five leading ports in the central and south part of the Adriatic east coast. The horizontal axis indicates each year of the research, while the vertical axis shows number of passenger movements in selected ports. Each port is marked with its coloured bar, and on the top of each bar is a figure indicating number of passenger movements in the selected port. Passenger movement trends follow the cruise ship call trends; the growth was constant with slight oscillation in 2018.

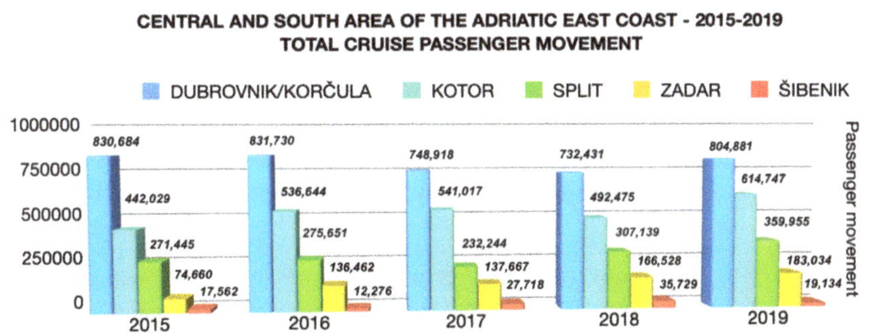


**Figure 5.** Total cruise passenger movement from 2015 to 2019 in five leading ports in the central and south part of the Adriatic east coast.

The number of total passenger movements in the central and south part of the Adriatic east coast increased by 46.07% in the period from 2015 to 2019 compared to 20.90% in the Adriatic and only 14.49% in the Mediterranean (Figure 4) [10].

Data on cruise ship calls and passenger movement show that the central and south part of the Adriatic east coast is the fastest growing region in the Adriatic Sea as well as in the Mediterranean.

#### 1.2.1. Dubrovnik

Dubrovnik is the largest cruise port in Croatia and the second largest in the Adriatic, with 16% of cruise passenger traffic in the Adriatic [14]. The port of Dubrovnik is one of the most perspective Croatian ports and it has great potential to become one of Europe's leading cruise ports [16]. Dubrovnik is located in a prominent Adriatic cruise route, between Venice in the north and Kotor in the south. The growth of cruising tourism is very evident in the Croatian part of the Adriatic Sea, especially in Dubrovnik and Korčula [17]. Dubrovnik/Korčula had 475 cruise calls and 830,684 total cruise passenger movements in 2015 and 622 cruise calls and 804,881 total cruise passenger movements in 2019 (Figures 3 and 4). Due to the attractiveness and growing popularity of Dubrovnik Old Town on one side and its very limited area on the other, Dubrovnik has experienced the problem of over-tourism. A large number of cruise ships, cruising passengers and independent tourists at the same time causes congestion in Dubrovnik Old Town [16]. In order to solve the problem of congestion and to reduce navigational, environmental and social risks, cruising traffic have to be well planned and controlled. The Dubrovnik Port Authority has implemented measures which limited the number of cruise passengers up to 5000 per day and the number of cruise ships alongside up to two ships in passenger terminals and one ship at the anchorage at the same time [18]. The effects of the imposed measures are visible from the stable number of cruise calls and passenger movement trends in the period between 2015 and 2019 (Figures 3 and 4).

#### 1.2.2. Korčula

Korčula is a relatively new destination in the cruising industry. In 2019, Korčula had 136 cruise calls and 35,957 cruise passengers (Figures 3 and 4). Although cruise traffic is still relatively low compared to the leading Adriatic cruise destinations, cruising trends are positive and have prospects for further growth [18]. Its geographical location near established cruise ports of Dubrovnik, Kotor and Split makes Korčula an ideal cruise transit port [19]. Korčula Old Town is one of the best examples of a fortified medieval town in the Mediterranean and it is listed in UNESCO's tentative list of outstanding world heritage sites [20]. Taking into consideration all the advantages of Korčula and future infrastructure investments in terminal Polačište, Korčula has a promising perspective as a desirable cruising destination [21].

#### 1.2.3. Kotor

Kotor is the largest cruise destination in Montenegro and the third largest cruise port in the Adriatic Sea [1]. Kotor is located in Boka Kotorska Bay (fjord). The natural beauty and historical value of Kotor has been recognized by the UNESCO. In 2019, Kotor had 464 cruise calls and 614,747 cruise passengers' visits (Figures 3 and 4) [10]. Kotor can accommodate three to four cruise ships (up to 250 m length) and has a river berth for smaller cruise ships and three anchorages [22]. Investment in a modern cruise terminal and passenger facilities is required [23] in order to make port of Kotor safe in terms of navigational safety as well as in the terms of sustainability in the full meaning of the phrase sustainable development.

#### 1.2.4. Split

The port of Split is the largest Croatian passenger port and the second largest Croatian cruise port after Dubrovnik [14]. Its geographical location in the vicinity of prominent cruising destinations Dubrovnik and Kotor and rich historical heritage recognized by the UNESCO makes a Split competitive and perspective destination in the cruising market. Split has experienced remarkable touristic success and has evolved from a transit ferry destination to one of the leading touristic and cruising destinations in the Adriatic Sea. In 2019, the total passenger turnover was over 5.6 million passengers and 829,594 vehicles [24]. Split had 261 cruise calls and 251,455 total cruise passenger movements in 2015 and 282 cruise ship calls and 359,955 cruise passenger movements in 2019 (Figures 3 and 4) [10]. Cruise tourism in Split started in 2002, when 82 cruise ships with 20,616 passengers visited Split [24]. In the period between 2007 and 2016, cruise traffic almost tripled, while in the period between 2014 and 2016, it doubled in spite of the lack of cruise vessel berths and adequate passengers' facilities [14]. In 2017, the extension of the port operational quay and two external cruise berths with infrastructure (265 m long berth 26 and 245 m long berth 27) were completed as a part of a port development project. The new berths are capable to simultaneously accommodate two cruise ships of 320 and 270 m. The berths are equipped with border crossing points, sanitary facilities, an access road and other supporting infrastructure that offers high-quality services to passengers [14]. The extension of the quays and the construction of new external berths improved the maritime security and road traffic inside the port and have directly increased port service quality and competitiveness among other cruise ports in the region [25]. Split has the potential to become one of the Adriatic homeports, especially because the construction of a new terminal building has been included in short-term plans and better land access to the port is under consideration [25].

#### 1.2.5. Zadar

Zadar is the second largest ferry port and the third largest cruise port in Croatia [14]. Zadar is located in the central part of the Adriatic east coast, and it is an important Croatian traffic node where continental traffic corridors meet the Adriatic Sea [16]. Its ideal location between notable cruise ships ports, Venice to the north and Split, Dubrovnik and Kotor to the south, makes Zadar an ideal cruise ship port option. Zadar has the potential to become an important destination in the cruise market due to its rich heritage and the vicinity of UNESCO sites and national parks but also due to its developed infrastructure, highway and railway connections and the vicinity of the international airport. In 2019, the total passenger turnover was 2,390,575 passengers and 484,690 vehicles [26]. Zadar had 96 cruise calls and 74,660 total cruise passenger movements in 2015, while in 2016, Zadar's cruising tourism exploded, with 114 cruise ship calls and 136,452 cruise passenger movements (Figures 3 and 4). In 2019, Zadar had 131 cruise ship calls and 183,034 cruise passenger movements (Figures 3 and 4) [10]. Zadar's cruise traffic and maritime traffic boom was caused by the completion of a new spacious port operation and handling area with four quays and a new passenger terminal in Gaženica. The new terminal can simultaneously accommodate seven ships in domestic traffic, two ships in international traffic and three cruise ships [26]. The construction of the new terminal gives Zadar a chance to become a homeport and a competitive cruise destination in the Adriatic region.

#### 1.2.6. Šibenik

Šibenik is predominately a passenger ferry port with an approximate passenger turnover of 250,000 per year [27]. Šibenik is a relatively new cruising destination in the cruise market. The town is situated in the central Adriatic east coast near prominent cruising destinations such as Split, Dubrovnik and Kotor, and it has numerous national and natural parks in the vicinity. Šibenik had 31 cruise calls and 17,562 total cruise passenger movements in 2015 and 103 cruise ship calls and 19,134 cruise passenger movements in 2019 (Figures 3 and 4). The port of Šibenik is situated in a flooded estuary of the river Krka, and it is well sheltered from winds and waves. The entrance to the port is through St. Anthony channel, and only ships up to 50,000 GT and up to 260 m in length can enter the port. The port has two designated cruise ship berths that can simultaneously accommodate two ships of 260 m length and 200 m length [27]. In order to make Šibenik an important cruise destination, investment in a modern passenger terminal with full cruise passenger service is required.

## 2. Materials and Methods

Research into cruise ships' navigational practices and cruise ships' trends in the central and south part of the Adriatic east coast is carried out since the region records the highest growth in cruise ship calls and passenger movement in the Mediterranean. Taking this and the fact that the strong expansion of cruise ship traffic has not been equally followed by the development of new navigational corridors and the implementation of improved protected measures into consideration, this makes the region ideal for cruise traffic monitoring and analysis.

The methodology of the research was based on the analysis of cruise industry trends in the Mediterranean region. The results brought the dynamic expansion of the central and south part of the Adriatic east coast to our attention. Strong cruise ship passenger movements and an increment in cruise ship calls in the region which has not had dense maritime traffic before offered valid grounds to identify newly established cruise ship routes as well as their influence on navigational and environmental safety. In addition, the research will determine cruise ships' traffic density in the vicinity of marine protected areas and will identify areas of high navigational and environmental risks. The obtained results and traffic comparison provide information regarding how often cruise ships use the Central Adriatic Separation Scheme. The methodology of the paper is presented in the diagram, in which all levels of the research are chronologically presented (Figure 6).

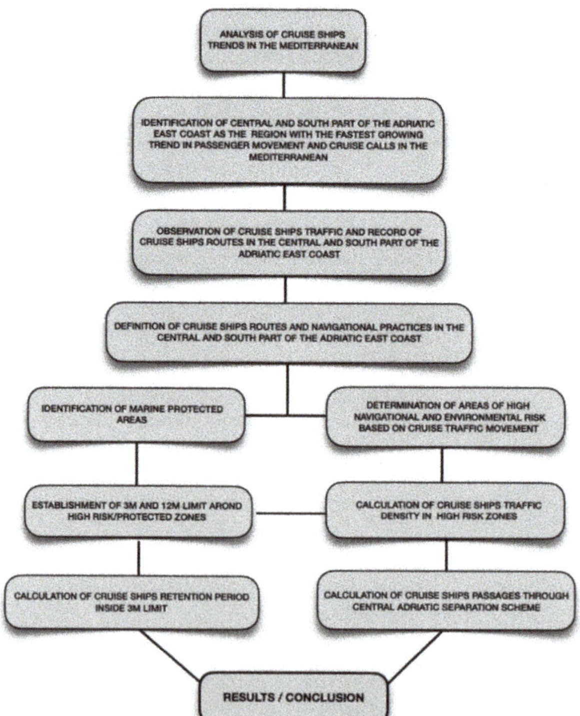


**Figure 6.** Research methodology diagram.

The analyses of leading cruise destinations trends are based on four-year (from 2015 to 2019) cruise passenger movement and cruise calls data. The data of cruise ships' movement (ships over 50,000 GT) were recorded on a daily basis in the period from August 2014 to

July 2015 using the Marine Traffic application. The data obtained from Perić (2016) [28] were processed and analyzed in the research. During the research, the routes of every cruise ship in the Adriatic east coast were recorded. The collection and analysis of all cruise routes offered real and detailed insight into cruise ships' navigational practices in the Adriatic east coast, in particular the central and south part of the Adriatic east coast.

### 2.1. Cruise Ships' Routes Presentation

Figure 7a–e show real cruise ship traffic movement in the researched period of time among leading cruise destinations in the east Adriatic region. Each figure represents the movement of one cruise ship. They are selected samples among all analyzed routes and represent general traffic flow in the east Adriatic coast. The presented figures are maps downloaded from the Marine Traffic application which are adjusted and edited with research data. Each figure shows different cruise ships' route patterns. Destinations are named, blue lines show the cruise ship route, black squares indicate separation schemes and the light blue line is the Croatian and Italian territorial waters border.

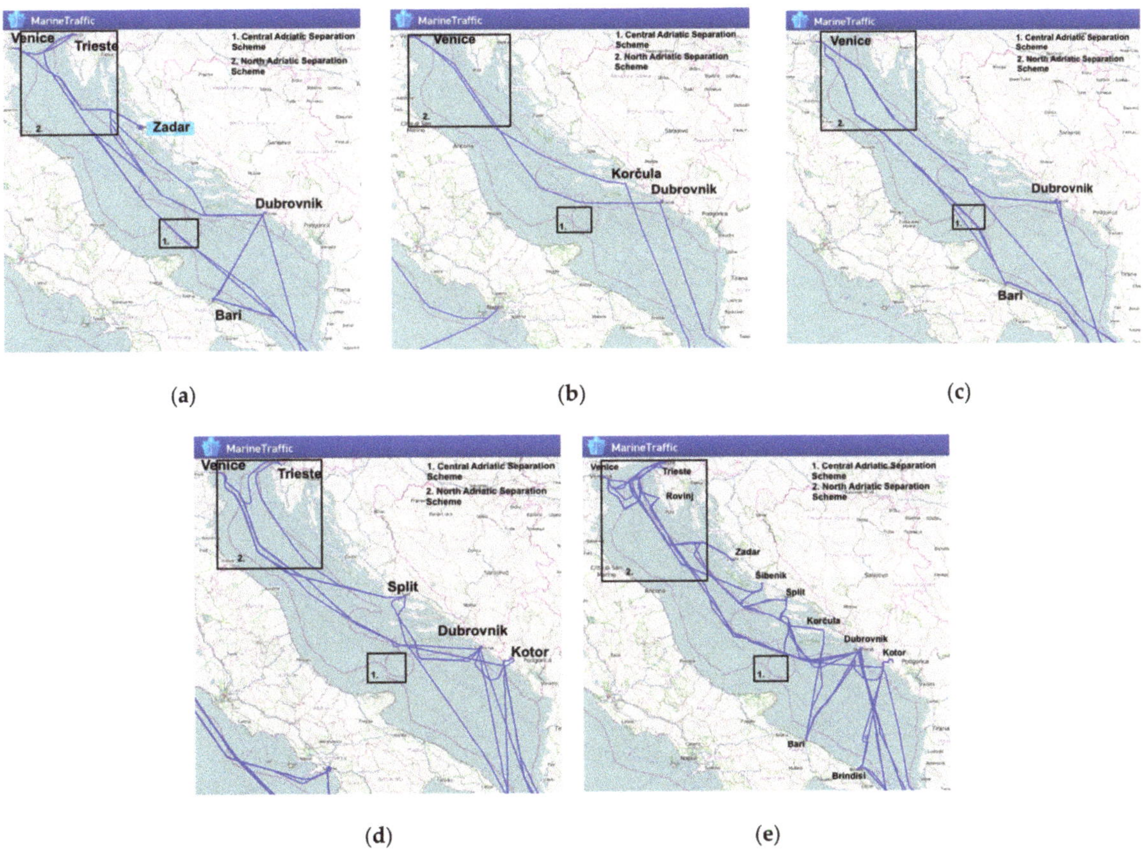


**Figure 7.** Real cruise ship traffic movement in the Adriatic east coast; (**a**) longitudinal cruise traffic corridor; (**b**) longitudinal cruise traffic corridor via port of Korčula; (**c**) international longitudinal cruise ships corridor through Croatian territorial waters in coastal navigation; (**d**) transversal cruise traffic corridor; (**e**) interaction between transversal and horizontal routes.

Figure 7a shows the routes of one selected cruise ship during a one-month period. The routes are part of two different circular itineraries. First itinerary: Venice—Trieste—Zadar—Dubrovnik—Bari—Mediterranean port—Venice. Second itinerary: Venice—Dubrovnik—Mediterranean ports—Venice. The figure emphasizes the longitudinal cruise traffic corridor from the north Adriatic ports to the central and south Adriatic east coast ports and the corridor from the Mediterranean ports to the north Adriatic ports. Cruise ships heading from the north Adriatic ports to the central and southern Adriatic destinations, after leaving the North Adriatic Separation Scheme, stay in coastal navigation. They proceed along the other island ridge or in inland navigation heading to the south Adriatic regions. Cruise ships do not use the Central Adriatic Separation Scheme on the way to the south Adriatic east-coast destinations. The Central Adriatic Separation Scheme is only in use on direct routes, from Mediterranean ports or from the south Adriatic west-coast ports to the north Adriatic ports and vice versa.

Figure 7b shows the routes of one selected cruise ship's itinerary: Venice—Korčula—Mediterranean port—Dubrovnik—Venice. The figure emphasizes the longitudinal cruise traffic corridor from the north Adriatic ports to a Mediterranean port via the port of Korčula. Cruise ships follow the North Adriatic Separation Scheme, and when leaving the North Separation Scheme, cruise ships proceed in inland navigation through the Vis and Hvar channel to the port of Korčula. Departing Korčula cruise ships proceed through the channel between Lastovo and Mljet island heading to the Otranto strait.

Figure 7c shows the routes of one selected cruise ship during a one-month period. The routes are part of two different circular itineraries. First itinerary: Venice—Dubrovnik—Mediterranean port—Venice. Second itinerary: Venice—Bari—Mediterranean port—Venice. This figure emphasizes irregular navigational practice where cruise ships on the international longitudinal route from Mediterranean ports to the north Adriatic ports proceed in coastal navigation inside Croatian territorial waters. The cruise ship did not use the Central Adriatic Separation Scheme, and the route was not planned for the shortest stay inside Croatian territorial waters. On the contrary, the route proceeded in coastal navigation between the islands on the way to the north Adriatic Italian ports. The Central Adriatic Separation Scheme was in use only on itineraries from the south Adriatic west coast ports to the north Adriatic ports and vice versa.

Figure 7d shows the routes of one selected cruise ship during a two-month period. The routes are part of four different circular itineraries. First itinerary: Venice—Dubrovnik—Kotor—Mediterranean port—Venice. Second itinerary: Venice—Split—Mediterranean port—Venice. Third itinerary: Venice—Kotor—Mediterranean port—Venice. Forth itinerary: Venice—Dubrovnik—Mediterranean port—Venice. The figure emphasizes the transversal navigational corridor that passes the west coast of Sušac island and between Sušac island and Lastovo island.

Figure 7e shows the routes of one selected cruise ship during a two-month period. The routes are part of different itineraries among: Venice—Trieste—Rovinj—Zadar—Šibenik—Split—Korčula—Dubrovnik—Kotor—Brindisi—Bari. The figure emphasizes the interaction between horizontal and vertical routes. It shows vertical routes that pass west of Sušac island and between Sušac and Lastovo island and routes between Lastovo island and Mljet island.

### 2.2. *Calculation of Traffic Density and Distance Cruise Ships Pass from the Island Shores*

The calculation of traffic density and the distance cruise ships pass from the island shores is based on traffic analysis during the busiest five-month period (August–October 2014 and June–July 2015). In order to identify density and determine the distance cruise ships traffic pass from the island shores and marine protected areas, regions of high navigational and environmental risk were selected. Regions of high navigational and environmental risk were selected based on the traffic monitoring results, geographical particularities, and environmental sensitivities of the selected area. Regions of high interest were Svetac, Biševo and Sušac and Lastovo islands, as well as the Biševo–Vis passage, Vis channel, Hvar channel and the Sušac–Lastovo and Lastovo–Mljet passages.

On the chart, around the selected islands, 3 M diameter black circles are set. A distance of 3 M was set as a safe navigational reference and was in reference to International Convention for the Prevention of Pollution from Ships Annex IV (MARPOL), which stipulates a 3 M limit for ships' sanitary water (grey water) discharge. The second reference was the 12 M territorial water limit marked with a light blue line. The distance of 12 M was also chosen in reference to International Convention for the Prevention of Pollution from Ships Annex IV (MARPOL), which stipulates a 12 M limit for ships' untreated wastewater (black water) discharge. Protected areas on the chart are marked with a green line (Figure 8a–c).

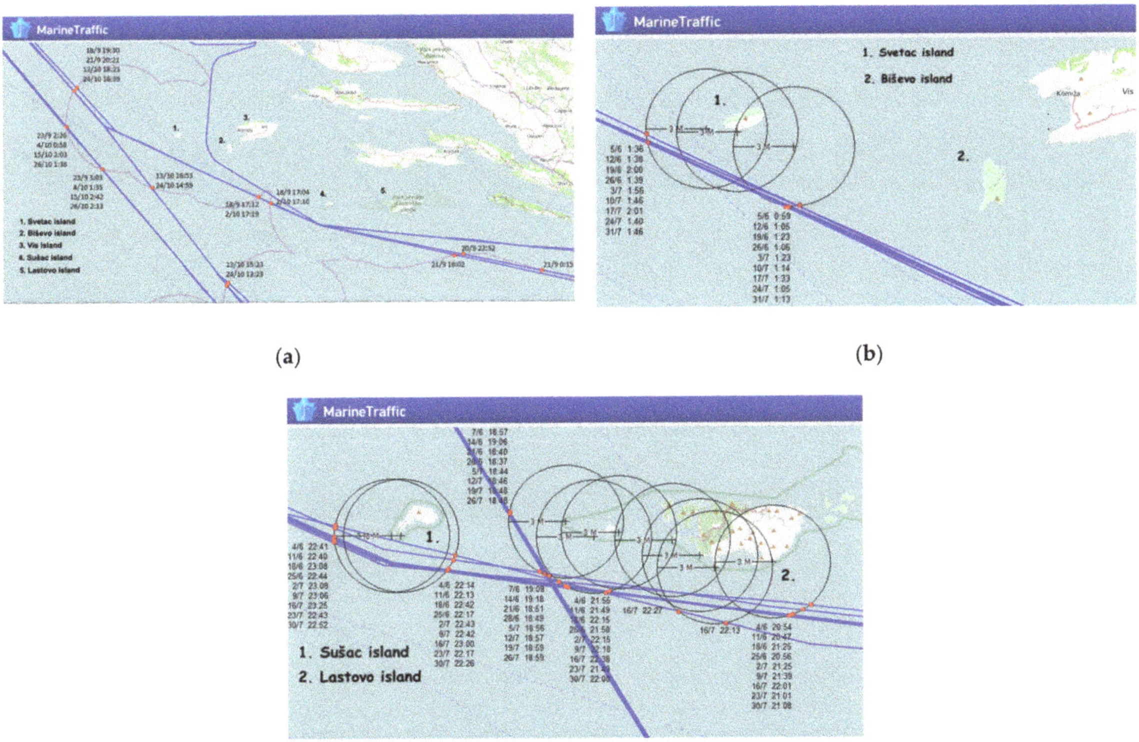


(**a**) (**b**)

(**c**)

**Figure 8.** (**a**) Presentation of cruise traffic distance from the islands' shores and display of >12 M retention calculation in the central and south Adriatic east coast; (**b**) presentation of cruise traffic distance from the islands' shore and display of >3 M retention calculation with focus on the islands of Svetac and Biševo; (**c**) presentation of cruise traffic distance from the islands' shores and display of >3 M retention calculation with focus on the islands of Sušac and Lastovo.

Figure 8a–c are routing samples in the central and south Adriatic east coast, selected among all monitored cruise ship routes. Each of them represents the movement of one cruise ship in the researched period of time. Areas of interest are numbered and named, the blue lines show the cruise ship routes, the black circles indicate 3 M distance from the shore, dates and times indicate when a ship entered and left a 3 M or 12 M limit and the light blue lines show the Croatian territorial waters border (12 M limit). Analyzing Figure 8a–c and observing the times and dates of ships' arrivals and departures from 3 M or 12 M limits, it is evident that cruise routes are repetitive and have become navigational routine in the central and south part of the Adriatic east coast.Cruise ships' density in the researched areas was calculated following each cruise ship's route. In the process, the time of each ship's entrance and exit to 3 M and 12 M limits was recorded. Pre-set 3 M diameter circles and 12 M territorial waters limits were used as the main references in Figure 8a–c. Cruise ships that crossed the 3 M circle limit were considered to pass inside 3 M from the shore, cruise ship routes which passed close to the 3 M circle limit were considered to pass with a distance 3 M to 6 M from the shore, and open sea routes closer to the 12 M territorial water border were considered to pass more than 6 M from the shore. Each cruise ship was counted and put into one of three categories according to the distance they passed from the islands shore: (1) <3 M from shore, (2) 3 M to 6 M from shore and (3) >6 M from shore. Cruise ships retention period inside 3 M from shore is also calculated. The time difference between the entrance and exit from the 3 M diameter was taken into consideration, and it was calculated and expressed in minutes. The collected data related to the distance cruise ships pass from the islands' shores and traffic density are grouped in four tables (Tables A1–A4). Each table is related to one high-risk navigational and environmental region. Information is given for each month (August–October 2014 and June–July 2015) and cumulatively. Data available from each table are the number of cruise ships that pass <3 M from shore, 3–6 M from shore and >6 M from shore, the retention period <3 M from shore expressed in minutes and the average retention period expressed in minutes. Cumulative data are expressed as totals and give a summary of cruise ships' impact on the designated area.

The number of cruise ship transits through Svetac—Biševo, Sušac—Lastovo and Lastovo–Mljet passages were calculated separately, as well as number of transits through the Central Adriatic Separation Scheme (Tables A5–A8). Information is given for each month of the research and cumulatively. The comparison between routes that used the Central Adriatic Separation Scheme and routes that kept north of the Central Adriatic Separation Scheme offers an important conclusion regarding cruise ships' navigational practice in the central and south Adriatic east coast. The number of cruise ships that used the Central Adriatic Separation Scheme were counted from cruise traffic monitoring. In order to make a comparison, the number of routes that passed north of the Central Adriatic Separation Scheme were calculated. Sušac island was taken as a reference; it is the island located north of the Central Adriatic Separation Scheme, it is positioned on the southern point of the longitudinal corridor that cruise ships frequently use, and it is the closest land from the Central Adriatic Separation Scheme. Palagruža island, being a remote island at the immediate border to the Central Adriatic Scheme, does not offer reliable data and was not taken as the reference.

The calculation was carried out as follows: From Table A2, cumulative information related to the number of cruise ships that passed 3–6 M (265 cruise ships) and <6 M (30 cruise ships) from Sušac island were added (data < 3 M were not taken into consideration, as cruise ships that passed <3 M from the shore passed 3 M–6 M from the shore too). From the result, the number of vertical routes that transited through the Sušac–Lastovo passage (58) were subtracted (Table A6), because for the comparison, only horizontal routes had to be taken in consideration. The final result was: 237 cruise ships kept north of the Central Adriatic Separation Scheme passing in the vicinity of Sušac island, while 187 cruise ships used the Central Adriatic Separation Scheme.

## 3. Results and Discussion

Cruise traffic expansion in the central and south part of the Adriatic east coast has established new frequently used routes as navigational options to the Central Adriatic Separation Scheme. New insight into cruise ships' navigational practices was obtained by processing and analyzing data from a one-year survey of cruise ships in the Adriatic Sea from Perić (2016). The data acquired are charted in two integrated charts (Figure 9a,b). Figure 7a–e justify the data in Figure 9a,b.

Figure 9a defines the longitudinal and transversal traffic flow of cruise ships over 50,000 GT in the Adriatic east coast. The figure shows cruise ships' routes (black lines), the well-defined North Adriatic Separation Scheme (red square 1) and the shorter, less defined Central Adriatic Separation Scheme (red square 2). The North Adriatic Separation Scheme directs the cruise traffic on steady routes without any oscillation, while the Central Adriatic Separation Scheme offers more routing options, and therefore, cruise routes are more dispersed than in the north Adriatic.

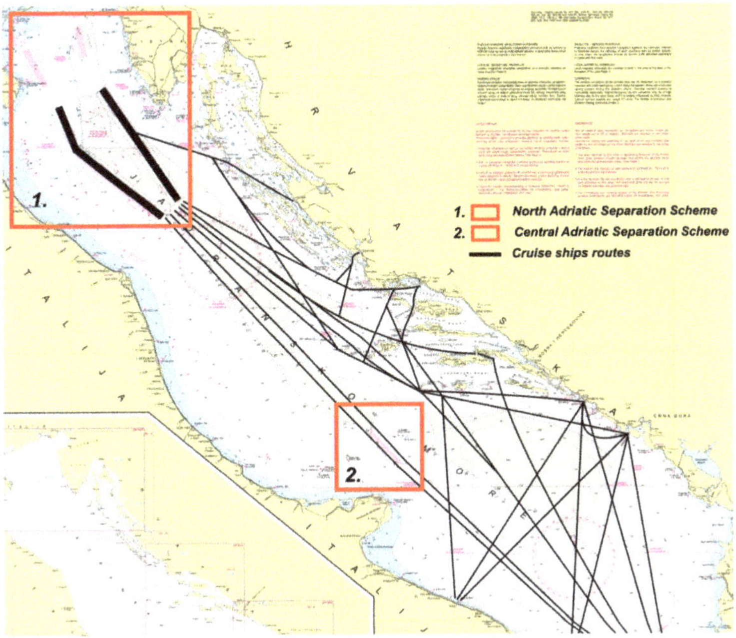


(a)

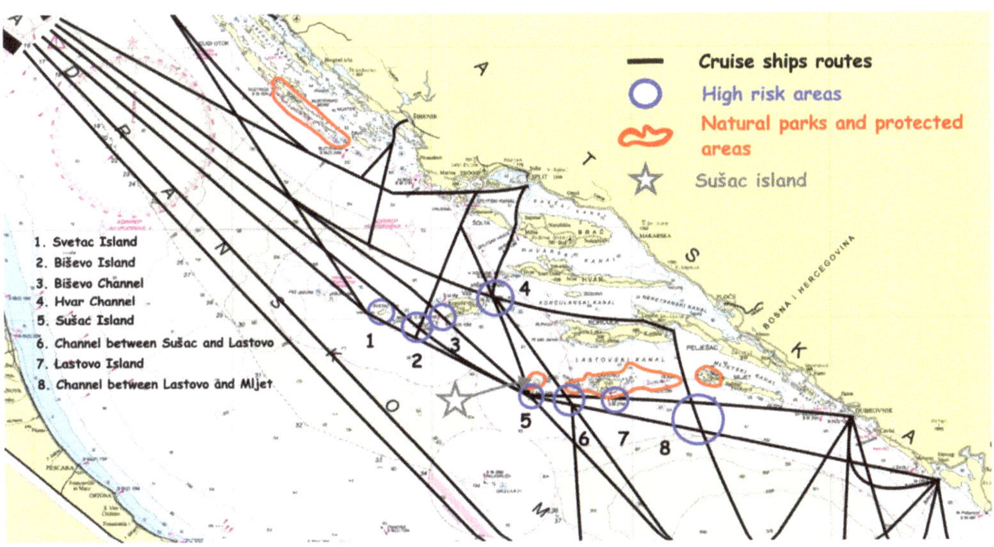


(b)

**Figure 9.** (**a**) Cruise traffic in the Adriatic east coast [7]. Black lines—cruise ships' routes; red squares—red square 1: the North Adriatic Separation Scheme; red square 2: the Central Adriatic Separation Scheme. (**b**) Cruise ships' routes and high-risk zones in the central and south part of the east Adriatic. Black lines—cruise ships routes; red squares—red square 1: the North Adriatic Separation Scheme; red square 2: the Central Adriatic Separation Scheme; red curves—marine protected areas; grey star—the island of Sušac; blue circles—areas of high navigational risks.

According to the traffic monitoring and route analysis in the east Adriatic, cruise ships' movements can be defined in three standard patterns:

1. From the north Adriatic ports to the central and south destination of the Adriatic east coast. Cruise ships departing from the north Adriatic ports proceed on southeasterly courses using the North Adriatic Separation Scheme. Once they leave the North Adriatic Separation Scheme, they follow the route along the outer skirts of the islands along the east Adriatic coast. They do not use the Central Adriatic Separation Scheme (Figure 7a–e).
2. From the north Adriatic ports to the Otranto strait and vice versa. Cruise ships departing from the north Adriatic ports proceed on southeasterly courses using the North Adriatic and Central Adriatic Separation scheme, heading to the Otranto strait. On northwesterly courses, cruise ships proceed from the Otranto strait to the north Adriatic ports using the Central and North Adriatic Separation Scheme (Figure 7a,c).
3. From the Otranto strait to the northern Adriatic via south Adriatic east coast destinations. Cruise ships arriving from Mediterranean ports through the Otranto strait proceed directly to the south Adriatic east coast destinations. Departing from the South Adriatic east coast destinations, cruise ships proceed in coastal navigation along the outer skirts of islands and keep out of the Central Adriatic Separation Scheme. In northwesterly courses, they join the North Adriatic Separation Scheme on the way to north Adriatic ports (Figure 7a–e).

Cruise ships' traffic monitoring has shown irregular navigational practice where cruise ships on international voyages in transit through Croatian territorial waters (on routes: Montenegro—Italy or Mediterranean port to Italy) do not use the Central Adriatic Separation Scheme, nor are the routes planned to enable the shortest stays inside Croatian territorial waters. Cruise ships tend to choose coastal navigation between the islands in Croatian territorial waters until they reach the North Adriatic Separation Scheme (Figure 7c). The Central Adriatic Separation Scheme is only used on direct voyages from the north Adriatic ports to the Otranto strait and from the Otranto strait to north Adriatic ports.

### 3.1. High-Risk Zones Created by Cruise Ships' Routes

Figure 9b is a fragment of Figure 9a; it shows high-risk zones created by cruise ships' routes. Figure 9b shows cruise ships' traffic in the central and south part of the Adriatic east coast (black lines), the position of four marine protected areas (red curves) and eight areas of high navigational and environmental risks (blue circles). Sušac island was selected as a reference point being the closest to the Central Adriatic Separation Scheme (grey star). Four marine protected areas are Kornati National Park, Mljet National Park, Lastovo Nature Park and Sušac Island Nature Park. Eight areas of high navigational and environmental risks were selected based on cruise ships' routes data (the vicinity that cruise ship routes pass from the islands' shores and marine protected areas and cruise routes' mutual interaction). Designated areas of high navigational risks are the islands of Svetac, Biševo, Sušac and Lastovo, as well as the Svetac–Biševo passage, Hvar channel and the Sušac–Lastovo and Lastovo–Mljet passages.

The comparison of cruise traffic density in the Central Adriatic Separation Scheme and cruise traffic density in the vicinity of the island of Sušac provide valuable information on cruise ships' navigational principles in newly established and navigationally less defined cruise regions. The research showed that cruise ships used the Central Adriatic Separation Scheme 187 times, while cruise ships kept north of the Central Adriatic Separation Scheme and passed south of Sušac island 237 times.

Figure 9b shows that cruise ships frequently pass through areas of high navigational and environmental risk which are geographically restricted, navigationally challenging and environmentally sensitive areas. Cruise ships' routes that pass very close to outer island shores as well as between the islands have become standard navigational practice for cruise ships. Additionally, longitudinal and transversal routes' interaction in coastal navigation puts cruise ships in crossing, head on and overtaking situations, which increase the risk of collision grounding and pollution, especially in areas of high navigational risks.

During the research period, cruise ships' traffic inside designated high-risk areas was thoroughly analyzed. The results are presented for each high-risk zone separately.

#### 3.1.1. Svetac Island

Svetac island is part of the northwesterly and southeasterly longitudinal corridor. Cruise ships departing from south Adriatic east-coast destinations to the northern Adriatic ports on northwesterly routes do not use the Central Adriatic Separation Scheme. The Central Adriatic Separation Scheme is also not in used in southeasterly routes from the northern Adriatic ports to the south Adriatic east-coast destinations. Cruise ships proceed along the outer islands route, passing south or north of Svetac island's shores.

Traffic monitoring showed that 76 cruise ships passed less than 3 M from the shores of Svetac island with average retention periods of 16.1 min. A total of 155 cruise ships passed with 3 M to 6 M distances, and 47 cruise ships kept 6 M or more from the island's shores (Table A3). Traffic monitoring showed that two cruise ships met less than 3 M from the island shores six times in a 90 min period. On one occasion, during a 90 min period, three cruise ships met less than 3 M from the island's shore.

#### 3.1.2. Biševo Island

Biševo island is part of the longitudinal and transversal corridor that cruise ships use. Longitudinal southeasterly routes proceed to the south Adriatic east-coast destinations, while northwesterly routes head to north and central Adriatic ports. Transversal routes in northerly and southerly directions connect the port of Split with southern Adriatic destinations and the Italian coast.

During the research period, 21 cruise ships passed less than 3 M from Biševo island with average retention periods of 19.23 min, 51 cruise ships passed with 3 M to 6 M distances from Biševo island, 6 cruise ships kept 6 M and more from Biševo island and 43 cruise ships passed through the Svetac and Biševo passage (Table A4). Traffic monitoring showed that two cruise ships met less than 3 M from Biševo island's shores five times in a 90 min period.

#### 3.1.3. Biševo Channel

Cruise ships' routes through the Biševo channel are not usual; however, despite restrictions, cruise ships' passage through the Biševo channel was detected. During the monitoring period, cruise ships passed through the Biševo channel five times with average retention periods of 25.2 min less than 3 M from the shore. The passage was carried out by identical cruise ships, which shows that the route has become a standard navigational routine for monitored cruise ships. The observed cruise route puts cruise ships in immediate grounding and collision danger.

#### 3.1.4. Hvar Channel

Hvar channel is situated between the island of Hvar and Vis in the central Adriatic region. It is located in an intersection of the longitudinal and transversal routes. The longitudinal inland cruising route connects the northern and central Adriatic region with southern Adriatic ports. The transversal route connects Split with the south Adriatic coast and Italy.

The main navigational risk is the crossing of transversal and longitudinal routes and head on situations in the restricted area of Hvar channel. During the research period, longitudinal inland cruise ships' traffic was not constant; on the other hand, transversal traffic was frequent. In addition to cruise traffic, there was a frequent ferry and catamaran connection from Vis and Lastovo islands to Split and vice versa. In addition to that, the area of Hvar channel is touristy and very popular with developed nautical tourism and dense leisure craft traffic. Taking the above into consideration, navigation in Hvar channel has to be taken with precaution, since the risk of collision and grounding is elevated.

#### 3.1.5. Sušac Island

Longitudinal northwesterly and southwesterly routes as well as transversal southerly and northerly routes often pass along the coast of Sušac island. Cruise ships on northwesterly and southeasterly longitudinal routes connecting south Adriatic east-coast destinations and north Adriatic destinations (and vice versa) do not use the Central Adriatic Separation Scheme. They proceed along outer island routes passing the south shores of Sušac island. Transversal routes pass the eastern and western shores of Sušac island. Cruise ships on southern transversal routes depart from northern Adriatic ports or the port of Split and proceed to the western Adriatic coast or join the longitudinal corridor on the way to Dubrovink, Kotor or the Otranto strait. Northerly transversal routes connect western Adriatic ports, the Otranto strait, Dubrovnik or Kotor to central Adriatic ports or northern Adriatic ports.

During the research period, 101 cruise ships passed less than 3 M from Sušac island with average retention periods of 13.84 min, 265 cruise ships passed with 3 M to 6 M distances from the south shores of Lastovo, while 30 cruise ships kept distances of 6 M or more from the south island shores (Table A2). Traffic monitoring showed that two cruise ships met less than 3 M from the island shores seventeen times in a 90 min period.

#### 3.1.6. Sušac–Lastovo Passage

The passage between Sušac island and Lastovo island is part of a transversal route that cruise ships use on northerly and southerly courses. Cruise ships on southerly courses proceed through the Sušac–Lastovo passage to the western Adriatic coast or they join the longitudinal corridor on the way to Dubrovnik, Kotor or the Otranto strait. Meanwhile, cruise ships on northerly courses proceed through the Sušac–Lastovo passage from western Adriatic shores or from the longitudinal corridor from Dubrovnik or Kotor on the way to the central and north Adriatic ports.

During the research period, 58 cruise ships passed through the Sušac–Lastovo passage (Table A6). Traffic monitoring showed that two cruise ships met in the passage within a 3 M distance from the shore six times in a 90 min period. On one occasion, four cruise ships met inside the passage within 3 M from the shore in a 90 min period.

#### 3.1.7. Lastovo Island

Longitudinal northwesterly and southeasterly cruise ship routes often pass along the south coast of Lastovo island. Cruise ships on a northeasterly longitudinal route from Kotor and Dubrovnik to northern Adriatic ports do not use the Central Adriatic Separation Scheme. They proceed in the east coast outer island route, heading along the south coast of Lastovo island and Sušac island until they reach the Northern Adriatic Separation Scheme. Cruise ships on southeasterly routes from the north Adriatic ports, after leaving the Northern Adriatic Separation Scheme, proceed in east coast outer island route along Lastovo south coast to Dubrovnik and Kotor.

During the research period, 113 cruise ships passed less than 3 M from the south shores of the island with average retention periods of 21.2 min, 281 cruise ships passed with 3 M to 6 M distances from the south shores of Lastovo, while 19 cruise ships kept distances of 6 M or more from the south island shores (Table A1). Traffic monitoring showed that two cruise ships met within 3 M from the shore six times in a 90 min period.

#### 3.1.8. Lastovo–Mljet Passage

The passage between Lastovo island and Mljet island is precisely between Glavat island and Mljet island, which cruise ships use on northerly courses and on southerly courses when arriving/departing to or from Korčula island. During the summer season, there is dense sailing boat, yacht and leisure boat traffic, which often interacts with cruise ships' transit through the passage.

During the research, 30 ships passed through the Lastovo–Mljet passage (Table A7). Traffic monitoring showed that two cruise ships met in the passage within a 3 M distance from the shore two times in a 90 min period. On six occasions, two cruise ships met in a crossing course 3 M to 6 M from the shore in a 90 min period.

The results of the research give a new conclusion and new information from the paper 'Main sailing routes in the Adriatic' [10], in which one of conclusions states that maritime flow in the central Adriatic region is mostly directed through the Central Adriatic Separation Scheme and that maritime accidents are rare, which indicates good maritime coordination. Additionally, in a statement in the paper 'Analysis of the maritime traffic in central part of the Adriatic' [9], one of the conclusions states that the greater part of the longitudinal sailing route extends in the area of sufficient depth and width where there is no significant danger to navigation with the exception of the danger of collision with opposite and transverse traffic and the danger of grounding in the broader area of Palagruža island.

The study brought elevated navigational, safety and environmental risks in the navigationally less defined region of the central and south Adriatic east coast to our attention. The present cruise routing practice presents serious environmental and safety risks to protected areas of the Lastovo Natural Park, Sušac island and the Mljet National Park because the preserved natural ambience and ecosystems give these locations high natural, national and international value.

The analysis of cruise ship movement shows that high-risk routes are repetitive and have become navigational standard routine for cruise ships. This practice questions efficient maritime coordination in the researched area. It is of high importance that cruise traffic expansion is well controlled and equally complemented with investment, the development of routing systems, the implementation of restricted areas, efficient traffic control and straightened maritime regulations in order to ensure the sustainable development of the central and south Adriatic east coast.

This research showed the standard cruise shipping navigational practices in coastal navigation in areas that have been recently discovered by the cruise industry. The observed cruise ship routing practice can be related to any costal region where strong cruise traffic expansion has not been equally supported by investment, the development of routing systems, the implementation of restricted areas and efficient marine traffic coordination. The results are not only related to the central and south part of the Adriatic east coast; on the contrary, they show cruise ships' navigational practice that can be associated globally to any developing cruise region.

## 4. Limitations of the Study

The monitoring of cruise ships' traffic was carried out in period from August 2014 to June 2015, and although the period of the research does not display the current status, the results showed cruise ships' navigational routines where selected cruise routes are constant as cruise ships operate on circular itineraries. The defined navigational practice in costal navigation has become standard operational procedure for cruise ships. The results are not closely related to the time period; on the contrary, they reveal established cruise ships' navigational practice in coastal navigation.

Cruise industry trends were analyzed for the period from 2015 to 2019. Further research was not possible due to the COVID pandemic and the subsequent inactivity in the cruise ship industry. However, the obtained results directed research to the central and south part of the Adriatic east coast. The uniqueness of the region gave solid ground for the detailed analysis of cruise ships' routing practice in coastal navigation. Global cruise industry trends and cruise ships' routing practices in costal navigation raised the question of the sustainable development of coastal regions that have not been involved in cruise tourism before.

The research covered sizes of cruise ships from 50,000 GT and more. The size limit was placed in order to only take large cruise ships with higher environmental impacts and elevated safety risks into consideration. However, smaller cruise ships under 50,000 GT should be taken into consideration in future research, as they are also important factors of navigational safety and the sustainable development of coastal areas.

## 5. Conclusions

Research on leading cruise destination trends in the Adriatic region for the period from 2015 to 2019 showed that the central and south part of the Adriatic east coast is the region with the highest growth in cruise ship calls and passenger movement in the Adriatic and in the entire Mediterranean region. The strong expansion of cruise tourism has created an impact on sustainable development in already established and popular destinations in the Adriatic region such as Venice and Dubrovnik. As a result of imposed restrictions in Venice and Dubrovnik and the development of the cruise industry in general, the cruise industry has discovered new destinations in the Adriatic. The central and south part of the Adriatic east coast has taken advantage of the present circumstances and became an important factor in the Mediterranean cruise market. Destinations such as Zadar, Šibenik, Split, Korčula and Kotor together with established Dubrovnik have become recognizable cruise ports with solid perspectives for continuing growth.

New cruise destinations in the central and south part of the Adriatic east coast have created new cruise ship routes in a region that did not have dense maritime traffic before. In addition, the region is of important national and natural value due to its unspoiled beauty and natural protected areas. The aim of the paper was to define individually implemented cruise ship routing practices in costal navigation, to determine cruise ships' traffic density in the vicinity of maritime protected areas and to identify areas of high safety, navigational and environmental risk.

The results of this research show that cruise ships do not use the Central Adriatic Separation Scheme on the way to the south Adriatic ports; they keep north of the Central Adriatic Separation Scheme and proceed in coastal navigation along outer island shores. Navigation close to the outer island shores along outer island ridges as well as between the islands has become standard navigational practice for cruise ships. Additionally, longitudinal and transversal routes' interaction in coastal navigation puts cruise ships in crossing, head on and overtaking situations, which increases the risk of collision, grounding and pollution, especially in environmentally sensitive, naturally preserved, navigationally challenging and restricted areas. The execution of high-risk cruise ship routes, which have become standard navigational practice, shows a lack of maritime regulation and coordination. The Adriatic region has not had major maritime accidents recently; however, with the present navigational practice, maritime accidents with serious safety, environmental and economic consequences can easily occur.

The study has brought to our attention how efficient traffic coordination is of high importance for regulated and safe maritime traffic. With that in mind, it is of high importance that cruise traffic expansion is well controlled and equally complemented with investment and the implementation and development of routing systems, efficient traffic control and maritime regulation.

The results of the study gave perspectives for further research into cruise ships' practices in coastal navigation with a focus on cruise ships' environmental impact and preventive measures. Cruise ships' practices under 50,000 GT and their impact on coastal areas should be also taken into consideration in future research. In addition to environmental impacts, the results of the study offered valid grounds to plan the modification and extension of the Central Adriatic Separation Scheme and the implementation of restricted areas around high-risk and marine protected areas of the central and south part of the Adriatic east coast.

The obtained results offer a general overview of cruise ships' navigational practices in coastal navigation not only in the central and south Adriatic east coast region but in any coastal region in the world.

**Author Contributions:** Conceptualization, J.D.; methodology, J.D.; validation, J.D.; formal analysis, J.D.; investigation, J.D.; resources, J.D. and T.P.; data curation, T.P. and J.D.; writing—original draft preparation, J.D.; writing—review and editing, J.D., T.P. and G.J.M.; visualization, J.D.; supervision, J.D., T.P. and G.J.M. All authors have read and agreed to the published version of the manuscript.

**Funding:** This research received no external funding.

**Institutional Review Board Statement:** The study did not require ethical approval, nor was involved in animal, human studies.

**Informed Consent Statement:** The study is not involving humans.

**Data Availability Statement:** Data used for the analyses will be available upon request from the corresponding author.

**Conflicts of Interest:** The authors declare no conflict of interest.

## Abbreviations

| | |
|---|---|
| GDP | Gross Domestic Product |
| IMO | International Maritime Organisation |
| COVID | Coronavirus disease |
| UNESCO | The United Nations Educational, Scientific and Cultural Organization |
| GT | Gross Tonnage |
| M | Nautical mile (1852 m) |
| MARPOL | International Convention for the Prevention of Pollution from Ships |

## Appendix A

**Table A1.** Lastovo island—density, distance and retention period of cruise ships' traffic in selected high-risk navigational and environmental region.

| Month | <3 M | 3–6 M | >6 M | Retention Period <3 M (min) | Average (min) |
|---|---|---|---|---|---|
| August 2014 | 17 | 49 | 1 | 319 | 18.8 |
| September 2014 | 20 | 64 | 8 | 356 | 17.8 |
| October 2014 | 16 | 39 | 7 | 269 | 16.8 |
| June 2015 | 31 | 61 | 3 | 712 | 23.0 |
| July 2015 | 29 | 68 | 0 | 854 | 29.4 |
| Total | 113 | 281 | 19 | 2510 | 21.16 |

**Table A2.** Sušac island—density, distance and retention period of cruise ships' traffic in selected high-risk navigational and environmental region.

| Month | <3 M | 3–6 M | >6 M | Retention Period <3 M (min) | Average (min) |
|---|---|---|---|---|---|
| August 2014 | 15 | 38 | 4 | 200 | 13.3 |
| September 2014 | 17 | 60 | 9 | 140 | 8.2 |
| October 2014 | 19 | 41 | 7 | 243 | 12.8 |
| June 2015 | 26 | 60 | 7 | 451 | 17.3 |
| July 2015 | 24 | 66 | 3 | 423 | 17.6 |
| Total | 101 | 265 | 30 | 1457 | 13.84 |

**Table A3.** Svetac island—density, distance and retention period of cruise ships' traffic in selected high-risk navigational and environmental region.

| Month | <3 M | 3–6 M | >6 M | Retention Period <3 M (min) | Average (min) |
|---|---|---|---|---|---|
| August 2014 | 13 | 26 | 3 | 255 | 17.3 |
| September 2014 | 15 | 35 | 18 | 143 | 9.5 |
| October 2014 | 17 | 25 | 8 | 233 | 13.7 |
| June 2015 | 18 | 40 | 7 | 369 | 20.5 |
| July 2015 | 13 | 29 | 11 | 745 | 19.7 |
| Total | 76 | 155 | 30 | 1745 | 16.14 |

**Table A4.** Biševo island—density, distance and retention period of cruise ships' traffic in selected high-risk navigational and environmental region.

| Month | <3 M | 3–6 M | >6 M | Retention Period <3 M (min) | Average (min) |
|---|---|---|---|---|---|
| August 2014 | 10 | 11 | 1 | 156 | 15.6 |
| September 2014 | 4 | 15 | 1 | 66 | 16.5 |
| October 2014 | 3 | 9 | 0 | 61 | 20.3 |
| June 2015 | 0 | 6 | 1 | 0 | 0 |
| July 2015 | 4 | 10 | 3 | 98 | 24.5 |
| Total | 21 | 51 | 6 | 381 | 19.23 |

**Table A5.** Svetac–Biševo passage—number of cruise ships' transits in selected high-risk navigational and environmental locations (passages).

| Month | Number of Transits |
|---|---|
| August 2014 | 8 |
| September 2014 | 14 |
| October 2014 | 6 |
| June 2015 | 6 |
| July 2015 | 9 |
| Total | 43 |

**Table A6.** Sušac–Lastovo passage—number of cruise ships' transits in selected high-risk navigational and environmental locations (passages).

| Month | Number of Transits |
|---|---|
| August 2014 | 5 |
| September 2014 | 14 |
| October 2014 | 7 |
| June 2015 | 13 |
| July 2015 | 19 |
| Total | 58 |

**Table A7.** Lastovo–Mljet passage—number of cruise ships' transits in selected high-risk navigational and environmental locations (passages).

| Month | Number of Transits |
|---|---|
| August 2014 | 7 |
| September 2014 | 9 |
| October 2014 | 2 |
| June 2015 | 9 |
| July 2015 | 3 |
| Total | 30 |

**Table A8.** Central Adriatic Separation Scheme—number of cruise ships' transits.

| Month | Number of Transits |
|---|---|
| August 2014 | 24 |
| September 2014 | 42 |
| October 2014 | 37 |
| June 2015 | 42 |
| July 2015 | 42 |
| Total | 187 |

## References


1. Kovačić, M.; Silveira, L. Cruise tourism: Implications and impacts on the destinations of Croatia and Portugal. *Sci. J. Marit. Res.* **2020**, *34*, 40–47. [CrossRef]
2. Peričić, D. Analysis of the World Cruise Industry. In Proceedings of the 4th Dubrovnik International Economic Meeting, DIEM 2019, Dubrovnik, Croatia, 27–28 September 2019; Available online: https://www.researchgate.net/publication/348276595_Analysis_of_the_World_Cruise_Industry (accessed on 20 April 2022).
3. Perić, T. Wastewater Pollution from Cruise Ships in the Adriatic Sea. *Promet-Traffic Transp.* **2016**, *28*, 425–433. [CrossRef]
4. Orsini, K.; Ostojić, V. Economic brief 036 I March 2018 Croatia's Tourism Industry—Beyond the Sun and Sea. Available online: https://ec.europa.eu/info/sites/info/files/economy-finance/eb036_en.pdf (accessed on 22 November 2021).
5. Europa Economy and Finance Forecast. 2020. Available online: https://ec.europa.eu/economy_finance/forecasts/2020/spring/ecfin_forecast_spring_2020_me_en.pdf (accessed on 16 April 2022).
6. Dorigatti, J.; Perić, T.; Jelić Mrčelić, G. Sustainability in Maritime Transport: Impact of Cruise Ships' Routing in Coastal Navigation. In Proceedings of the Kotor International Maritime Conference, Kotor, Montenegro, 26–27 November 2021.
7. Luković, T.; Kovačić, M. Seasonality of World and Croatian Cruising, New Technologies, New Challenges—Conference proceedings. In Proceedings of the 28th International Conference on Organisational Science Development, Portorož, Slovenia, 25–27 March 2009.
8. Lušić, Z.; Pušić, D.; Medić, D. Analysis of the maritime traffic in the central part of the Adriatic. In *International Congress on Transport Infrastructure and Systems (TIS 2017)*; Dell'sAcqua, G., Wegman, F., Eds.; CRC Press/Balkerma: Rome, Italy, 2017; pp. 1013–1020. ISBN 978-1-138-03009-1. Available online: https://www.researchgate.net/publication/321034646_Analysis_of_the_maritime_traffic_in_the_central_part_of_the_Adriatic (accessed on 11 November 2021).
9. Lušić, Z.; Kos, S. Glavni plovidbeni putovi na Jadranu—The main sailing routes in the Adriatic. *Naše More* **2006**, *53*, 198–205. Available online: https://www.researchgate.net/publication/236667819_The_Main_Sailing_Routes_in_the_Adriatic (accessed on 12 November 2021).
10. MedCruise Report 2019 Statistics Cruise Activities in Med Ports. Available online: https://www.medcruise.com/2019-statistics-cruise-activities-in-medcruise-ports (accessed on 5 October 2021).
11. UNESCO World Heritage Convention. Available online: https://whc.unesco.org/en/convention/ (accessed on 8 March 2022).
12. Zec, D.; Frančić, V.; Rudan, I.; Malgić, L.; Lušić, Z.; Đurđević-Tomaš, I.; Brajović, M.; Vukuć, M. Prometno—Plovidbena Studija Plovnog Područja Split, Ploče i Dubrovnik, Pomorski Fakultet Rijeka. 2014. Available online: https://mmpi.gov.hr/UserDocsImages/arhiva/MPPI%20PRZ%20PROM-PLOV%20studija%2019-11_14.pdf (accessed on 20 February 2021).
13. Cruise Industry News 2020–2021 Annual Report. Available online: https://www.cruiseindustrynews.com/annual-cruise-industry-report.html (accessed on 22 May 2022).
14. Zanne, M.; Beškovnik, B. Assesing Home Port Potential of Selected Adriatic Ports. *TOMS Trans. Marit. Sci.* **2008**, *7*, 143–153. [CrossRef]
15. Rodrigue, J.-P.; Notteboom, T. The geography of cruises: Itineraries, not destinations. *Appl. Geogr.* **2013**, *38*, 31–42. [CrossRef]
16. Milić Beran, I.; Draganić, N. An analysis of the dynamics of cruise traffic in the port of Dubrovnik. *DIEM* **2020**, *5*, 65–72.
17. Ban, I.; Peručić, D.; Vrtiprah, V. Izazovi razvoja crusing turizma u Dubrovačko-neretvanskoj županiji. *Zb. Sveučilišta U Dubrov.* **2014**, *1*, 1–33.
18. Dubrovnik Port Authority. Available online: https://www.portdubrovnik.hr/dubrovnik-cruise-destination (accessed on 20 December 2021).
19. Vojvodić, K. Cruise port positioning—The case of Korčula, Pozicioniranje luka na tržištu pomorskih krstarenja-primjer Korčule—Pregledni članak. *Naše More* **2003**. Available online: https://www.researchgate.net/publication/27193815_Pozicioniranje_luka_na_trzistu_pomorskih_krstarenja_-_primjer_Korcule (accessed on 20 December 2021).
20. Korčula Tourist Board. Available online: https://www.visitkorcula.eu (accessed on 5 April 2022).
21. Ministarstvo Mora Prometa i Infrastrukture. Available online: http://mppi.hr/default.aspx?id=9733 (accessed on 6 April 2022).
22. Nikičević, J. Strengthening the role of local government to ensure sustainable development of the cruise sector: The case of Kotor. *Mar. Policy* **2019**. Available online: https://www.sciencedirect.com/science/article/pii/S0308597X1930332X (accessed on 16 April 2022). [CrossRef]

23. Kofjač, D.; Škurić, M.; Dragović, B.; Škraba, A. Traffic Modelling and Performance Evaluation in the Kotor Cruise Port. *Stroj. Vestn.* **2013**, *59*, 526–535. Available online: https://www.researchgate.net/publication/256492372_Traffic_Modelling_and_Performance_Evaluation_in_the_Kotor_Cruise_Port (accessed on 16 April 2022). [CrossRef]
24. Split Port Authority. Available online: https://portsplit.hr/en/cruising/ (accessed on 17 December 2021).
25. Perić, T.; Mihanović, V.; Golub Medvešek, I. Analysis of cruise ship traffic in the port of Split. *Istraz. I Proj. Za Privredu* **2019**, *17*, 304–310. Available online: https://www.researchgate.net/publication/336432724_Analysis_of_cruise_ship_traffic_in_the_Port_of_Split (accessed on 4 March 2022). [CrossRef]
26. Zadar Port Authority. Available online: https://www.port-authority-zadar.hr/centar-za-korisnike/statistike/ (accessed on 18 December 2021).
27. Port of Šibenik Authority Bulletin. Available online: https://www.portauthority-sibenik.hr/en/23-2/ (accessed on 20 December 2021).
28. Perić, T. Evaluation Model of Sanitary Waste Water Pollution from Cruise Ships in the Adriatic Sea. Ph.D. Thesis, University of Rijeka, Rijeka, Croatia, 2016.

*applied sciences*

MDPI

*Article*

# Hydrogen vs. Batteries: Comparative Safety Assessments for a High-Speed Passenger Ferry

**Foivos Mylonopoulos, Evangelos Boulougouris *, Nikoletta L. Trivyza, Alexandros Priftis, Michail Cheliotis, Haibin Wang and Guangyu Shi**

Maritime Safety Research Centre, Department of Naval Architecture, Ocean and Marine Engineering, University of Strathclyde, Glasgow G4 0LZ, UK; fivos.mylonopoulos.2017@uni.strath.ac.uk (F.M.); nikoletta.trivyza@strath.ac.uk (N.L.T.); alexandros.priftis@strath.ac.uk (A.P.); michail.cheliotis@strath.ac.uk (M.C.); haibin.wang.100@strath.ac.uk (H.W.); guangyu.shi@strath.ac.uk (G.S.)
* Correspondence: evangelos.boulougouris@strath.ac.uk

**Abstract:** Batteries and hydrogen constitute two of the most promising solutions for decarbonising international shipping. This paper presents the comparison between a battery and a proton-exchange membrane hydrogen fuel cell version of a high-speed catamaran ferry with a main focus on safety. The systems required for each version are properly sized and fitted according to the applicable rules, and their impact on the overall design is discussed. Hazards for both designs were identified; frequency and consequence indexes for them were input qualitatively, following Novel Technology Qualification and SOLAS Alternative Designs and Arrangements, while certain risk control options were proposed in order to reduce the risks of the most concerned accidental events. The highest ranked risks were analysed by quantitative risk assessments in PyroSim software. The gas dispersion analysis performed for the hydrogen version indicated that it is crucial for the leakage in the fuel cell room to be stopped within 1 s after being detected to prevent the formation of explosive masses under full pipe rupture of 33 mm diameter, even with 120 air changes per hour. For the battery version, the smoke/fire simulation in the battery room indicated that the firefighting system could achieve a 30% reduction in fire duration, with firedoors closed and ventilation shut, compared to the scenario without a firefighting system.

**Keywords:** liquefied hydrogen; batteries; high-speed ferry; safety; hazard identification; quantitative risk assessments

**Citation:** Mylonopoulos, F.; Boulougouris, E.; Trivyza, N.L.; Priftis, A.; Cheliotis, M.; Wang, H.; Shi, G. Hydrogen vs. Batteries: Comparative Safety Assessments for a High-Speed Passenger Ferry. *Appl. Sci.* **2022**, *12*, 2919. https://doi.org/10.3390/app12062919

Academic Editor: José A. Orosa

Received: 9 February 2022
Accepted: 10 March 2022
Published: 12 March 2022



## 1. Introduction

The International Maritime Organisation (IMO) requires a 50% reduction in greenhouse gas (GHG) emissions by 2050 compared to 2008 levels, which renders the need for utilising alternative fuels in the maritime industry mandatory. Hydrogen and proton exchange membrane (PEM) fuel cells constitute a zero-emission alternative under the prerequisite that hydrogen is produced by renewable sources. Several boats sail at rivers or lakes utilising hydrogen fuel cells, with hydrogen stored in compressed gas form [1–3]. PEM fuel cells receive hydrogen and air, and through electrochemical reactions, electricity and hot water are produced without any carbon emissions. They are light, producing insignificant noise and low vibrations, as well as having a high efficiency (50–60%), especially if combined with a waste heat recovery system, where efficiency can reach even higher levels.

The battery-powered vessel is also a solution to fulfil the demand of GHG reduction from IMO. This type of vessel can store and use electricity supplied from the power grid on shore in battery racks and eliminate all the exhaust gas emissions from the burning of fossil fuels in internal combustion engines (ICE). The European Union H2020 Project-TrAM has investigated and developed battery-powered vessels using such a concept to implement the emission control strategy in short-sea shipping.

Both emerging zero-emission technologies could be a potential solution for high-speed passenger ferries; however, the use of these novel propulsion methods and green technologies introduces new safety concerns and challenges in the design of ships, including weight limitations and internal arrangement restrictions. Regarding the safety issues, hydrogen gas leakage can lead to explosions under certain concentrations, while batteries are associated with fire risk. As far as the design of zero-emission high-speed vessels is concerned, substantial work has been carried out by the authors, within the TrAM project [4] and on alternative fuels [5], which is being extended through the work presented in this paper. Design optimisation has been beneficial for the performance improvement of various kinds of vessels [6,7], including high-speed catamarans [8]. Recent studies on the hydrodynamic performance of high-speed catamarans in various operational conditions [9,10] aim at reducing the energy requirements, resulting in more economical designs while mitigating the aforementioned challenges related to this kind of vessel.

A brief overview of the existing literature, involving batteries and hydrogen applications on ships, is presented in Section 2. In Section 3, the proposed designs are discussed, including the considered systems for both versions. In Section 4, the most severe hazards and the most effective risk control options (RCOs) are presented after a hazard identification (HAZID) analysis was performed for the hydrogen and the battery versions of the ferry, following the formal safety assessment (FSA). In Section 5, materials and methods are presented, including quantitative risk assessments for both designs, using the PyroSim software. For the hydrogen version, a gas dispersion analysis in the fuel cell room is presented, while for the battery version, a smoke/fire simulation is performed in the battery room. In Section 6, the results are presented and discussed for various scenarios for both designs. The concluding remarks are presented in Section 7.

## 2. Background

Hydrogen is an abundant, non-toxic, zero-emission fuel from well to wake if it is produced by electrolysis, but with a highly flammable and explosive nature. It is more gravimetrically and volumetrically efficient to be stored in liquefied form ($LH_2$) at −253 degrees Celsius, compared to high-pressure compressed gas. The fuel parameters of liquefied hydrogen are presented in Table 1. Current experience with liquefied hydrogen is limited in the marine industry, and regulations are still under development.

**Table 1.** Liquefied hydrogen properties.

| Parameter | Value |
|---|---|
| Storage temperature (°C) | −253 |
| Storage pressure (bar) | 1 |
| Gas constant (J/kgK) | 4124 |
| Volumetric energy density (GJ/m$^3$) | 8.5 |
| Autoignition temperature (°C) | 585 |
| Minimum ignition energy (mJ) | 0.019 |
| Energy density LHV (MJ/kg) | 119.96 |
| Flammability range in air (%) | 4–75 |
| Explosive range in air (%) | 18–59 |

Regarding the hydrogen systems' applications on ships, Ervin and Dincer [11] presented a thermodynamic analysis for an integrated solid oxide fuel cell system onboard a liquefied hydrogen-fuelled ship to assess the overall energy and exergy efficiencies. Cavo et al. [12] presented a model-based dynamic analysis onboard a zero-emission ship focusing on the coupling of PEM fuel cells and metal hybrids (MH). MHs provided promising results in terms of fuel storage and supply of hydrogen to the fuel cells. Sari et al. [13] investigated the environmental impact for auxiliary powered systems of a hydrogen fuel cell-powered chemical tanker until 2050, using the reference energy system concept. Diesel generators were replaced with fuel cells in 2030, and this resulted in a reduction of around 7000 tonnes of $CO_2$ compared to 2029.

Another alternative option to reduce gaseous emissions is by utilising batteries. They are energy storage systems but with high fire risk, which can negatively affect their applicability. Regarding the battery systems' applications on ships, Wang et al. [14] presented a life cycle analysis, with a focus on environmental footprint, and cost assessment of a battery ferry, and compared the results with a conventional ferry. It was demonstrated that when grid mix electricity was supplied in 2019, a 30% reduction of GHG emissions could be achieved, along with a 15% reduction of lifecycle costs when battery-powered systems are used. Lindstad et al. [15] studied the conversion of conventional offshore support vessels to hybrid by retrofitting of batteries, with the main focus on environmental and economic aspects. The results demonstrated a 40–45% reduction of annual global warming potential in Arctic regions when batteries and ICEs are combined. However, a significant payback period of 12.5 years was obtained, indicating that for existing old vessels, hybridisation with batteries will not be beneficial. Vanem et al. [16] discussed the various data-driven states of health modelling approaches that can be used to estimate the available energy stored in marine battery systems on the basis of sensor data during the operational phase. This is a crucial parameter to avoid loss of propulsion, which can lead to collision or grounding.

Currently, in the literature, there are a few safety assessments performed for hydrogen-powered ships. Mao et al. [17] used ANSYS Fluent software to analyse the overpressure and high-temperature damages induced by an explosion due to hydrogen leakage and ignition in the fuel cell room, control room, and passenger area. An explosion in the fuel cell room exhibited the greater brisance, imposing the most severe damages in ship structure. Yuan et al. [18] assessed the effect of fine water mist on suppressing jet fires around the hydrogen storage tank at the upper deck of a passenger ship. It was demonstrated that jet fires caused by hydrogen leakage could not be extinguished but fire field temperature could be reduced by setting the appropriate spray velocity and droplet size values, preventing fire spread and damage to adjacent equipment/structures. Aarskog et al. [19] presented a consequence assessment using the FLACS CFD model to estimate the fatality risks during operation and at night, related to various hydrogen systems, including high-pressure piping, vent mast, and high-pressure storage tanks, onboard a hydrogen-fuelled high-speed ferry. It was demonstrated that the design was equivalent in terms of safety with conventionally fuelled ferries. Pratt and Klebanoff [20] presented a preliminary HAZID risk assessment for another concept design, the SF-Breeze high-speed catamaran, with hydrogen stored in liquefied form. In this assessment, bunkering was also considered, and it was demonstrated that collision during operation and fuel spill during bunkering were the most severe hazards (highest risk indexes). Another safety analysis was conducted to assess the hydrogen diffusion using ANSYS Fluent software in the SF-Breeze by Li et al. [21], but it was assumed that hydrogen is stored in compressed gas form at 200 bar. It was assessed how hydrogen was concentrated in space depending on the leakage position for different ventilation and hydrogen detection systems' arrangements. The highest hydrogen concentrations were observed in the corners of the fuel cell rooms.

Similarly, there are very few research studies related to the risk assessments on marine battery power plants. One study carried out by Jeong et al. [22] developed a multi-criteria decision-making approach for a hybrid battery-engine system focusing on cost–environment–risk issues. In their study, the qualitative risk assessment conducted could be further expanded to a quantitative risk assessment. There are also classification societies' guidelines providing risk assessment and safety design for the maritime batteries' application [23,24]. In the TrAM project, a safety level evaluation was carried out including HAZID, fault tree analysis (FTA), event tree analysis (ETA), and cost–benefit assessment on RCOs [25].

Currently, most studies are focused on the design/efficiency, as well as the cost or emission analysis, of hybrid arrangements including both batteries and fuel cells for the propulsion [26–29]. In this work, two zero carbon emission designs are proposed. For the design of the hydrogen ferry, hydrogen is stored as a cryogenic liquid ($LH_2$), since it is more gravimetrically and volumetrically effective, enabling larger amounts of fuel to be stored onboard and used for propulsion, without requiring frequent refuelling [20]. $LH_2$ is used as the only fuel source for the catamaran ferry. Batteries are used only for emergency situations in case of severe failure of hydrogen systems. In the battery version, the ferry is fully electric, powered solely by electricity. The novelty of this paper is twofold. First, two novel zero-emission solutions are proposed and compared for a catamaran ferry, showcasing the possibility of a pure hydrogen solution, with hydrogen stored in liquefied form, and a pure electric alternative. Secondly, a detailed safety comparison is performed for these zero-emission solutions for the high-speed passenger vessel including HAZID analysis but also gas and smoke dispersions in the fuel cell and battery rooms, respectively. As was derived from the literature review, this is the first work to perform such an in-depth safety assessment for these decarbonising solutions for passenger ferries.

## 3. Proposed Designs

### 3.1. Case Study

The case ship selected for the comparative assessment between the battery and the hydrogen version was the Stavanger demonstrator, for which the main particulars and ferry details are presented in Table 2.

**Table 2.** Case ship specifications.

| Main Particulars and Ferry Details | | | |
|---|---|---|---|
| Length overall (m) | 30.6 | route length | 23 nm |
| Length waterline (m) | 29.32 | service hours/day | Up to 20 |
| Draft (m) | 1.26 | round trips/day | Up to 15 |
| Breadth (m) | 9 | electric motors | 2 × 550 kW |
| Demihull breadth waterline (m) | 2.44 | service speed | 23 knots |
| Demihull spacing (m) | 6.56 | crew | 3 |
| Displacement (c.m.) | 80 | passengers | 147 |

The battery-powered catamaran ferry, which is expected to be operational in 2022–2023, will operate in a multi-stop route in the Stavanger area in Norway up to 20 h per day with each round trip including up to 12 stops upon passenger's request. This ferry is part of the TrAM project, which is funded by the European Union [30]. Two more studies were carried out in London and Belgium for the same type of vessel.

In this study, for the design of hydrogen systems and their arrangement onboard, the IGF code is mainly used [31], along with the guidelines provided by the recently published handbook for hydrogen-fuelled vessels by DNV [32]. Even though IGF rules are mostly applicable for LNG, they can also be used as a basis for $LH_2$ systems as well, considering their similar properties [20]. However, for novel projects, an equivalent level of safety with a referenced design should be demonstrated for the vessel through an alternative design approach [33,34].

For a maritime application, the design of a battery-powered system has different criteria, mainly focusing on the performance and safety levels. Regarding the performance of the battery-powered system, the biggest challenges are energy density, power density, charging duration, life span, cost, and sustainability [35]. DNV has also published technical guidelines to support the design of such vessels and the evaluation of safety and risk levels [36].

### 3.2. Systems and Equipment Onboard

#### 3.2.1. Hydrogen Version

The connections from the hydrogen tank at the upper deck to the fuel cells at the aft of the main deck are shown in Figure 1. The catamaran's design provides sufficient stability, enabling the placement of the hydrogen tank, piping, and vaporisers at the top deck. The type C double-walled $LH_2$ tank has a pressure transmitter that will measure the pressure levels in the system. If there is a rapid pressure increase in the tank, there are two pressure relief valves mounted on top of it, as required by regulations, which will open immediately. The gas vents from these valves are led to the vent mast, which is located above tank connection space (TCS) adjacent to the tank, through gas vent piping. In this ventilated enclosure of TCS in which there are hydrogen sensors, fire detectors, vents, pipes, safety valves, vaporisers, etc., at least a 30 air changes per hour (ACH) ventilation rate is required. There is also a pressure building unit (PBU) that passively warms the liquid derived from the bottom of the tank and delivers it at its top in the cold gas space. Then, hydrogen is delivered to the vaporisers which convert cold hydrogen gas to room temperature hydrogen required for the fuel cells. After the vaporisers, there are pressure regulators to measure the pressure levels in the pipes, and if the pressure exceeds 10 bar, the pressure relief valves will open and gas will be vented to the vent mast. Then, there are three-way valves so that hydrogen is directed to the required fuel cell room even in case of damage/leakage in one of the pipes. The numerous master-gas valves, also called emergency shut-down (ESD) valves, throughout the whole gas distribution system should quickly shut off the flow of hydrogen once leakage is detected to avoid hazardous conditions by large accumulations of gas in the air. This arrangement with the cross-connected pipes and dual vaporisers was also utilised in the SF-Breeze high-speed ferry concept design and it provides redundancy of equipment in terms of safety [20]. The fuel pipes should either be gas-tight ventilated ducts or double-walled pipes, and they should not be led directly through control or accommodation spaces as required by regulations.

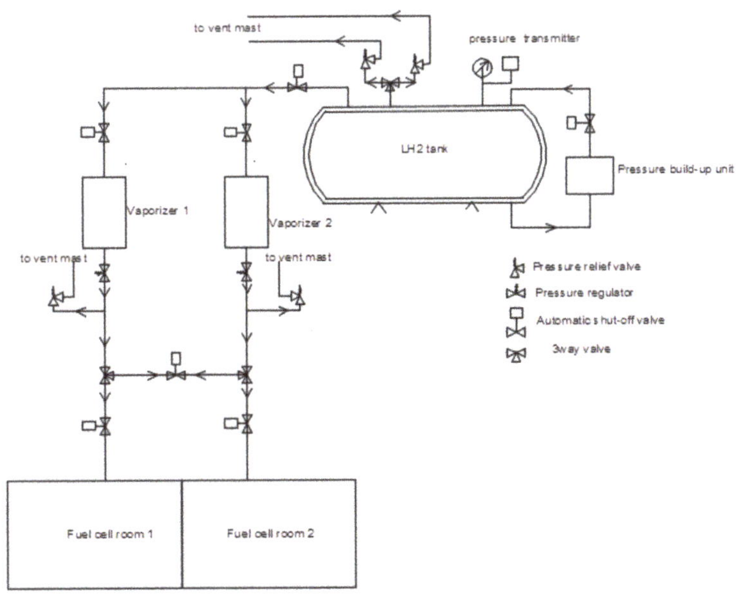


**Figure 1.** Onboard hydrogen systems.

Once hydrogen enters the fuel cell room through the gas supply piping, there are double block-and-bleed valves for each of the fuel cells, before hydrogen enters the stack (Figure 2). Double block and bleed valves are used to stop (block) the flow of hydrogen immediately after the hydrogen leakage is detected. Delay time for leakage to stop once detected should in general be around 0.5–2 s. The bleed valve will open to release ("bleed") any pressure that remains in the pipes and hydrogen gas will flow through the vent pipe to the vent mast. According to regulations inside each stack (gas consumer), there should also be double block and bleed valves.

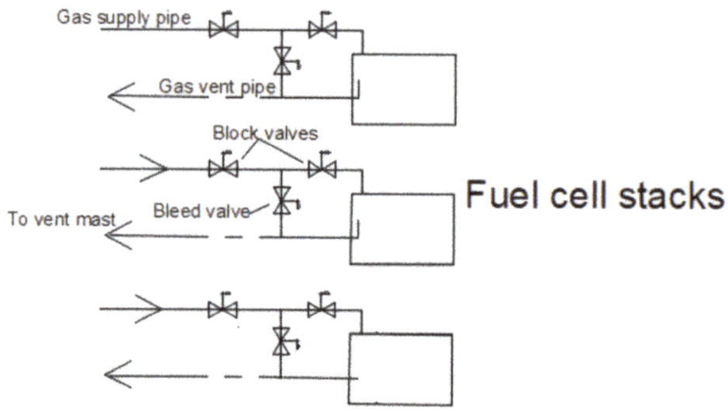


**Figure 2.** Gas and vent piping in the fuel cell room.

Fuel cell stacks used are Powercell MS-100, which are low-temperature PEM fuel cells [37]. They are placed at the aft of the main deck, behind the passengers' area as far away from the crew cabin as possible, in two separated spaces at the port and starboard within gastight enclosures as required by regulations. The technical data of the fuel cell stacks are presented in Table 3. Hybridisation with batteries is not required since MS-100 stacks have a very fast response time with a minimum operational lifetime of 20,000 h (high durability). Hence, in this study, batteries are installed onboard only for emergency purposes in case of severe damage in hydrogen systems to provide a safe return to port. Battery packs are placed at the bottom deck in demihulls.

**Table 3.** Technical data of MS-100 stacks.

| Parameter | Value |
|---|---|
| Rated power (kW) of each stack | 100 |
| System efficiency @ rated power | 50% |
| Dimensions of each stack: H × D × W (m) | 0.75 × 0.75 × 0.25 |
| Weight of each stack (kg) | 150 |
| Fuel inlet pressure (bar) | 8 |
| Fuel inlet temperature (°C) | 10 |
| Response time (s): off-stanby | 10 |
| Response time (s): standby-run | 10 |
| Minimum operational lifetime (h) | 20,000 |

The control room with all the DC/DC converters and DC/AC inverters is adjacent to the fuel cell room in the main deck. Each DC/DC converter controls two fuel cell stacks in series and delivers DC power to the DC/AC inverter which converts it to AC, so that electricity is delivered to e-motors that drive the propellers. The e-motors are located at the bottom deck, one in each demihull.

3.2.2. Battery Version

Figure 3 presents the battery power system on the battery version of the ferry. The two battery racks are connected to two DC hubs separately. In each DC hub, DC/AC converters were applied to enable the availability of power to motors and other energy feeders. There is also a shore connection within the hub which not only provides energy to the switchboard while docking but also charges the batteries. After the motors, the electricity will be converted into mechanical energy which drives water pumps of a waterjet. The waterjet will eventually drive the ferry. Both hydrogen and battery versions of the ferry satisfy the stability requirements for high-speed crafts (HSC). More details about the arrangement of the battery systems onboard the battery version of the high-speed ferry are presented by Boulougouris et al. [4].

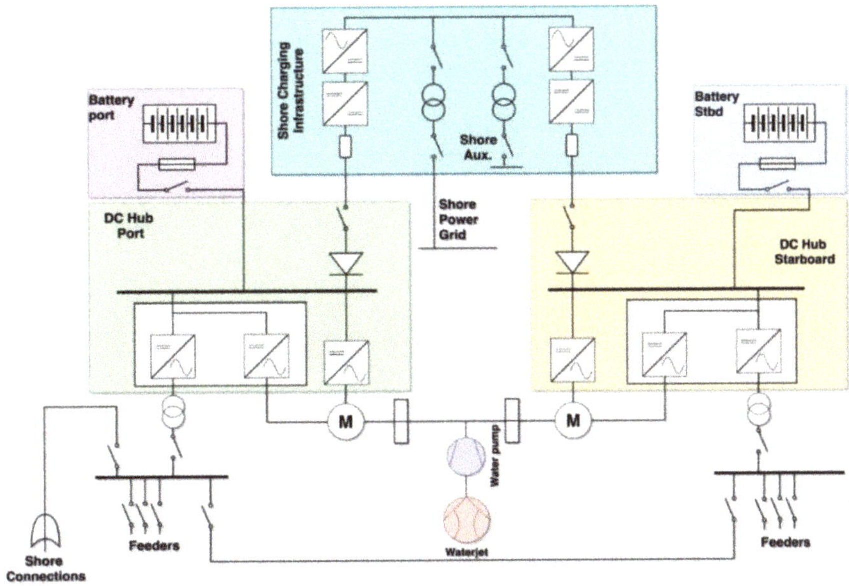


**Figure 3.** Battery power system.

## 4. HAZID

A HAZID analysis was performed for both designs to identify the most concerning hazards. Then, for each of them, the causes that can lead to the critical (accidental) event and the subsequent consequences were analysed. The determination of frequency index and consequence index that follows is based on the IMO's FSA guideline [38]. The frequency index indicates the probability of an event to occur, and the consequence index is related to the severity of the event and the subsequent repercussions on human safety and ship structure. Since there are no past accident statistics from similar ships considering the novelty of the designs, the hazards' rankings were discussed with experts for verification of results. The risk matrix is obtained by utilising the logarithmic scales of frequency and consequence as presented in Equation (1) on the basis of [25]:

$$\begin{gathered} \text{Risk} = \text{Frequency} \times \text{Consequence} \\ \log(\text{Risk}) = \log(\text{Frequency}) + \log(\text{Consequence}) \\ \text{Risk Index (RI)} = \text{Frequency Index (FI)} + \text{Consequence Index (SI)} \end{gathered} \quad (1)$$

Hence, the risk index is obtained by adding the frequency index, which is also called probability index (PI), and the consequence (severity) index (SI). Certain RCOs need to be proposed to mitigate the risks of the most concerning hazards. A simplified flowchart of the FSA procedure followed is shown in Figure 4.

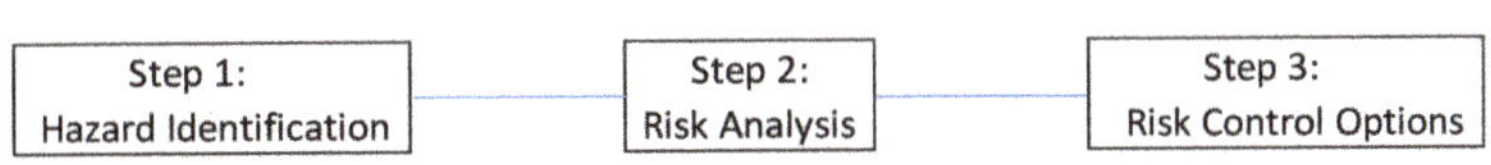


**Figure 4.** FSA methodology.

### 4.1. Hydrogen Version

There are four categories that were included in the HAZID analysis for the hydrogen version. All hazards should be either low or medium risk after the proper risk reduction measures are applied. Hazards are caused by improper installation, malfunction of equipment, or systems belonging to the design/construction/installation category, while accidental events can potentially occur during operations belonging to the operation category. The rest of the hazards can potentially occur during bunkering (refuelling) or emergency situations. The total number of hazards considered in this study is 35. The most severe hazards for the hydrogen version are presented in Table 4.

**Table 4.** The most severe hazards of the hydrogen-fuelled version from HAZID.

| No. | Initial Accidental Event |
|---|---|
| 1 | Leakage at FC room |
| 2 | Fire/explosion at FC room |
| 3 | Fire/explosion in the control room |
| 4 | Fire/explosion at the upper deck |
| 5 | Collision during operation |
| 6 | Fuel spill during bunkering |
| 7 | Fire propagation indoors (emergency) |

The most effective RCOs identified after the HAZID analysis for the hydrogen version of the ferry were the following:

- Placement of $LH_2$ tank and tank connection space at the upper deck.
- Proper alarm/firefighting equipment.
- Redundancy and proper arrangement of ventilation, hydrogen gas detection equipment, and safety valves.
- Use equipment of proven usage and test it prior to use.

### 4.2. Battery Version

Following a similar approach, a HAZID was conducted for the battery version of the ferry [25]. A total of 55 hazards with frequencies and consequences were evaluated. The following number of hazards was identified for three categories:

- Design, construction, installation (21 hazards);
- Operation (25 hazards);
- Emergency (9 hazards).

In Table 5, the most severe hazards associated with the battery-powered systems are presented; thus, in Section 5.2, the quantitative risk assessment is conducted to evaluate the safety of the vessel in detail.

**Table 5.** The most severe hazards of the battery-powered vessel from HAZID.

| No. | Initial Accidental Event |
|---|---|
| 1 | Battery breach/punctures during construction and installation |
| 2 | Fire and explosion in battery room during construction and installation |
| 3 | Battery breach while in operation |
| 4 | Internal cell failure/thermal runaway while in operation |
| 5 | Battery on fire while in operation |
| 6 | Fire and explosion while in operation |
| 7 | Fire propagation during an emergency |
| 8 | Evacuation failed during an emergency |

The most effective RCOs identified after the HAZID analysis for the battery version of the ferry were the following:

- Movement of batteries in the main deck to reduce the fire risk;
- Proper alarm/firefighting system;
- Testing of equipment prior to use;
- Regular inspection and maintenance of battery-related equipment.

## 5. Gas and Smoke Dispersion Analyses

A gas dispersion analysis was conducted for the hydrogen version and a smoke/fire simulation for the battery version. Gas and smoke paths are presented in this section.

Hydrogen has a highly flammable and explosive nature. Maximum hydrogen concentrations and durations that flammable masses remain in the fuel cell room were measured. The effect of different ventilation conditions (30–120 ACH) was assessed. The leakage from different piping diameters (3–33 mm) at the inlet of the fuel cell stacks was stopped within 1 s after being detected by any of the sensors at the ceiling.

Batteries are associated with high fire risk. The smoke dispersion phenomenon was analysed in the battery room. The effects of firedoors (open or shut), FFS (on or off), and wind towards the bow of the ferry (no wind, wind velocities of 38mph and 6.9 mph) on the smoke paths was assessed.

### *5.1. Hydrogen Version—Gas Dispersion Analysis*

The most severe hazard according to HAZID results is the leakage in the fuel cell room due to piping damage, which was analysed with a quantitative approach in this section using CFD software called PyroSim, provided by Thunderhead Engineering. These simulations aimed to gain insight and understanding of the hydrogen dispersion in cases of potential leakage due to pipe damage under certain vent and sensor arrangements in the room. Leakage probability was not assessed in this study.

A mesh size of 7920 cells was used to perform the simulation in the fuel cell room. It was assumed for the analysis that the leakage due to the piping damage was at the inlet of the bottom stacks, which was 0.2 m above ground, but for simplicity of the simulations, the leakage position was assumed at a ground level. Fuel cell stacks are depicted with grey colour in Figure 5. Each of the six stacks provided a net power of 100 kW. This modular arrangement provides easy access and sufficient space for repairs and maintenance.

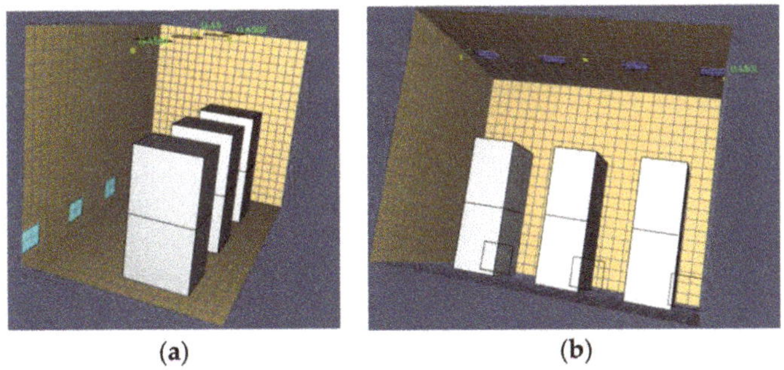

**Figure 5.** Vent and sensor arrangement in the fuel cell room: (**a**) inlet vents; (**b**) exhaust vents and hydrogen sensors in the ceiling.

There were three natural (inlet) vents, depicted with light blue colour in Figure 5a, for the supply of air in the room from the lower part as it is required by regulations for gases lighter than air. These are ducts that were routed upwards to the upper deck, and not just holes (openings), to reduce the possibility of flooding under excessive trim by stern since the fuel cell room was at the aft of the main deck. There were also four mechanically driven exhaust air fans at the ceiling of the room (Figure 5b), considering the highly buoyant nature of hydrogen gas. Backflow of gas from the vent outlet to the vent inlet should be prevented by providing sufficient distance between them. For an ESD, protected machinery space at least 30 ACH ventilation rate is required [31]. There were also three sensors at the ceiling to detect hydrogen concentrations above 0.4% and trigger the automatic valves to stop the flow of hydrogen in the room within 1–2 s.

The double block and bleed valves in the fuel cell room stop the supply of hydrogen in the fuel cells and bleed any remaining hydrogen in the fuel pipes to the vent mast at the upper deck through vent pipes. Hydrogen supply in the room where leakage or ventilation failure has been detected should be stopped by the master-gas valves outside the fuel space enclosure.

The hydrogen release rate ($m^3/s$) was calculated on the basis of [39] and [40], as shown in Equation (2).

$$Q_{gas} = \frac{C_d \times A \times \sqrt{\rho \times P \times \gamma \times \left(\frac{2}{\gamma+1}\right)^{\left(\frac{\gamma+1}{\gamma-1}\right)}}}{\rho}\ m^3/s \quad (2)$$

where Cd is the discharge coefficient for circular bore which is equal to 1 [17]. A is the hole area, which is $\pi d^2/4$, where d is the leakage diameter (meters). In these simulations, leakage diameters considered were 3, 16, and 33 mm. The ratio of specific heats ($\gamma$) was assumed to be constant and equal to 1.4

The density ($\rho$) of hydrogen can be calculated as shown in Equation (3):

$$\rho = P/RT\ kg/m^3 \quad (3)$$

where P is the pressure at the inlet of the stack and was $8 \times 10^5\ N/m^2$ and temperature (T) at the same position was 10 degrees Celsius so 283.15 K, according to the Powercell MS-100 data sheet [37]. The gas constant of hydrogen (R) was 4124 J/kgK, considering universal gas constant and hydrogen's molecular weight. On the basis of these data and Equation (3), the density was obtained: $\rho = 0.68\ kg/m^3$.

Only the area of leakage varied for the different simulations since the pressure and temperature were the same at the inlets of fuel cell stacks, regardless of position. The hydrogen release rate ($Q_{gas}$) could then be calculated on the basis of Equation (2) as follows:

$$Q_{gas} = \frac{1 \times A \times \sqrt{0.68 \times 8 \times 100{,}000 \times 1.4 \times \left(\frac{2}{1.4+1}\right)^{\left(\frac{1.4+1}{1.4-1}\right)}}}{0.68} = 742.69 \times A = 742.69 \times (\pi \times d^2/4)$$

### *5.2. Battery Version-Smoke/Fire Simulation*

In this section, a quantitative risk assessment is presented using a fire dynamic simulator (FDS) on the battery-powered version of the high-speed ferry to indicate the smoke and fire paths under various scenarios. In the fire simulation, the key is to model the heat release rate per area (HRRPA), which is related to the heat release rate and surface area of the burner as shown in Equation (4).

$$HRRPA = HRR/S \quad (4)$$

where HRRPA is the heat release rate per area ($kW/m^2$), HRR is the heat release rate from the burning (kW), and S is the surface area of the burner ($m^2$).

For the firefighting system (FFS), there are two key factors, flow rate and control strategy. The flow rate of the system is calculated as shown in Equation (5).

$$\dot{v} = m/\rho \cdot t \quad (5)$$

where $\dot{v}$ is the volume flow rate of the FFS chemical agent ($m^3/s$), m is the total mass release of the agent (kg); ρ is the agent density ($kg/m^3$), and t is the total release time duration (s). For the control strategy, the FFS is triggered when the temperature reaches 68 °C.

The main deck of the ferry includes both the battery room and the passenger area. Following the provided geometry from the shipyard, the 3D geometry model was developed, as shown in Figure 6. The blue plate is the main deck with passengers and the red blocks are the battery packs. The dark yellow panels are the walls on the main deck; the brown objects in the passenger area are chairs.

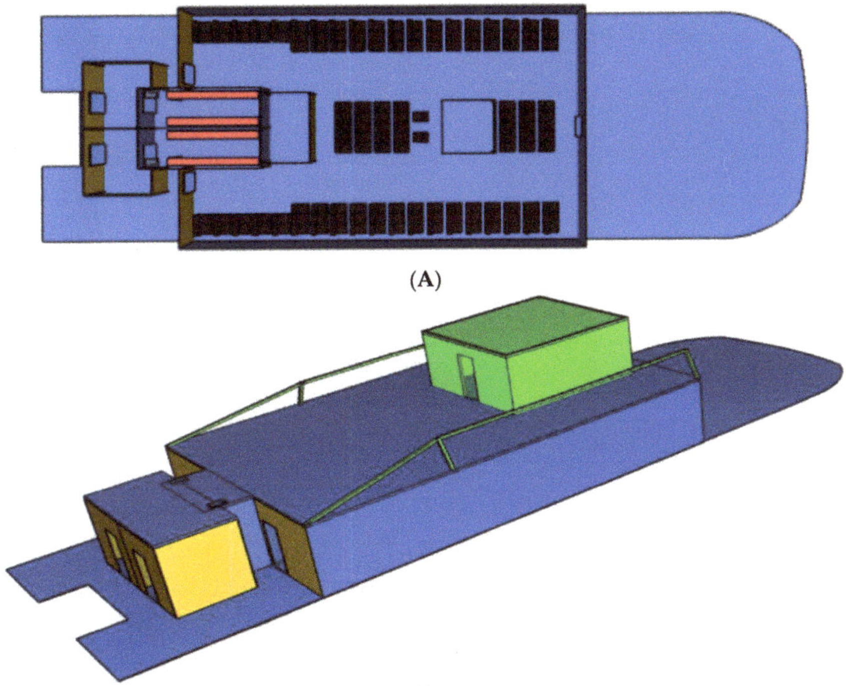

**(A)**

**(B)**

**Figure 6.** Three-dimensional geometry model of the ferry in PyroSim ((**A**) battery room deck; (**B**) with superstructure).

The battery room onboard the ferry includes the battery room walls, the battery packs, the sprayers of the FFS, and the inlet and outlet of the air ventilation system. The battery room was modelled and developed in PyroSim, and Figure 7 presents the developed model from the views from the top, bottom, and transparent inside:

1. Two air outlets were designed on the ceiling and are shown in the top view.
2. Another two air inlets are shown on the floor of the battery room, which can be seen in the bottom view.
3. Four battery packs (red blocks in inside details) are equipped onboard the ferry.
4. Two sprayers are installed on the ceiling, and they are marked as SPRK01 and SPRK02.

The battery fire heat released was provided by the manufacturer with a value of around 15,833 kJ/kWh obtained from a 14 min battery fire laboratory experiment; hence, the determined HRR was about 24,503 kW/kWh [41]. The dimension of the battery was also provided by the manufacturer: the width is 4 m, the depth is 0.5 m, and the height is 2.37 m [42]; hence, the surface area of the battery pack was determined as 25.33 $m^2$. With the heat release rate and surface area, the HRRPA can be derived on the basis of Equation (4): HRPA = 242 $kW/m^2$. The technical data for the batteries are shown in Table 6.

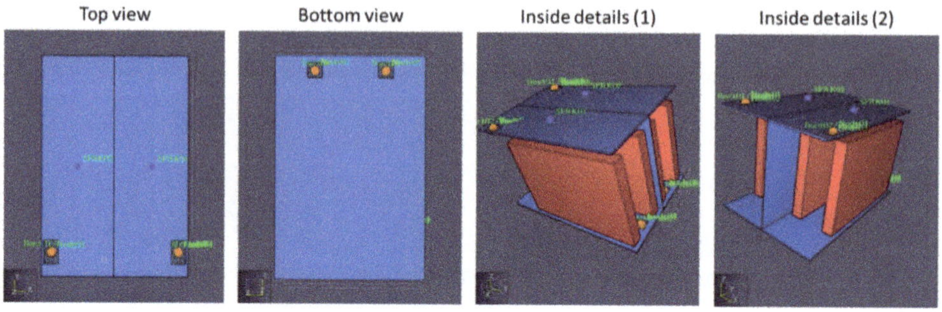


**Figure 7.** Battery room design (top view, bottom view, and inside details).

**Table 6.** Technical data of battery package (one set).

| Parameter | Value |
|---|---|
| Energy (kwh) | 650 |
| Dimensions: H × W × D (m) | 2.38 × 7.86 × 0.5 |
| Weight (ton) | 5.2 |
| Volume (m$^3$) | 9.35 |
| C-rate—Continuous (discharge) | 2.2 |
| C-rate—Continuous (charge) | 1.6 |
| Single module size (kwh) | 7.8 |
| Nominal voltage (V) | 805 |

The FFS system onboard the ferry was Novec 1230 [43]. The data of the FFS were provided [44]; hence, the volume flow rate was able to be obtained on the basis of Equation (5): $\dot{v}$ =192 litre/min.

The evacuation plan is summarised in Figure 8. The green arrows are the path of evacuation, and the green fishnet square is the evacuation exit.

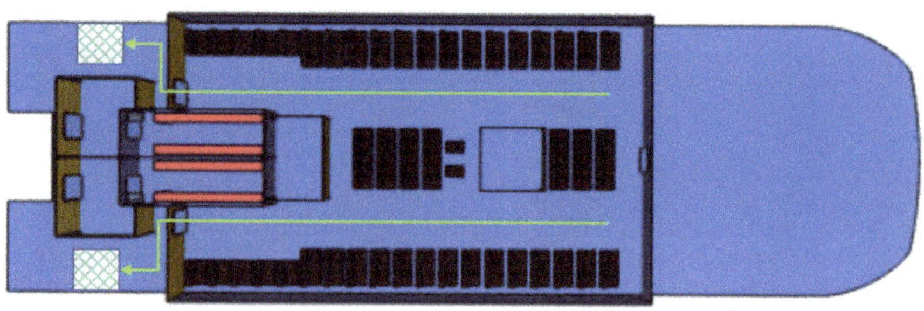

**Figure 8.** Ferry evacuation plan.

The wind profile targets the worst scenario on the River Thames according to the historical data from Met Office in the United Kingdom [45]. The strongest wind along the river was 38 mph (west) on 4 January 2018, and the average wind speed since 2011 is about 6.9 mph. These two data points were used for the analysis in the PyroSim simulation.

## 6. Results and Discussion

### 6.1. Hydrogen Version

In this section, scenarios of hydrogen leakage from different piping diameters at the inlet of the bottom right stack are presented. The results for leakages at the inlets of the bottom middle and left fuel cell stacks were similar in terms of maximum hydrogen concentrations and the time it takes for the release of flammable masses. This indicates the minor impact that the change of position of leakage has in the simulation results for the same hydrogen release rate and exhaust vent flow rate, with the designed vent and sensor arrangement in the room (Figure 5).

Different vent and sensor arrangements (numbers, positions) were considered, but the one depicted in Figure 5 was considered the most beneficial in terms of risk minimisation in the fuel cell room. Redundancy of sensors inside the room is crucial so that in cases of potential malfunction of one of the components, leakage can be detected shortly after by any of the other two sensors. Fuel pipes are double-walled, with inert gas between the two concentric pipes and an outer diameter of 33 mm [46].

6.1.1. Scenario 1: Leakage Diameter 3 mm at the Inlet of Bottom Right Stack

The hydrogen release rate ($Q_{gas}$) for a leakage of 3 mm was calculated on the basis of Equation (2): $Q_{gas} = 5.25 \times 10^{-3}$ $m^3/s$. Leakages of small piping diameters (3 mm or less) were considered to be the most likely failure scenarios for the hydrogen pipes.

The total volume flow rate of air is 0.16 $m^3/s$ to achieve 30ACH, considering that the volume of the room was 19.19 cubic meters. Hence, for each vent, the exhaust flow rate was 0.04 $m^3/s$.

Hydrogen leakage was detected first by the middle sensor, 5.6 s after the release, as can be observed in Figure 9a, and 1 s later, hydrogen flow was stopped in the room. With 30 ACH, flammable mass concentrations can be avoided since the maximum percentage is 1% = 0.01 mol/mol, and the flammability range for hydrogen is 4–75%. Hence, there was no ignition risk (Figure 9b,c), even in the presence of ignition sources, as long as the leakage was stopped 1 s after detection. Leakage and potential malfunction of multiple systems (two or three sensors) can delay the detection of gas and potentially create a hazardous atmosphere with 30 ACH ventilation rate. However, it is considered a highly unlikely scenario.

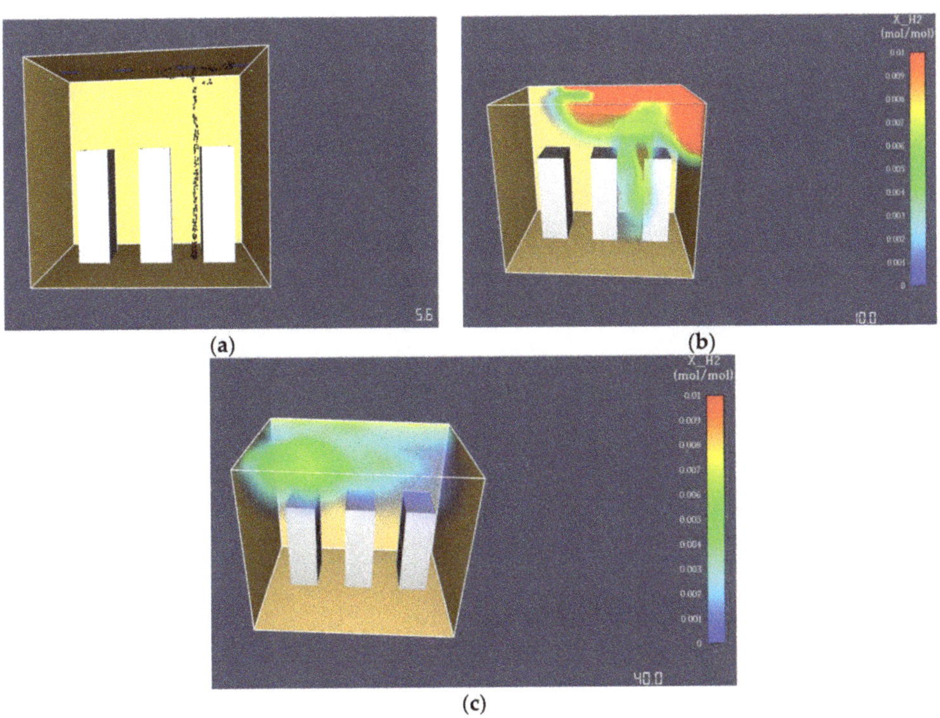


**Figure 9.** Leakage diameter 3 mm at the inlet of the bottom right stack: (**a**) detection at 5.6 s; (**b**) concentrations at 10 s; (**c**) concentrations at 40 s.

6.1.2. Scenario 2: Leakage Diameter 16 mm at the Inlet of the Bottom Right Stack

The hydrogen release rate ($Q_{gas}$) for a leakage of 16 mm was calculated on the basis of Equation (2): $Q_{gas} = 0.1493\ m^3/s$.

To achieve 30 ACH, the total exhaust flow rate needs to be $0.16\ m^3/s$. Leakage was detected at 1.9 s by the middle sensor and stopped 1 s later (Figure 10a). Maximum hydrogen concentration levels reached 7.5%, which was within the flammability range (Figure 10b). Flammable masses were mostly concentrated near the ceiling, due to hydrogen's highly buoyant nature, and released after 30 s with 30 ACH (Figure 10c,d). Since flammable masses were near the top for a few seconds, the fire risk was lower, considering that there were fewer ignition sources near the ceiling and higher ignition probability near the stacks. However, the minimum ignition energy of hydrogen is very low, and therefore in case of ignition, ventilation systems need to be shut off and a gaseous fire suppressant (Novec1230) should be used. The deflagration to detonation transition (DDT) phenomenon was considered unlikely since there was a limited run-up distance in the confined fuel cell room, which had a volume of $19.19\ m^3$ [20].

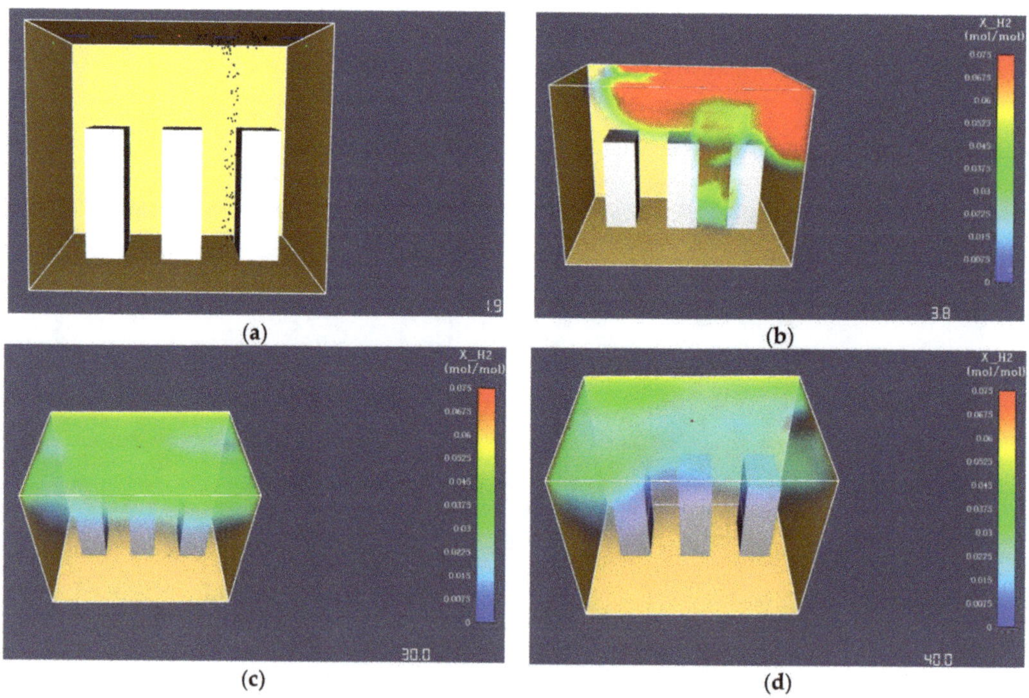


**Figure 10.** Leakage diameter 16 mm at the inlet of the bottom right stack: (**a**) detection at 1.9 s; (**b**) concentrations at 3.8 s; (**c**) concentrations at 30 s; (**d**) concentrations at 40 s.

By increasing the ventilation rate from 30 to 60ACH, flammable masses were released much faster at 18 s, resulting in lower ignition risk (Figure 11). Maximum hydrogen concentration was also reduced from 7.5 to 7%.

**Figure 11.** Leakage diameter 16 mm at the inlet of the bottom right stack (60 ACH).

6.1.3. Scenario 3. Full Pipe Rupture (33 mm) at the Inlet of Bottom Right Stack

The hydrogen release rate ($Q_{gas}$) for a leakage of 33 mm was calculated on the basis of Equation (2): $Q_{gas} = 0.635\ m^3/s$.

Full pipe rupture is the worst scenario, but it is also considered to be highly unlikely since pipes are double-walled, well-protected, away from the sides and of proven usage. Initially, a ventilation rate of 30ACH was assumed. Leakage was detected at 1.3 s by the middle sensor (Figure 12a). Hydrogen concentration levels reached 15%, which was below the lower explosion limit (LEL) of 18% (Figure 12b–d). With 30ACH, flammable masses (4–9%) were mostly concentrated near the ceiling, but they were not completely released outside, even after 40 s (Figure 12e). Hence, higher ventilation rates than 30ACH are suggested for safety under the worst-case scenario of full pipe rupture.

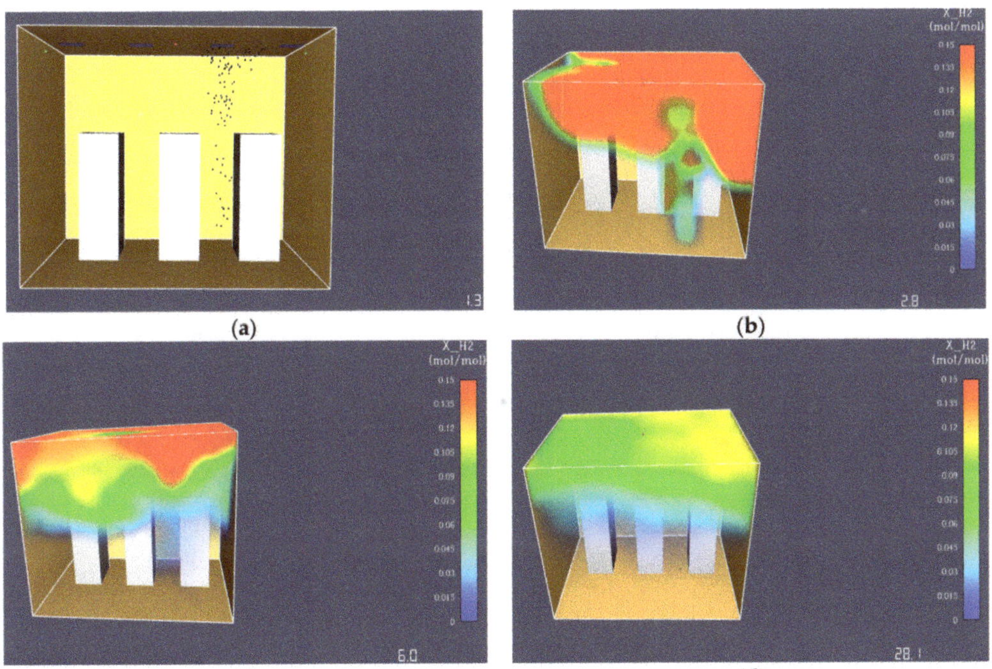


(a) (b) (c) (d)

**Figure 12.** *Cont.*

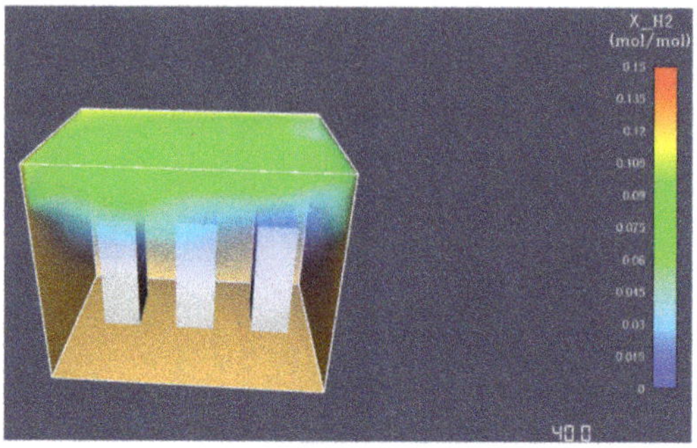


(e)

**Figure 12.** Full pipe rupture at the inlet of the bottom right stack with 30ACH: (**a**) detection at 1.3 s; (**b**) concentrations at 2.8 s; (**c**) concentrations at 6 s; (**d**) concentrations at 28.1 s; (**e**) concentrations at 40 s.

If 120ACH ventilation rate is used (Figure 13), the total volume flow rate of air needs to be 0.64 $m^3/s$. Hence, flow rate of each exhaust vent was set at 0.16 $m^3/s$. With 120ACH, flammable masses were released after 18 s, and at the end of the simulation (40 s), the room was completely emptied of hydrogen (100% air). Maximum concentrations remained at around 15% for the different ventilation rates considered, under full pipe rupture at the inlet of the bottom right stack. Unrealistically high ventilation rates (>200 ACH) would be required to avoid the flammability range completely during the entire simulation time for large leakages.

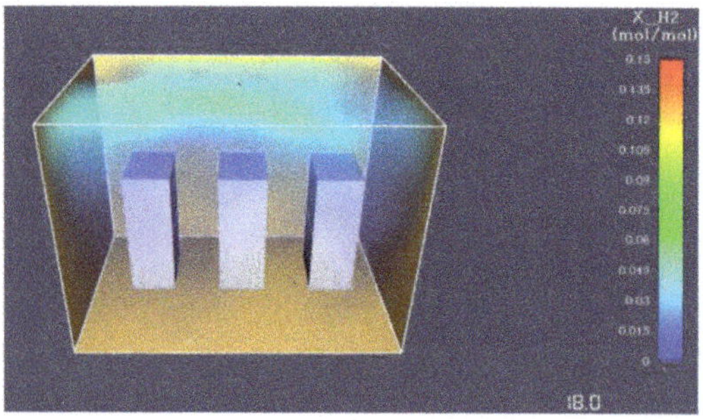


(a)

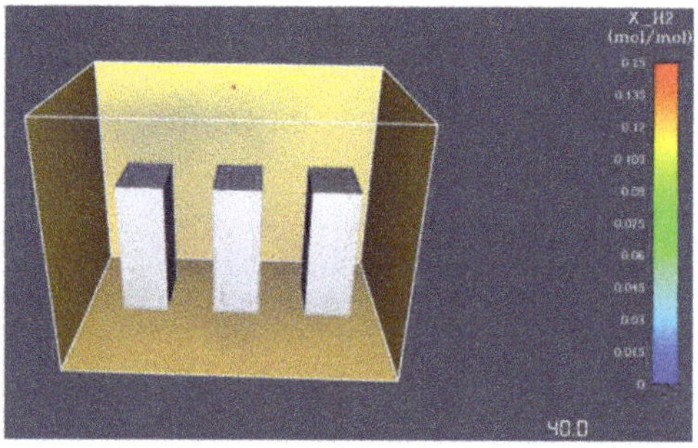


(b)

**Figure 13.** Full pipe rupture at the inlet of the bottom right stack with 120 ACH: (**a**) concentrations at 18 s; (**b**) concentrations at 40 s.

If the flow of hydrogen was stopped 2 s after leakage was detected instead of 1 s as in all the previous scenarios, maximum concentrations reached 20%, which was within the explosion range (Figure 14). Flammable masses were released after 25 s, and thus 7 s later compared to the corresponding scenario where the leakage was stopped 1 s after being detected (Figure 13a). This indicates that the leakage must be stopped within 1 s after being detected.

**Figure 14.** Full pipe rupture with 120ACH and 2 s delay.

A denser mesh with more cells included in the domain of the analysis resulted in higher accuracy. Hence, we attempted to increase the mesh from the original size of 7920 cells up to 16,000 cells to assess if the accuracy of simulations was enhanced. However, minor changes were observed in the maximum hydrogen concentrations and release time of flammable gases outside the fuel cell room. Hence, to save computational cost, without significant simulation time required to obtain the results, the mesh size remained unchanged (7920 cells). The cells of the computational domain are shown in Figure 5.

### 6.2. Battery Version

To quantitatively evaluate the safety level of the battery ferry, six scenarios were developed on the basis of the situation and condition of doors and wind levels:

- Scenario 1 (S1): this is the default condition with the fire doors shut, the FFS off, and no wind effect;
- Scenario 2 (S2): the FFS is on while keeping the fire doors shut and no wind effect;
- Scenario 3 (S3): the fire doors are open while keeping the FFS off and no wind effect;
- Scenario 4 (S4): the fire doors keep open and no wind effect while the FFS is on;
- Scenario 5 (S5): the fire doors are open and the FFS is on while adding a wind at 38 mph (61 km/h) towards the ferry's bow;
- Scenario 6 (S6): keep doors opened, FFS on, and add a wind at 6.9 mph (11 km/h) towards the ferry's bow.

All these scenarios were simulated and discussed from the perspectives of fire and smoke paths. The smoke paths under different scenarios can indicate the time to reach the evacuation area.

Figures 15–20 show the smoke in S1–6 with different conditions of fire doors, FFS, and winds. Figures 15 and 16 show the smoke trapped inside the battery room, with the only difference being the FFS working (green particles in Figure 16). Hence, the risk for passengers is low as long as the smoke/fire does not escalate in adjacent places, considering that fire doors are kept closed, preventing potential further damage.

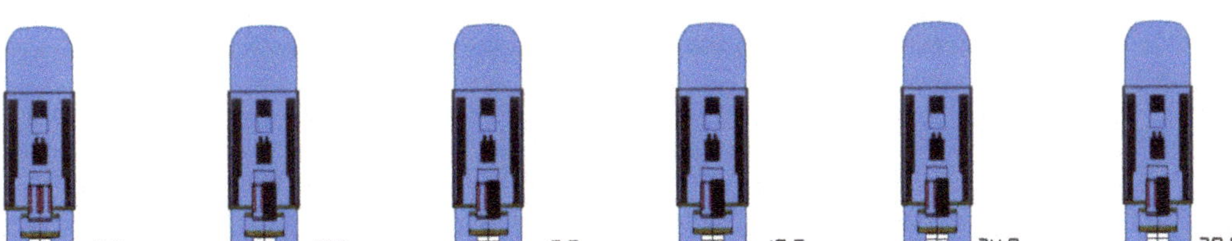


**Figure 15.** Simulation of scenario 1.

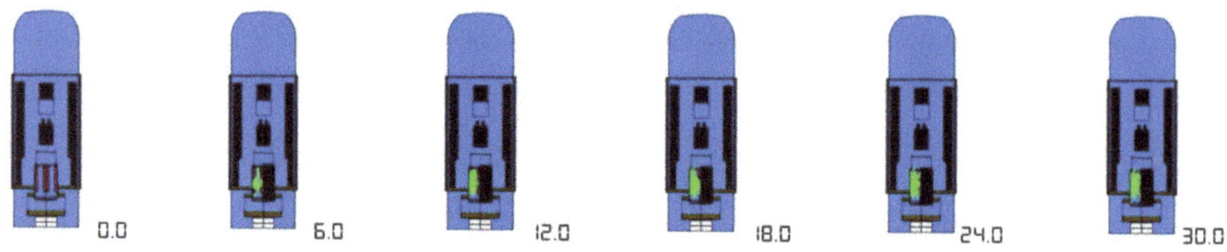


**Figure 16.** Simulation of scenario 2.

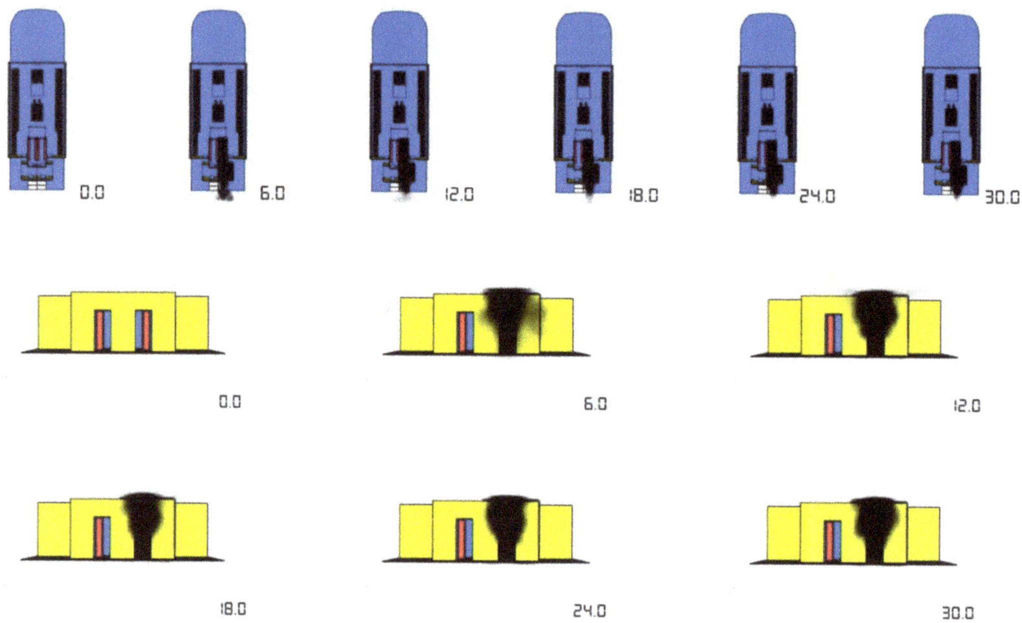


**Figure 17.** Simulation of scenario 3.

In S3 and S4, the situation of the fire door open was simulated, and the smoke paths are presented in Figures 17 and 18. Smoke was mostly concentrated at the aft of the main deck, but in S3, smoke was not blocking the evacuation exit, since it was rising and released into the atmosphere. It took around 6–8 s for the smoke to reach the aft extent of the main deck in S3, and until the end of the simulations, there was no significant difference in the accumulation of smoke at the back of the main deck.

In the last two scenarios, two wind speeds were integrated into the model to see the smoke movement under extreme and average wind conditions (Figures 19 and 20). It is apparent in S4 and S6 that the smoke accumulated in the evacuation region. The smoke reached the evacuation region in about 13 s in S4 and it took about 4 s to blow the smoke to the evacuation region; following this, the smoke was blown away and dispersed until accumulated again in 12 s. In S5, due to the extremely strong wind, there was no similar accumulation phenomenon. It was observed with the mixture of the agent from FFS in S4, S5, and S6 that the mixed smoke had a higher density than air, and therefore the mixture moved downward and accumulated, different from S3, in which the smoke rose and was released into the atmosphere.

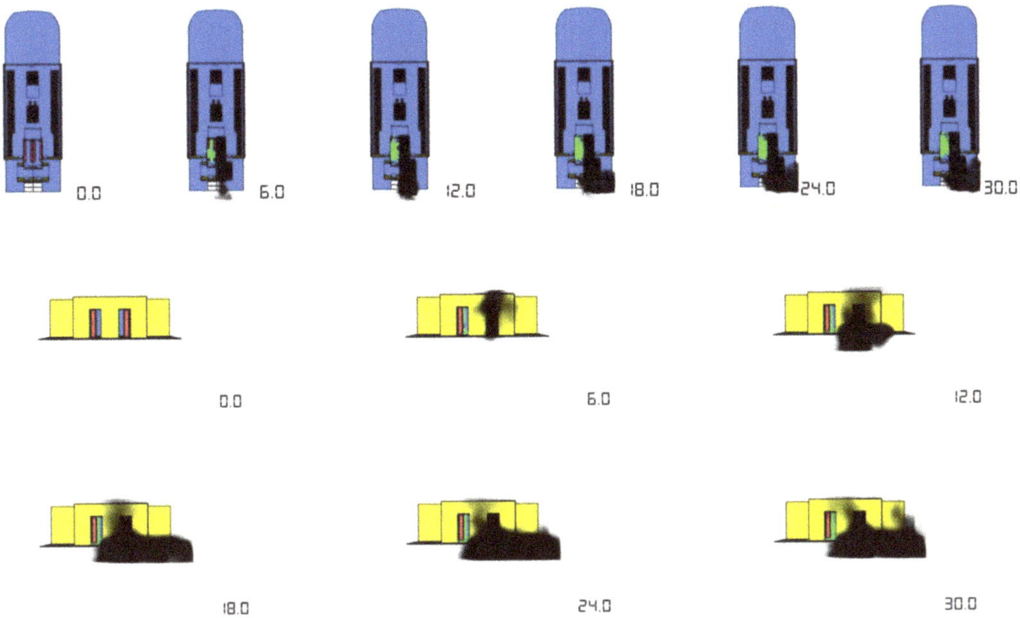


**Figure 18.** Simulation of scenario 4.

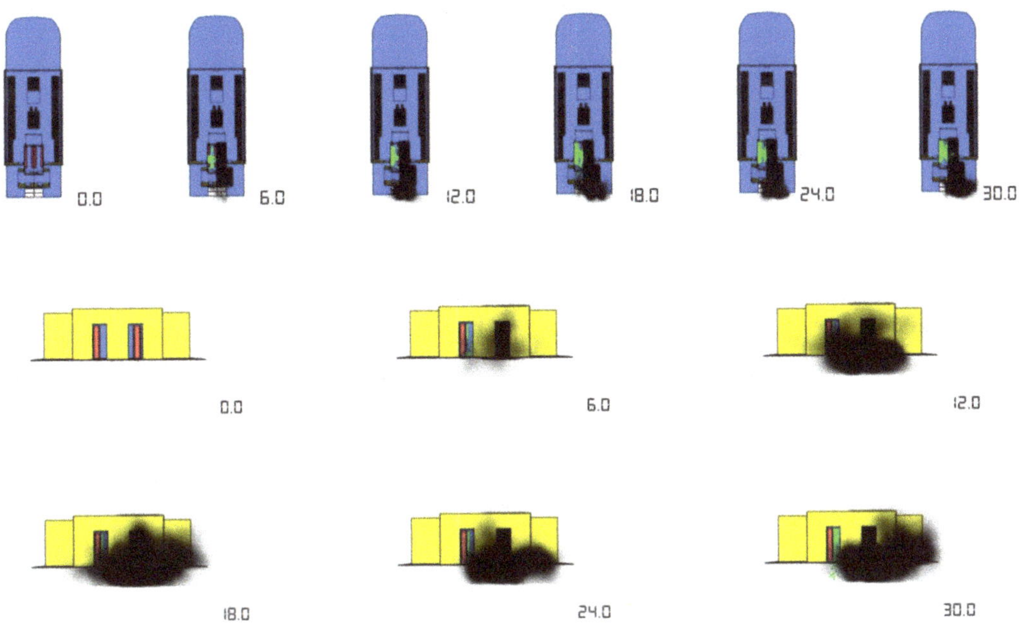


**Figure 19.** Simulation of scenario 5.

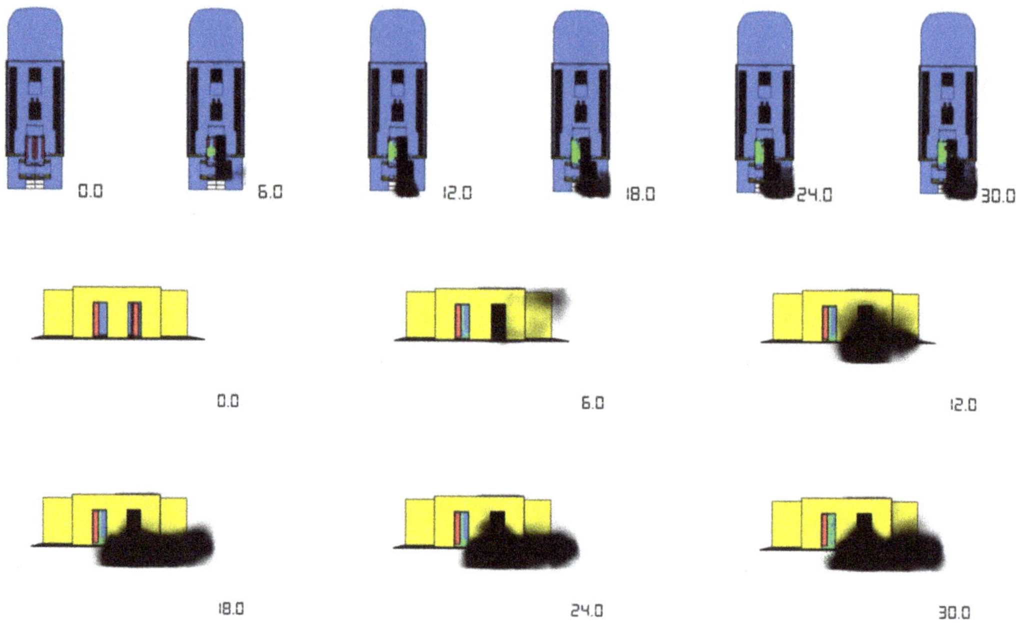


**Figure 20.** Simulation of scenario 6.

To identify the fire conditions under different scenarios, the heat release rates were monitored, as shown in Figure 21. The observation from S1 and S2 is that the FFS reduced the fire time by about 3 s. This means that the selected FFS system can delay the fire by 30%. In all other cases, all the fire doors were open, and the air was continuously circulated for the fire; hence, the HRRs fluctuated significantly. When there was a high wind effect with a velocity of 61 km/h towards the bow of the ferry, with FFS on and firedoors open (S5), the highest heat release rates were observed between 15 and 30 s, varying between 1600 and 2300 kW with a fluctuating nature (Figure 21).

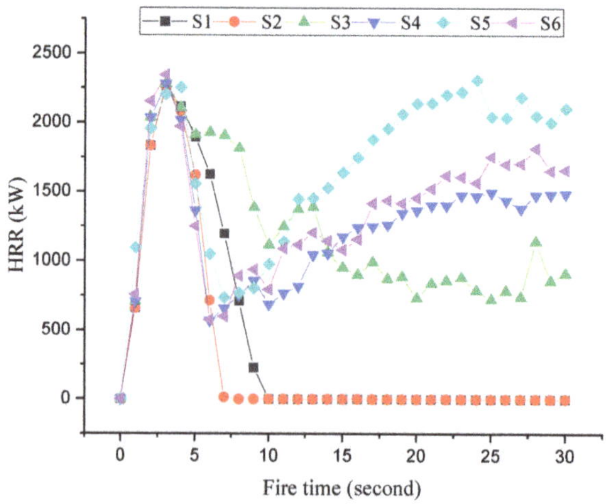


**Figure 21.** Heat release rate over time.

## 7. Conclusions

In this study, a comparative safety assessment of the battery and hydrogen version of the Stavanger high-speed catamaran ferry was presented.

Placement of the $LH_2$ tank at the upper deck and batteries at the aft of the main deck in the hydrogen and battery version, respectively, were considered the most suitable design solutions. In the hydrogen version, both fuel cell rooms were located within gastight enclosures in the main deck behind the passengers' area. The DC/AC converters were required to supply AC to e-motors in both hydrogen and battery versions. Both designs satisfied the HSC stability requirements.

A qualitative HAZID approach was conducted for both versions to determine the most concerning hazards. For both designs, proper installation of alarm and firefighting equipment, as well as testing of the equipment prior to its use, were considered to be amongst the most effective RCOs in terms of safety. Leakage in the fuel cell room due to piping damage at the inlet of the fuel cell stacks, as well as fire in the battery room during construction/installation or operation were considered to be the most severe hazards for the hydrogen and battery version, respectively, and they were analysed in quantitative assessments using PyroSim software.

A battery has more energy stored at any instant wherein an accident might occur compared to fuel energy in the hydrogen fuel cell, which constitutes an increased fire risk for battery applications. In other words, when hydrogen leakage is detected by sensors in the fuel cell stacks or the room, the supply can be automatically shut off by valves, reducing the fire risk, despite hydrogen's low ignition energy. In this study, in the case of hydrogen leakage, it was demonstrated that flow should be stopped within 1 s after gas concentrations higher than 0.4% are detected by any of the three sensors in the ceiling of the fuel cell room, in order to avoid explosive atmospheres under piping leakages from diameters of 16–33 mm. For the battery version, there is one sprayer of FFS in the ceiling of each battery room to reduce the smoke concentration and the probability of fire. The sprayer of FFS is activated when the temperature reaches 68 degrees Celsius, spraying with a volume flow rate of 192 litre/min. Minimising ignition sources in fuel cell and battery rooms is crucial for both designs so that potential gas or smoke released will not lead to fires. In the case of a fire, Novec1230 should be used as the firefighting system in fuel cell and battery rooms. Gastight bulkheads and firedoors are essential for hydrogen and battery versions, respectively, in order to mitigate the escalation risk to adjacent places and especially in the passenger area in the middle part of the main deck. Ventilation can be effective before any ignition takes place in both designs, but in the case of a fire, vents need to be shut off.

For the hydrogen version, leakages of 3 mm diameter did not constitute an ignition risk if leakage was stopped within 1 s after being detected. For 16 mm diameter of piping leakage, flammable masses, with 7.5% hydrogen concentration in air, were present in the room with 30ACH for around 30 s. However, they were mostly concentrated near the ceiling, away from the stacks, resulting in low ignition risk. In the case of full pipe rupture (33 mm diameter), increasing the ventilation rates from 30 to 120 ACH can result in a significant reduction of time duration that flammable masses remain in the room. In the case where leakage was stopped within 1 s after being detected, the flammable masses were completely released after 18 s with 120 ACH, whereas with 30ACH, flammable masses, with 4–9% hydrogen concentrations in air, remained near the ceiling, even after 40 s. In the case where leakage was stopped after 2 or more seconds, explosive atmospheres can be created with concentrations above 20% even with 120ACH.

For the battery version, it was demonstrated that the role of the FFS is crucial in cases of ignition. In ordinary conditions with fire door shut, FFS on, and air ventilation closed, the fire duration can be reduced by about 3 s, a 30% reduction compared to the condition without an FFS system. When the firedoors were open, without air ventilation, the use of FFS resulted in higher smoke concentrations near the ground, with the smoke accumulated in the evacuation region. The highest heat release rates over time were observed in the scenario in which there was a high wind velocity of 61 km/h towards the ferry's bow, with FFS on and firedoors open, with heat release rates varying between 1600 and 2300 kW from 15 to 30 s.

Overall, it seems that both hydrogen and battery designs can be feasible in terms of safety as long as all the proper RCOs are considered and there are sufficient safeguards in place to mitigate potential accident impacts. Gastight bulkheads and firedoors are mandatory in hydrogen and battery versions, respectively, and ignition sources should be minimised in both designs. Hydrogen design can be considered of higher risk because of the wide explosive range of the fuel. Appropriate measures, in terms of gas detection and ventilation, were taken to avoid the range of hydrogen concentrations between 18 and 59%, even if the flammability range (4–75%) could not be avoided. The DDT phenomenon was considered unlikely in a room with a volume of 19.19 $m^3$, which means that flame propagation speed could not be significantly increased to cause detonation. In the case of a fire in the fuel cell or battery rooms, the timely use of FFS (Novec 1230) can significantly reduce the fire duration and minimise the damage impact. Further research studies could include the cost–benefit analysis of the various components and arrangements, as well as the holistic optimisation of the designs including safety as an objective.

**Author Contributions:** Conceptualisation, F.M., E.B.; methodology, F.M., H.W.; formal analysis, F.M., H.W.; resources, E.B.; writing—original draft preparation, F.M.; writing—review and editing, F.M., E.B., N.L.T., A.P., M.C., H.W., G.S.; visualisation, F.M., H.W.; supervision, E.B., funding acquisition, E.B. All authors have read and agreed to the published version of the manuscript.

**Funding:** This research has been partially supported by the EC H2020 project TRAM with grant agreement 769303.

**Institutional Review Board Statement:** Not applicable.

**Informed Consent Statement:** Not applicable.

**Acknowledgments:** Boulougouris, Trivyza, Wang, Priftis, Guangyu, and Cheliotis greatly acknowledge the funding from DNV and the Royal Caribbean Group for the MSRC establishment and operation. The research herein has also been partially supported by EC H2020 project TRAM with grant agreement 769303. The opinions expressed herein are those of the authors and should not be construed to reflect the views of EC, DNV, or RCG.

**Conflicts of Interest:** The authors declare no conflict of interest.

## References


1. Choi, C.H.; Yu, S.; Han, I.S.; Kho, B.K.; Kang, D.G.; Lee, H.Y.; Seo, M.S.; Kong, J.W.; Kim, G.; Ahn, J.W.; et al. Development and demonstration of PEM fuel-cell-battery hybrid system for propulsion of tourist boat. *Int. J. Hydrog. Energy* **2016**, *41*, 3591–3599. [CrossRef]
2. Hydrogen Fuel Cell Boat Unveiled in Turkey News | SDG Knowledge Hub IISD. Available online: http://sdg.iisd.org/news/hydrogen-fuel-cell-boat-unveiled-in-turkey/ (accessed on 10 May 2021).
3. Chiche, A.; Andruetto, C.; Lagergren, C.; Lindbergh, G.; Stenius, I.; Peretti, L. Feasibility and impact of a Swedish fuel cell-powered rescue boat. *Ocean Eng.* **2021**, *234*, 109259. [CrossRef]
4. Boulougouris, E.; Papanikolaou, A.; Dahle, M.; Tolo, E.; Yan, X.K.; Jürgenhake, C.; Seidenberg, T.; Sachs, C.; Brown, C.; Jenset, F. Implementation of Zero Emission Fast Shortsea Shipping. In Proceedings of the SNAME Maritime Convention, Providence, RI, USA, 28 October 2021. [CrossRef]
5. Cheliotis, M.; Boulougouis, E.; Trivyza, N.L.; Theotokatos, G.; Livanos, G.; Mantalos, G.; Stubos, A.; Stamatakis, E.; Venetsanos, A. Review on the safe use of ammonia fuel cells in the maritime industry. *Energies* **2021**, *14*, 3023. [CrossRef]
6. Nikolopoulos, L.; Boulougouris, E. A Methodology for the Holistic, Simulation Driven Ship Design Optimization under uncertainty. In Proceedings of the International Marine Design Conference 2018, Helsinki, Finland, 1 June 2018. Available online: https://www.researchgate.net/publication/325735524 (accessed on 10 May 2021).
7. Priftis, A.; Boulougouris, E.; Turan, O.; Atzampos, G. Multi-objective robust early stage ship design optimisation under uncertainty utilising surrogate models. *Ocean Eng.* **2020**, *197*, 106850. [CrossRef]
8. Papanikolaou, A.; Xing-Kaeding, Y.; Strobel, J.; Kanellopoulou, A.; Zaraphonitis, G.; Tolo, E. Numerical and experimental optimization study on a fast, zero emission catamaran. *J. Mar. Sci. Eng.* **2020**, *8*, 657. [CrossRef]
9. Shi, G.; Priftis, A.; Xing-Kaeding, Y.; Boulougouris, E.; Papanikolaou, A.D.; Wang, H.; Symonds, G. Numerical investigation of the resistance of a zero-emission full-scale fast catamaran in shallow water. *J. Mar. Sci. Eng.* **2021**, *9*, 563. [CrossRef]

10. Xing-Kaeding, Y.; Papanikolaou, A. Optimization of the propulsive efficiency of a fast catamaran. *J. Mar. Sci. Eng.* **2021**, *9*, 492. [CrossRef]
11. Evrin, R.A.; Dincer, I. Thermodynamic analysis and assessment of an integrated hydrogen fuel cell system for ships. *Int. J. Hydrog. Energy* **2019**, *44*, 6919–6928. [CrossRef]
12. Cavo, M.; Gadducci, E.; Rattazzi, D.; Rivarolo, M.; Magistri, L. Dynamic analysis of PEM fuel cells and metal hydrides on a zero-emission ship: A model-based approach. *Int. J. Hydrog. Energy* **2021**, *46*, 32630–32644. [CrossRef]
13. Sarı, A.; Sulukan, E.; Özkan, D.; Sıdkı Uyar, T. Environmental impact assessment of hydrogen-based auxiliary power system onboard. *Int. J. Hydrog. Energy* **2021**, *46*, 29680–29693. [CrossRef]
14. Wang, H.; Boulougouris, E.; Theotokatos, G.; Zhou, P.; Priftis, A.; Shi, G. Life cycle analysis and cost assessment of a battery powered ferry. *Ocean Eng.* **2021**, *241*, 110029. [CrossRef]
15. Lindstad, H.E.; Eskeland, G.S.; Rialland, A. Batteries in offshore support vessels—Pollution, climate impact and economics. *Transp. Res. Part D Transp. Environ.* **2017**, *50*, 409–417. [CrossRef]
16. Vanem, E.; Salucci, C.B.; Bakdi, A.; Sheim Alnes, Ø.Å. Data-driven state of health modelling—A review of state of the art and reflections on applications for maritime battery systems. *J. Energy Storage* **2021**, *43*, 103158. [CrossRef]
17. Mao, X.; Ying, R.; Yuan, Y.; Li, F.; Shen, B. Simulation and analysis of hydrogen leakage and explosion behaviors in various compartments on a hydrogen fuel cell ship. *Int. J. Hydrog. Energy* **2021**, *46*, 6857–6872. [CrossRef]
18. Yuan, Y.; Wu, S.; Shen, B. A numerical simulation of the suppression of hydrogen jet fires on hydrogen fuel cell ships using a fine water mist. *Int. J. Hydrog. Energy* **2021**, *46*, 13353–13364. [CrossRef]
19. Aarskog, F.G.; Hansen, O.R.; Strømgren, T.; Ulleberg, Ø. Concept risk assessment of a hydrogen driven high speed passenger ferry. *Int. J. Hydrog. Energy* **2020**, *45*, 1359–1372. [CrossRef]
20. Pratt, J.W.; Klebanoff, L.E. *Feasibility of the SF-BREEZE: A Zero-Emission, Hydrogen Fuel Cell, High-Speed Passenger Ferry*; Sandia National Laboratories: Livermore, CA, USA, 2016. Available online: https://www.ebdg.com/wp-ebdg-content/uploads/2016/10/SF-BREEZE-SAND2016-9719.pdf (accessed on 18 March 2021).
21. Li, F.; Yuan, Y.; Yan, X.; Malekian, R.; Li, Z. A study on a numerical simulation of the leakage and diffusion of hydrogen in a fuel cell ship. *Renew. Sustain. Energy Rev.* **2018**, *97*, 177–185. [CrossRef]
22. Jeong, B.; Oguz, E.; Wang, H.; Zhou, P. Multi-criteria decision-making for marine propulsion: Hybrid, diesel electric and diesel mechanical systems from cost-environment-risk perspectives. *Appl. Energy* **2018**, *230*, 1065–1081. [CrossRef]
23. Andersson, P.; Wikman, J.; Arvidson, M.; Larsson, F.; Willstrand, O. *Safe Introduction of Battery Propulsion at Sea*; Safety and Transport Borås: Borås, Sweden, 2017. Available online: https://publications.lib.chalmers.se/records/fulltext/250628/250628.pdf (accessed on 15 June 2021).
24. DNV GL. *Rules for Classification Ships*; DNV GL AS: Oslo, Norway, 2019. Available online: https://rules.dnvgl.com/ServiceDocuments/dnvgl/#!/industry/1/Maritime/1/DNV%20GL%20rules%20for%20classification:%20Ships%20(RU-SHIP) (accessed on 1 July 2020).
25. Wang, H.; Boulougouris, E.; Theotokatos, G.; Priftis, A.; Shi, G.; Dahle, M.; Tolo, E. Risk assessment of a battery-powered high-speed ferry using formal safety assessment. *Safety* **2020**, *6*, 39. [CrossRef]
26. Lanssen, C.; Sandblost, T.; Lanssen, E. Battery/Fuel Cell Fast Ferry. Trondheim/Sandtorg. 2017. Available online: https://www.nho.no/siteassets/nox-fondet/rapporter/2018/nox-report---rev-8.doc-002.pdf (accessed on 1 July 2021).
27. Pivetta, D.; Dall'Armi, C.; Taccani, R. Multi-objective optimization of hybrid PEMFC/Li-ion battery propulsion systems for small and medium size ferries. *Int. J. Hydrog. Energy* **2021**, *46*, 35949–35960. [CrossRef]
28. Wu, P.; Partridge, J.; Bucknall, R. Cost-effective reinforcement learning energy management for plug-in hybrid fuel cell and battery ships. *Appl. Energy* **2020**, *275*, 115258. [CrossRef]
29. Klebanoff, L.E.; Caughlan, S.A.M.; Madsen, R.T.; Conard, C.J.; Leach, T.S.; Appelgate, T.B. Comparative study of a hybrid research vessel utilizing batteries or hydrogen fuel cells. *Int. J. Hydrog. Energy* **2021**, *46*, 38051–38072. [CrossRef]
30. Boulougouris, E.; Priftis, A.; Dahle, M.; Tolo, E.; Papanikolaou, A.; Xing-Kaeding, Y.; Jürgenhake, C.; Svendsen, T.; Bjelland, M.; Kanellopoulou, A.; et al. TrAM-Transport: Advanced and Modular. In Proceedings of the 8th Transport Research Arena TRA 2020, Helsinki, Finland, 27–30 April 2020. Available online: https://tramproject.eu/ (accessed on 11 September 2021).
31. IMO. *Adoption of the International Code of Safety for Ships Using Gases or Other Low-Flashpoint Fuels (IGF CODE)*; MSC 95/22/Add.1; IMO: London, UK, 2015.
32. DNV. *Handbook for Hydrogen Fueled Vessels, MarHySafe JDP Phase 1*, 1st ed.; DNV: Hovik, Norway, 2021.
33. IMO. *Guidelines for the Approval of Alternatives and Equivalents as Provided for in Various IMO Instruments*; MSC.1/Circ.1455; IMO: London, UK, 2013.
34. IMO. *Revised Guidelines on Alternative Design and Arrangements for SOLAS Chapters II-1 and III.*; MSC.1/Circ.1212/Rev.1; IMO: London, UK, 2019.
35. Comsol. Meeting the Challenges of Battery Design with Modeling and Simulation. IEEE Spectrum. Available online: https://spectrum.ieee.org/energy/renewables/meeting-the-challenges-of-battery-design-with-modeling-and-simulation (accessed on 15 September 2021).
36. DNV. *Rules for Classification: Ships—DNV-RU-SHIP Pt.6 Ch.2. Propulsion, Power Generation and Auxiliary Systems*; DNV: Hovik, Norway, 2019.

37. Powercell MS-100 Fuel Cells System Technical Description. Available online: https://solteknikk.no/wp-content/uploads/2020/12/PCS-MS-100-Technical-description-1.pdf (accessed on 15 March 2021).
38. IMO. *Revised Guidelines for Formal Safety Assessment (Fsa) for Use in the Imo Rule-Making Process*; IMO: London, UK, 2018.
39. Failure Frequency Guidance by DNV GL (Old Account)—Issuu. Available online: https://issuu.com/dnv.com/docs/failure_frequency_guidance_process_ (accessed on 25 May 2021).
40. Howard, G.W.; Tchouvelev, A.V.; Cheng, Z.; Agranat, V.M. Defining hazardous zones—Electrical classification distances. In Proceedings of the International Conference on Hydrogen Safety, Ontario, Canada, 8 September 2005.
41. Larsson, F.; Andersson, P.; Blomqvist, P.; Mellander, B.E. Toxic fluoride gas emissions from lithium-ion battery fires. *Sci. Rep.* **2017**, *7*, 10018. [CrossRef] [PubMed]
42. Corvus Dolphin Energy Datasheet. Available online: https://corvusenergy.com/wp-content/uploads/2019/04/CE_Datasheets_Corvus_Dolphin_Energy_pages-1.pdf (accessed on 15 May 2021).
43. 3M™ Novec™ 1230 Fire Protection Fluid. Available online: https://multimedia.3m.com/mws/media/124688O/3m-novec-1230-fire-protection-fluid.pdf (accessed on 20 May 2021).
44. Wang, Q.; Li, K.; Wang, Y.; Chen, H.; Duan, Q.; Sun, J. The efficiency of dodecafluoro-2-methylpentan-3-one on suppressing the lithium-ion battery fire. *J. Electrochem. Energy Convers. Storage* **2018**, *15*, 041001. [CrossRef]
45. Met Office River Thames—Blackfriars Railway Bridge Wind Forecast, Greater London—WillyWeather. Available online: https://wind.willyweather.co.uk/se/greater-london/river-thames----blackfriars-railway-bridge.html (accessed on 1 June 2021).
46. Blaylock, M.; Pratt, J.; Bran-Anleau, G.; Proctor, C. *Informing Hazardous Zones for On-Board Maritime Hydrogen Liquid and Gas Systems*; Sandia National Laboratories: Livermore, CA, USA, 2018. Available online: https://energy.sandia.gov/wp-content/uploads/2018/01/SFBreeze_SAND2018-0585_Blaylock.pdf (accessed on 15 July 2021).

applied sciences

MDPI

Article

# Green Shipping—Multifunctional Marine Scrubbers for Emission Control: Silencing Effect

**Giada Kyaw Oo D'Amore [1,*], Marco Biot [1], Francesco Mauro [2] and Jan Kašpar [3]**

[1] Department of Engineering and Architecture, University of Trieste, Via Alfonso Valerio 6, 34127 Trieste, Italy; biot@units.it
[2] Maritime Safety Research Centre, Department of Naval Architecture, Ocean and Marine Engineering, University of Strathclyde, Glasgow G4OLZ, UK; francesco.mauro@strath.ac.uk
[3] Department of Chemical and Pharmaceutical Sciences, University of Trieste, Via Licio Giorgieri 1, 34127 Trieste, Italy; kaspar@units.it
* Correspondence: giada.kyawood'amore@phd.units.it

**Featured Application: The presented investigation indicates the feasibility of an a priori numerical approach aimed at designing a multifunctional scrubber that includes muffler functionality within the product. The integration between this component along the exhaust line, for example scrubbers and silencers, allows for space to be saved onboard ships. The numerical investigation allows to study the optimized acoustic properties of marine scrubbers during the design phase while reducing the number of the prototypes that need to be constructed and tested. Considering the large dimensions of a scrubber–silencer design allows for time and money to be saved.**

**Abstract:** Scrubber systems abate the sulphur oxide emissions of engines when cheap fuel oils that are high in sulphur content are employed as combustibles. However, the ships with these voluminous devices installed on board is space demanding. This work analyses the feasibility of incorporating the acoustic abatement of the exhaust gas noise functionality into the scrubber design to provide a combined scrubber–silencer system. For this purpose, a finite element analysis is performed on a simple expansion chamber, which is assessed using both analytical and experimental data. The transmission loss is the acoustic parameter chosen in this work. The numerical model depicts a good correlation with the transmission loss measured on a model scale scrubber. Finally, scrubber geometry modifications alter the transmission loss, changing and/or enhancing its featuring. These abilities indicate the feasibility to confer to scrubber silencing effects.

**Keywords:** marine scrubber; muffler; FEA; experimental test; transmission loss

**Citation:** Kyaw Oo D'Amore, G.; Biot, M.; Mauro, F.; Kašpar, J. Green Shipping—Multifunctional Marine Scrubbers for Emission Control: Silencing Effect. *Appl. Sci.* **2021**, *11*, 9079. https://doi.org/10.3390/app11199079

Academic Editor: José A. Orosa

Received: 17 August 2021
Accepted: 25 September 2021
Published: 29 September 2021



## 1. Introduction

Although maritime transport accounts for approximately 10–15% of global sulphur (SOx) and nitrogen (NOx) oxide emissions, the volume of seaborne trade has increased by about 3% annually over the past 50 years or so, accounting for about 80–85% of world trade by volume [1]. This has led to increasing concerns about the global impact of maritime emissions, and, consistently, the IMO (International Maritime Organization) has restricted the limits imposed by MARPOL 73/78 Regulations [2,3] on ship emissions, specifically those of SOx [3] and NOx [2].

EGR (exhaust gas recirculation) and SCR (selective catalytic reduction) systems are recognized as effective technologies for the control of marine NOx emissions [1,4]. However, the former solution increases particulate matter emissions and fuel consumption by about 4% [5], making SCR the preferred option.

Different strategies allow for SOx emissions to be reduced [6]: alternative fuels (e.g., LNG—liquefied natural gas), alternative energy sources (e.g., fuel cells), and conventional fuels with low a sulphur content (e.g., VLSFO—very low sulphur fuel oil).

Since January 2020, for ships using conventional fuels, the above-quoted regulations have made the use of VLSFO mandatory, as its sulphur content is lower than 0.5 wt%. Moreover, to sail in ECAs (emission control areas), ULSFO (ultra low sulphur fuel oil) must be used, as the sulphur content limit in fuels has been pushed down to 0.1 wt% in these areas. Whereas the adoption of alternative fuels or onboard energy sources represents, besides the costs, the burden of entire propulsion system refitting, the use of low sulphur content combustibles significantly impacts the total operating costs of a ship, with over 60% of them being associated with the fuel. Despite COVID-19 resulting in an extreme reduction in oil prices, prices of USD 365, USD 472, USD 515, and USD 520 per ton were reported for HSFO (high sulphur fuel oil), VLSFO, ULSFO, and MGO (marine gas oil), respectively, at the Rotterdam/Antwerp hub on 22 February 2021 [7], indicating that use of HSFO may represent higher savings compared to other fuels. Accordingly, an alternative solution is to install a scrubber as an emission abatement device, ensuring compliance with SOx regulations. Scrubbers represent a viable solution [8] and show a lower climate impact than low sulphur fuels [9].

The installation of both SCR and scrubber systems onboard ships could represent a reliable technical solution to satisfy the limits imposed on both NOx and SOx emissions; however, this solution is hardly applicable because of space limitations, especially for existing ships [10]. Indeed, pollution control devices are usually installed in the funnel and are often large in size, so integrating such systems with other exhaust line components, such as silencers, becomes mandatory in order to save space [1].

What is also remarkable is the importance of controlling and minimizing exhausted gas noise. Several regulations [11,12] limit the perceived noise level at ship decks and have prescribed distances from vessels. As such, the integration of pollution control devices and silencers can represent a smart way to satisfy these regulations by saving space along the exhaust line.

Concerning the acoustic performances of pollution control devices such as scrubbers, in the literature, only a notable example of an SCR converter [1] is present, and a patent for a marine scrubber with acoustic properties is deposited [13]. On the contrary, the acoustic performance of mufflers has been deeply studied [14–18].

The present work is a part of a research project (Project ABE: see funding) oriented to integrate different functionalities within emission control systems and intending to develop a methodology to achieve compact, efficient, and cost-effective scrubber-based marine engine exhaust emission systems. In the first instance, the integration of silencing effects within scrubber technology is addressed, and a FEM (finite element method) model that can predict the efficiency of a scrubber–silencer is developed.

In this work, transmission loss (TL) was used as an acoustic parameter to evaluate the acoustic performance of a model-scale scrubber; TL is a property of the muffler only, and it can be easily evaluated with models available in the literature [15].

Here, the FEM model used to evaluate the TL is compared with both analytical and experimental data.

## 2. Materials and Methods

The following sections first outline the theoretical aspects of the experimental measurement techniques and numerical methods used to evaluate TL. Then, we present the experimental set-up used to measure the TL of both a simple expansion chamber and a model-scale scrubber; finally, we address the case studies analysed in this work.

### 2.1. Transmission Loss Measurements and Calculations

In general, the TL of a component can be calculated with the following equation when considering an anechoic termination [16]:

$$\mathrm{TL} = 10\log_{10}\frac{W_i}{W_t}, \tag{1}$$

where $W_i$ and $W_t$ are the sound power of the incident and the transmitted waves, respectively.

2.1.1. Experimental Techniques for Transmission Loss Measurements

Three distinct techniques allow the TL to be measured for mufflers using an impedance tube: the decomposition method, the two-source method, and the two-load method.

The decomposition method [19,20] properly measures the acoustic properties in ducts [21] using a measurement set-up as depicted in Figure 1, where 1, 2, and 3 represent microphone locations. $S_{ii}$, $S_{rr}$, and $S_{tt}$ represent the auto-spectra of the incident and the reflected and transmitted sound waves, respectively. The auto-spectra are defined as the product between a function and its complex conjugate; they provide the power distribution of the analyzed signal as a function of frequency.

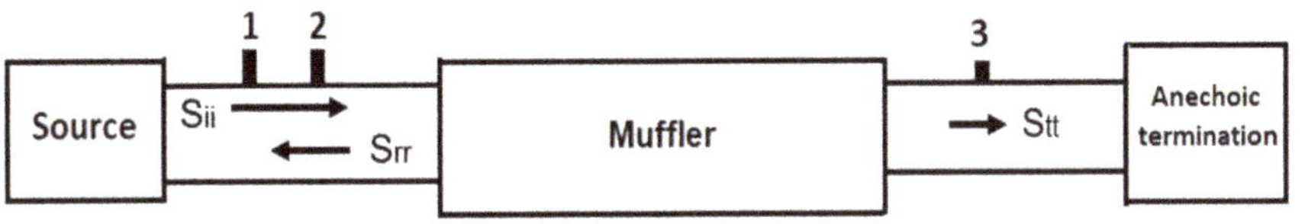


**Figure 1.** Measurement set-up according to the decomposition theory.

Considering a plane wave propagation and using two microphones at locations 1 and 2, the sound pressure can be decomposed into its incident and reflected waves. The microphone at location 3 directly measures the transmitted sound pressure. The decomposition theory provides an expression for the auto-spectrum $S_{ii}$ of the incident wave, so the sound power of the incident and transmitted waves can be derived, and the TL can be expressed as follows [15]:

$$\mathrm{TL} = 20\log_{10}\frac{p_i}{p_t} + 10\log_{10}\frac{S_i}{S_o}, \tag{2}$$

where $p_i$ and $p_t$ are the rms of the sound pressure of the incident and transmitted waves, respectively, and $S_i$ and $S_o$ are the cross-section area of the muffler inlet and outlet pipes, respectively.

The major drawback of the decomposition method is that a fully anechoic termination is difficult to reproduce in experiments, affecting the reliability of the TL measurement.

A muffler can also be modeled using the so-called four-pole parameters method [22]. The four parameters ($A$, $B$, $C$, and $D$) relate the inlet pressure ($p_i$) and velocity ($v_i$) to the respective outlet values ($p_o$, $v_o$), assuming a plane wave propagation.

$$\begin{bmatrix} p_i \\ v_i \end{bmatrix} = \begin{bmatrix} A & B \\ C & D \end{bmatrix} \begin{bmatrix} p_o \\ v_o \end{bmatrix}, \tag{3}$$

Two methods are available to calculate the four-pole parameters exploiting the transfer-matrix approach: the two-source method and the two-loads method [15].

When using the two-sources method, four microphones are needed. Consequently, the test has to be performed in two configurations to have the data required for TL parameter calculation according to Equation (4). In Figure 2 the two configurations ($a$ and $b$) are reported. They are implemented by changing the source position while keeping the other pipe termination open. The numbers 1, 2, 3, and 4 represent the microphone locations.

The muffler TL can be calculated as follows [16,23]:

$$\mathrm{TL} = 20\log_{10}\left\{\frac{1}{2}\left|A_{23} + \frac{B_{23}}{\rho c} + \rho c\, C_{23} + D_{23}\right|\right\} + 10\log_{10}\frac{S_i}{S_o}, \tag{4}$$

where $\rho$ is the fluid density, $c$ is the speed of sound in the fluid medium, and $A_{23}$, $B_{23}$, $C_{23}$, and $D_{23}$ are the four-pole parameters between microphones 2 and 3. The specific expressions for the parameters are given in [15].

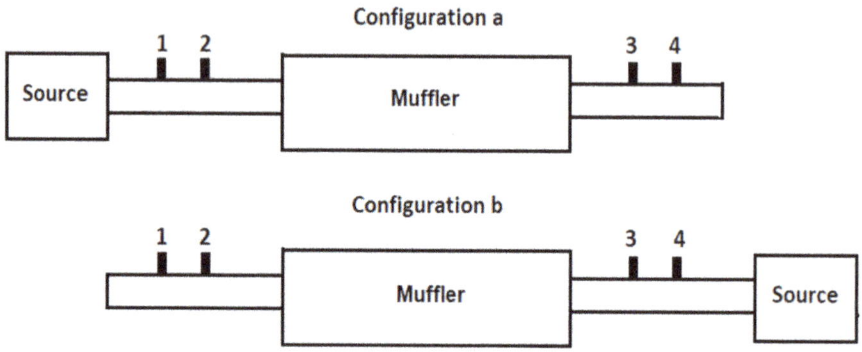


**Figure 2.** Measurement set-up according to the two-source method.

Figure 3 shows the set-up of the two-load method; it requires two configurations and leads to the same results as the two-source process by changing the outlet impedance (expressed as load a and b in the picture) instead of the source location, resulting in the TL being calculated with Equation (4) [14]. Additionally, in Figure 3, numbers 1, 2, 3, and 4 denote the microphones locations.

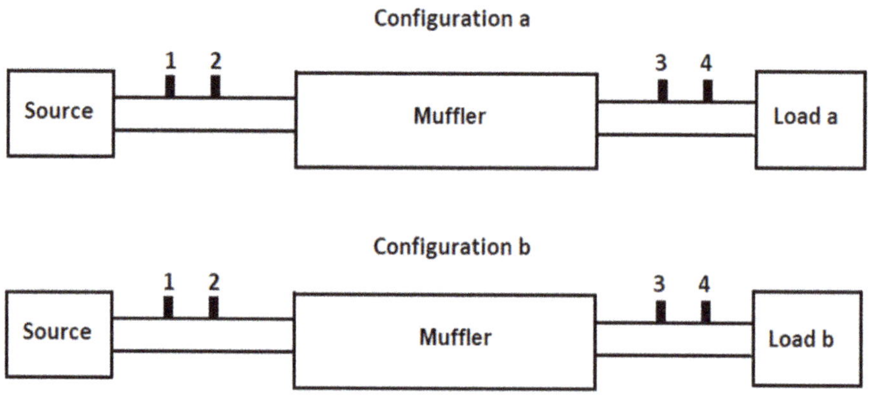


**Figure 3.** Measurement set-up according to the two-load method.

Different strategies allow the termination impedance to be changed: two terminations with different lengths, terminations with and without absorbing material, or even a closed and an open end. Of course, if the two loads are very similar, the results will be unstable.

In this work, the two-loads method suits the experimental layout, as a fully anechoic termination is difficult to build, and the impedance tube that used was designed to change the tube's end and not the source position.

2.1.2. Numerical Methods to Evaluate Transmission Loss

Two numerical methods for the estimation of the acoustic properties of a muffler are reported in the literature: the CFD (computational fluid dynamics) approach [24,25] and the FEM approach [26].

In the CFD approach, the pressure recorded at prescribed points during the simulation run determines the TL [25]. This procedure allows the influence of the flow conditions on the acoustic attenuation of the muffler to be considered. However, the CFD simulations may represent a limit for the computational costs. Moreover, especially in the earlier stage of the design process, the flow conditions at the inlet may be unknown, making difficult to set the inlet parameters.

The FEM approach, despite not considering the viscous flow, represents an easier and faster method to estimate the TL. It was for this reason that a FEM model was adopted in this study.

The FEM model uses the duct modes to simulate the incident and reflected pressure waves at the inlet and the transmitted pressure waves at the outlet. Then, the global pressure wave propagation inside the ducts results from a superposition of the duct modes (Figure 4) [22].

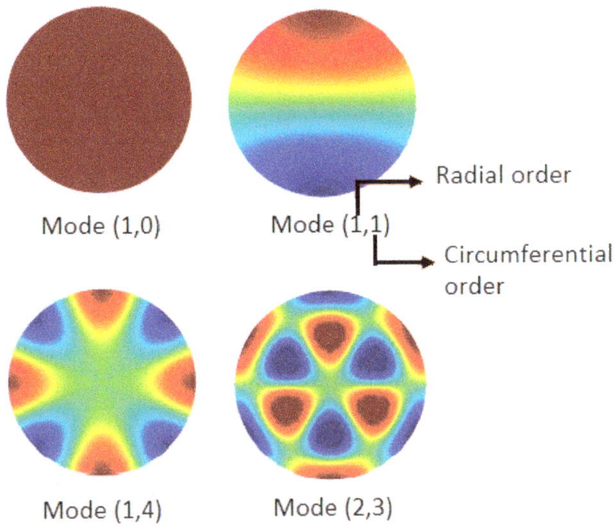


**Figure 4.** Example of acoustic mode shapes in circular ducts.

The duct modes follow an analytical representation of a semi-infinite duct (i.e., plane wave propagation with no reflection at the boundaries) that does not need to be meshed (Figure 5). This approach simulates the excitation at the inlet with imposed duct modes and the anechoic condition with free duct modes.

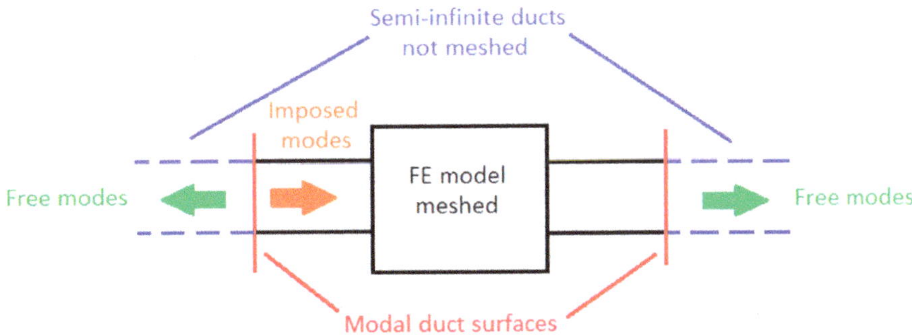


**Figure 5.** Schematization of the acoustic FEM model with duct modes.

### 2.2. The Case Studies: Experimental Tests and Numerical Analyses

In this work, two geometries (Figure 6) were analysed: a simple expansion chamber and a model-scale scrubber.

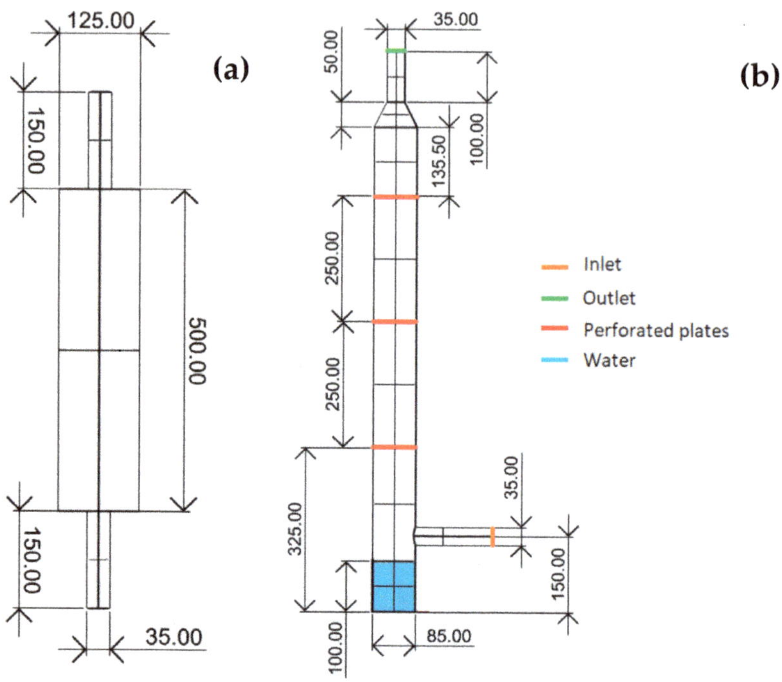


**Figure 6.** Geometry of (**a**) expansion chamber and (**b**) model-scale scrubber; dimensions in mm.

The expansion chamber represents one of the simpler reactive mufflers, and its TL can also be easily calculated analytically with the following equation:

$$\mathrm{TL} = 10\log_{10}\left(1 + \frac{1}{4}\left(h - \frac{1}{h}\right)^2 \sin^2(kl)\right) , \quad (5)$$

where $l$ is the expansion chamber length (500 mm in the present case), $k$ is the wavelength, and $h$ is the expansion ratio between the cross-sectional area of the chamber and the inlet/outlet pipe (12.7 in the present case), assuming the same diameter for the inlet and the outlet pipe [26]. Moreover, for a simple expansion chamber, the frequencies corresponding to the minima ($f_{min}$) and the maxima ($f_{max}$) of the TL curve can be calculated using the following equations, where $n$ is the number of the considered harmonics [27]:

$$f_{\min} = n\frac{c}{2l} \quad n = 0,\ 1,\ 2,\ \ldots, \quad (6)$$

$$f_{max} = \frac{c}{4l} + n\frac{c}{2l} \quad n = 0,\ 1,\ 2,\ \ldots, \quad (7)$$

The analytical expression was used to assess both the experimental set-up and the numerical model.

Then, a model-scale scrubber with perforated plates was tested. The perforated plates of the model-scale scrubber (Figure 6b) are 1 mm thick, with holes featuring a diameter of 6 mm. The spacing of the holes is 9 mm. The water at the bottom of the model-scale scrubber, which is generated by the water spraying on the perforated plate in the chemical processing of the exhaust gases to reduce emissions, was only considered numerically. The experimental set-up with the impedance tube was not designed to test the prototype with water inside.

During the experimental tests, the room temperature and the pressure were 15 °C and 101,325 Pa, respectively. The same environmental conditions apply to the numerical models for air inside of the chamber and the water at the scrubber bottom. Table 1 summarizes the characteristics of the fluids.

**Table 1.** Air and water characteristics used in the numerical model.

| | $c$ (m/s) | $\rho$ (kg/m$^3$) | $\mu$ (Pa/s) |
|---|---|---|---|
| Air | 340.0 | 1.225 | $1.78 \times 10^{-5}$ |
| Water | 1448.9 | 999.1 | $1.14 \times 10^{-3}$ |

2.2.1. Transmission Loss Experimental Measurements

The TL measurement was performed using an impedance tube featuring a diameter of 45 mm. The two-load technique explained in Section 2.1.1 was adopted, and the two loads were reproduced using a closed and an open termination (Figure 7).

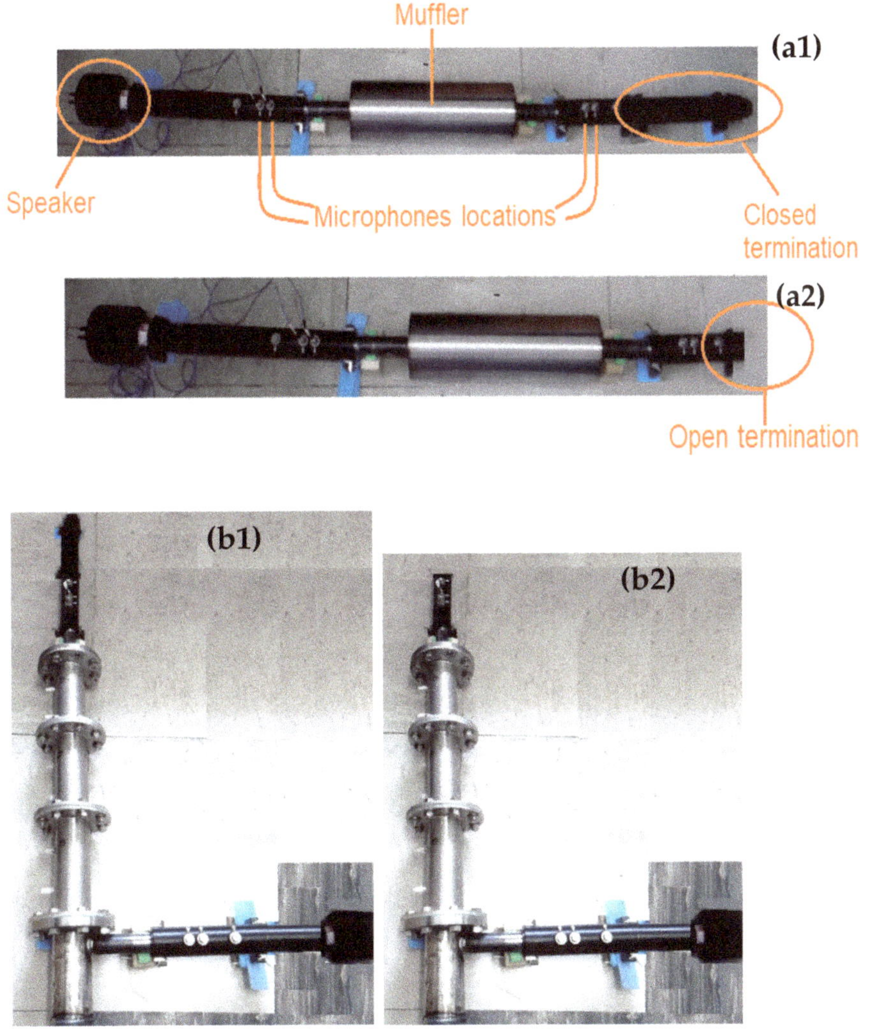


**Figure 7.** Experimental set-up of the (**a**) expansion chamber and the (**b**) model-scale scrubber with the closed (1) and the open (2) terminations.

The speaker emits a sine sweep with the following parameters: duration 10 s, frequencies range 50–5000 Hz, and variable amplitude between 0.05 V and 0.40 V.

The signals were acquired by two PCB Piezotronics 378C10 microphones connected to a data acquisition device NI USB 4431, National Instruments, Austin, TX, USA. Three measurements were performed for each set-up configuration, moving microphone 2 to the locations 3 and 4 while keeping microphone 1 in its location (see Figure 3).

The acquired traces were elaborated upon in order to obtain the TL values using the Main_TL software developed by Materiacustica s.r.l.

The distance between microphones 1–2 and 3–4 is equal to 30 mm, which is in accordance with the standard ISO 10534-2 [21], which fixes the following limit to the distance $s$ between the microphones:

$$s < 0.45\,\frac{c}{F}\,, \tag{8}$$

where $F$ is the maximum considered frequency.

Moreover, the distance between the sample and microphones 2 and 3 is greater than the 1–2 tube diameters suggested in the literature [28].

The ISO standard [21] also provides a frequency limit in order to ensure plane wave propagation:

$$f < 0.586\,\frac{c}{D}, \tag{9}$$

where $D$ is the diameter of the largest pipe in the structure. In our case, the frequency limit is about 1700 Hz and 2300 Hz for, respectively, for the expansion chamber and the model-scale scrubber.

The experimental tests include three repetitions for each geometry and microphone location; the results are reproducible, as the standard deviation between the tests' mean curve and the single experimental curve is less than 1%.

In this study, a bushing with O-rings (Figure 8) was realized to insert the pipes of the tested prototypes inside of the impedance tube. This solution ensures the airtightness of the system, facilitating the assembly of the setup but also creating discontinuity along the pipe that influences the measurements. The effects of this geometry setup on the TL will be analyzed using the FEM model in Section 3.1.2.

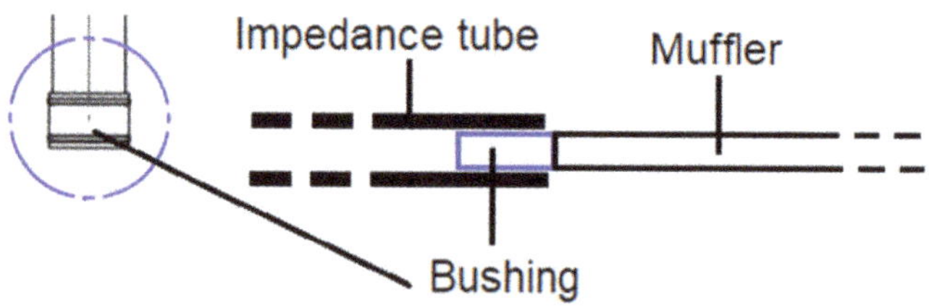


**Figure 8.** Details of the bushing used to connect the expansion chamber and the impedance tube.

### 2.2.2. FEM Simulations of Transmission Loss

The numerical simulations were performed using the software Actran VI [29].

The geometries were discretized with the Actran mesh generator. First, the surfaces of the studied geometries were meshed, then the volume mesh was generated starting from the meshed surface to discretize the interior volume. The guidelines [29–31] suggest using a minimum of 8–10 linear elements per wavelength: in this case, the base size is 20 mm and 10 mm with a deviation of 0.1 mm for the simple expansion chamber and the model-scale scrubber, respectively. A tetrahedral mesh was used due to its high flexibility with complex geometries (e.g., scrubber with perforated plates and pipes). A mesh sensitivity study was not performed since the mesh size was calculated on the basis of the maximum wavelength to be considered, so the number of elements is sufficient to adequately reconstruct it. Considering 10 elements per wavelength, as in this case, the error on the pressure estimation remains less than 3% [31].

In Table 2, the mesh parameters ensuring the mesh quality [29,32] are reported.

**Table 2.** Mesh quality parameters.

| | Max | Mean | Min |
|---|---|---|---|
| Jacobian | 1 | 1 | 1 |
| Aspect Ratio | 6.42 | 1.76 | 1.03 |

Figure 9 shows the discretized geometries. Two configurations are used to analyse the geometries: the simple one is analysed as it is (Figure 9(a1,b1)), and the other one considers the discontinuity introduced by the connection (bushing in Figure 8) with the impedance tube (Figure 9(a2,b2)). The bushing discontinuity was modelled by considering a sudden change in diameter from 35 mm (prototypes pipe diameter) to 45 mm (impedance tube diameter).

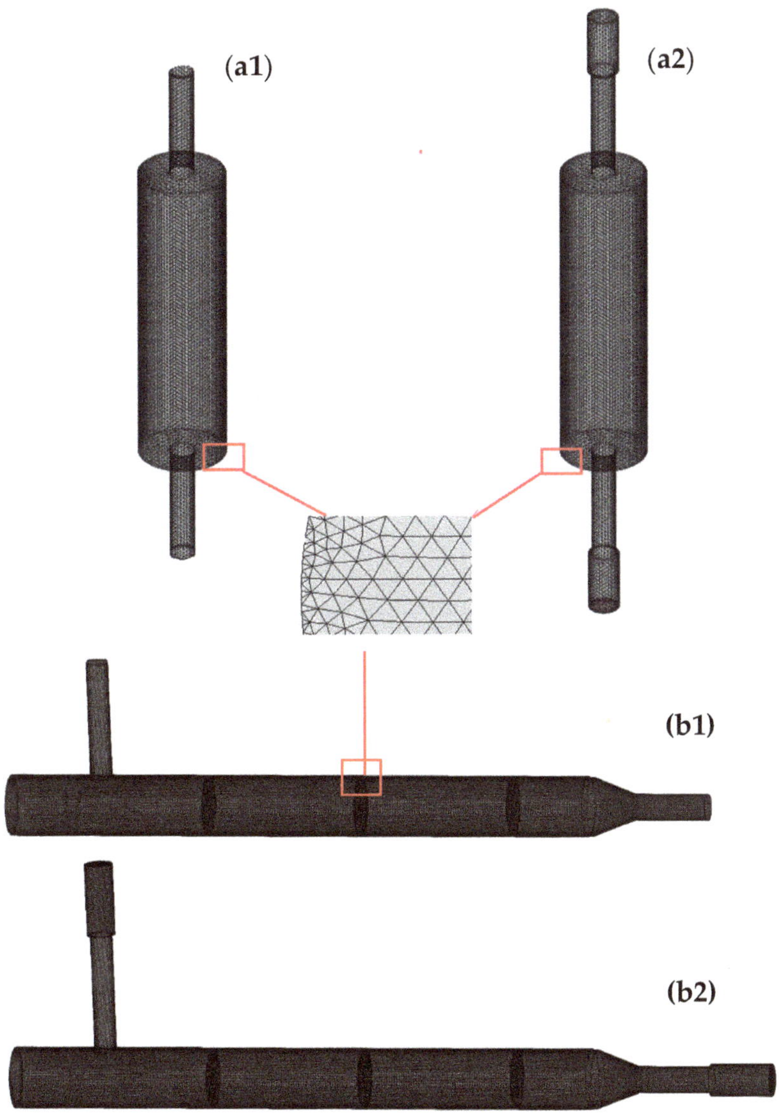


**Figure 9.** Discretized expansion chamber (**a1**) without and (**a2**) with the pipe discontinuity and model-scale scrubber (**b1**) without and (**b2**) with the pipe discontinuity.

The TL of the model-scale scrubber was evaluated with and without the presence of water at the bottom (Figure 6).

Perforated elements were modelled inside of the software environment considering their viscous dissipation. This was modelled by imposing a transfer admittance $A$ derived from the fluid characteristics and perforation geometry as follows:

$$A = \frac{1}{Z_p}, \tag{10}$$

$Z_p$ is the transfer impedance of the perforated element expressed as follows:

$$Z_p = \frac{Z}{\sigma}, \tag{11}$$

where $Z$ is the transfer impedance for a single hole expressed by Maa's formulation, and $\sigma$ is the porosity parameter reported in the following equations [33] according to a formulation that is valid for triangular meshes:

$$Z = \frac{8\mu L}{a^2}\sqrt{1+\frac{(k_s a)^2}{32}} + j\omega\rho l\left(1+\frac{1}{\sqrt{9+\frac{(k_s a)^2}{2}}}\right), \tag{12}$$

$$\sigma = \frac{2\pi a^2}{\sqrt{3}d^2}, \tag{13}$$

where $\mu$ is the dynamic viscosity of the fluid, $L$ is the thickness of the perforated shell, $a$ is the hole radius, $d$ is the hole spacing, $\omega$ is the angular frequency, $\rho$ is the fluid density, and $k_s$ the shear wavenumber, which can be expressed as follows:

$$k_s = \sqrt{\frac{\omega\rho}{\mu}}, \tag{14}$$

As discussed in Section 2.1.2, at the inlet, the duct modes were used to model the incident wave, imposing a plane wave propagation in a frequency range 0–1700 Hz and 0–2300 Hz for the expansion chamber and the model-scale scrubber, respectively. The upper limits were selected according to the ISO standard guidelines reported in Section 2.2 [21]. In order to avoid reflection, free mode propagation was set in the direction opposite to the excitation.

At the outlet, the anechoic condition was modelled using the duct modes and by setting free mode propagation.

Environmental conditions and fluid characteristics were set in the same way as they were during the experimental tests (Section 2.2.1, Table 1). The calculations assume the presence of a fluid inside the muffler but are not simulating the flow developing inside the geometry, as was the case during the tests using the impedance tube.

## 3. Results and Discussion

The comparison between the analytical and experimental TL of the simple expansion chamber is first addressed to assess the experimental setup. The accuracy of the numerical model is then evaluated. Finally, the assessed FEM model is used to estimate the TL of the model-scale scrubber, which was also measured experimentally, and, more importantly, to evaluate the influence of the modifications of the scrubber design on the TL.

In this work, to estimate the accuracy of the results, the guidelines suggested by Wärtsilä Italy S.p.A. [34] were used: a difference of less than 5% and 10% between the $f_{min}$ and $f_{max}$, respectively, and a difference of 5 dB between the TL amplitudes were considered as a criterion to assess the reliable fit of the TL curves. These are the same guidelines that used to simulate marine engine behaviour in the industrial environment.

It is important to stress that the accurate detection of a system's fundamental frequencies is of paramount importance, which corresponds to the $f_{min}$ of the TL (i.e., frequencies at which the minima of the TL curve occur). The $f_{min}$ are the so-called transparent frequencies, as they do not reflect the sound waves backwards, leading to sound attenuation at the outlet but allowing them to propagate unaltered through the component. As such, to reduce the exhausted gas noise, these frequencies must not coincide with the engine frequencies, producing a null sound attenuation.

Regarding the geometry modifications, ideally, TL values will increase. Higher TL values correspond to higher sound attenuation because the sound power transmitted through the component is less.

### 3.1. *Simple Expansion Chamber*

In the following sections, for the simple expansion chamber, the experimental and the numerical results are compared to the analytical TL to evaluate their accuracy. The FEM model is then used to investigate its influence on the TL of the discontinuity caused by the connection between the chamber and the impedance tube, as previously anticipated.

#### 3.1.1. Experimental Transmission Loss Assessment

As previously highlighted, different parameters influence the measurement setup, so the evaluation of its accuracy becomes mandatory.

Figure 10 compares the analytical (calculated with Equation (5)) and experimental TL values as a function of frequency for the simple expansion chamber (Figure 6a).

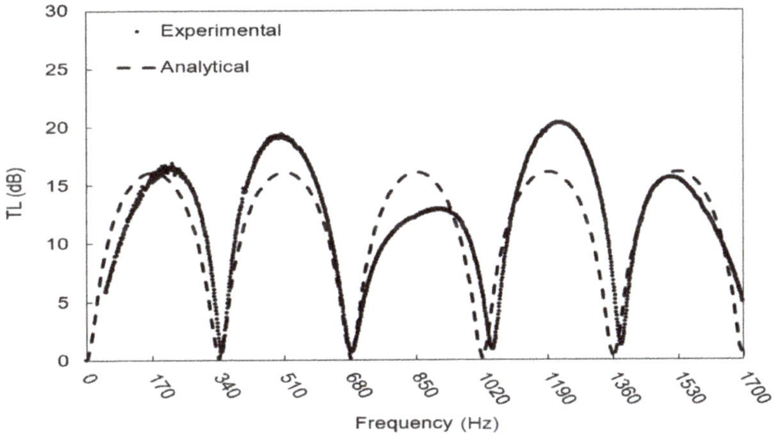


**Figure 10.** Comparison of analytical and experimental TL curves for the expansion chamber.

Table 3 compares the frequencies at which the minima ($f_{min}$) and the maxima ($f_{max}$) of the TL occur, as calculated with Equations (6) and (7), respectively. Delta represents the percentage difference between the analytical and experimental data.

**Table 3.** Comparison between analytical and experimental $f_{min}$ and $f_{max}$ of TL for the simple expansion chamber.

| $n$ | Analytical $f_{min}$ (Hz) | Experimental $f_{min}$ (Hz) | Delta $f_{min}$ (%) | Analytical $f_{max}$ (Hz) | Experimental $f_{max}$ (Hz) | Delta $f_{max}$ (%) |
|---|---|---|---|---|---|---|
| 0 | 0 | 0 | 0.0 | 170 | 220 | +29.4 |
| 1 | 340 | 348 | +2.4 | 510 | 508 | −0.4 |
| 2 | 680 | 685 | +0.7 | 850 | 919 | +8.1 |
| 3 | 1020 | 1050 | +2.9 | 1190 | 1226 | +3.0 |
| 4 | 1360 | 1382 | +1.6 | 1530 | 1520 | −0.7 |

Table 4 compares the amplitudes of the experimental and analytical TL. Delta represents the difference between the analytical and experimental data.

**Table 4.** Comparison between analytical and experimental TL amplitudes of simple expansion chamber.

| $n$ | Analytical Amplitudes (dB) | Experimental Amplitudes (dB) | Delta Amplitudes (dB) |
|---|---|---|---|
| 0 | 16.1 | 16.6 | +0.5 |
| 1 | 16.1 | 19.0 | +2.9 |
| 2 | 16.1 | 12.8 | −3.3 |
| 3 | 16.1 | 20.2 | +4.1 |
| 4 | 16.1 | 15.6 | −0.6 |

A perusal of the data reported in Tables 3 and 4 reveals differences of less than 3% between analytical and experimental $f_{min}$ and differences of less than 5 dB between the analytical and experimental amplitudes of TL; accordingly, the above reported fitting criterion [34] is satisfied.

A somewhat worse fit is observed for the $f_{max}$, which can be attributed to the irregular features of the TL peaks shown in Figure 10, which are reasonably caused by the discontinuity introduced along the pipe by the bushing (Figure 8) used to connect the expansion chamber and the impedance tube, as described in Section 2.2. The irregular feature of the experimental TL will be studied in the next section using the assessed FEM model.

However, considering the fundamental importance of the $f_{min}$ being measured correctly, as previously explained, the adopted experimental setup can be recognized as sufficiently accurate in the selected frequency range from an engineering point of view.

#### 3.1.2. FEM Transmission Loss Assessment

The accuracy of the adopted numerical model has to be evaluated to ensure the correctness of the settings illustrated in Section 2.2.2. Figure 11 compares the analytical (calculated with Equation (5)) and numerical TL values as a function of frequency for the simple expansion chamber (Figure 9(a1)).

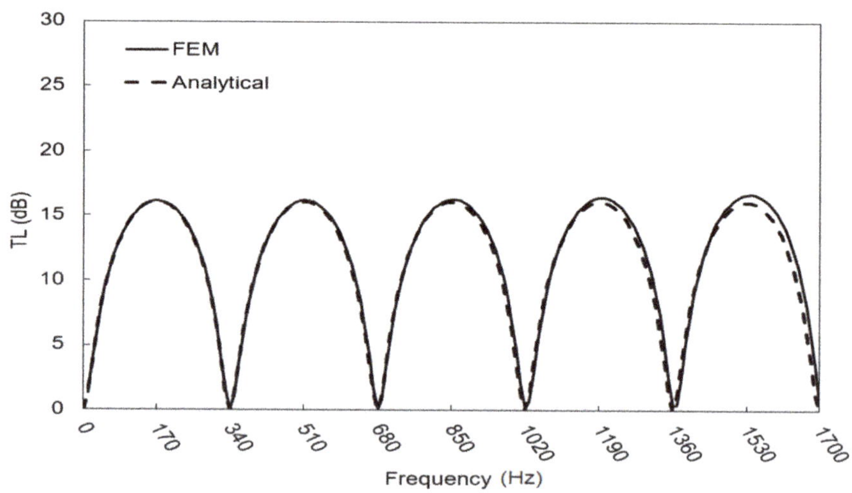


**Figure 11.** Comparison between analytical and FEM-simulated TL of the simple expansion chamber without discontinuity.

The analytical curve is nearly coincident with the FEM-simulated TL of the simple expansion chamber: frequency differences much lower than 1% and amplitude differences lower than 1 dB, as shown by the data comparison reported in Table 5. Delta represents the difference between the analytical and FEM-simulated TL of the simple expansion chamber.

The above-reported fitting criterion [34] is satisfied, clearly proving the reliability of the numerical model.

**Table 5.** Comparison between analytical and FEM data of simple expansion chamber without discontinuity.

| $n$ | FEM $f_{min}$ (Hz) | Delta $f_{min}$ (%) | FEM $f_{max}$ (Hz) | Delta $f_{max}$ (%) | FEM Amplitudes (dB) | Delta Amplitudes (dB) |
|---|---|---|---|---|---|---|
| 0 | 0 | 0.0 | 170 | 0.0 | 16.1 | 0.0 |
| 1 | 348 | 0.0 | 510 | 0.0 | 16.1 | 0.0 |
| 2 | 680 | 0.0 | 850 | 0.0 | 16.1 | 0.0 |
| 3 | 1040 | 0.0 | 1190 | 0.0 | 16.5 | +0.4 |
| 4 | 1370 | 0.0 | 1530 | 0.0 | 16.7 | +0.6 |

After comparing the numerical model against the analytical data, the discontinuity along the tube caused by the bushing was modelled considering a variation in diameter (Figure 9(a2)), as previously discussed. The tube discontinuity significantly affects the FEM-simulated TL of the expansion chamber (Figure 12), reinforcing the above-reported suggestion that the irregular features of the amplitudes are caused by the connection between the chamber and the impedance tube (Figure 8).

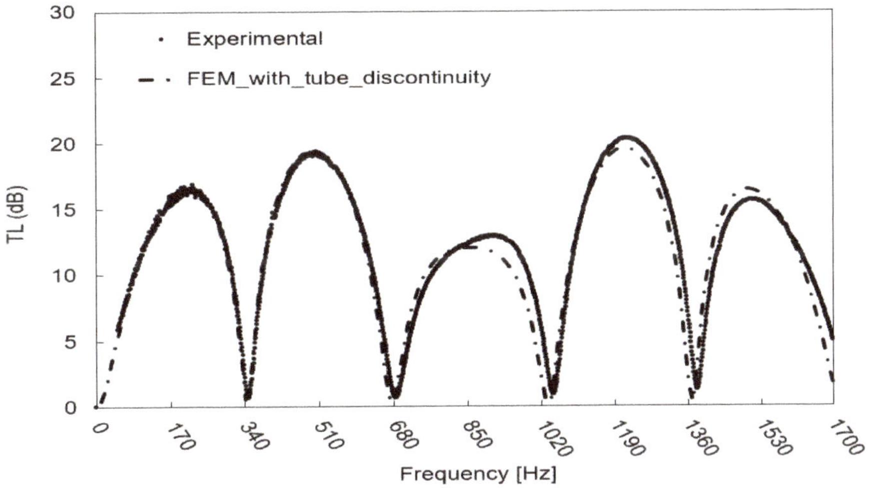


**Figure 12.** Comparison between experimental and FEM-simulated TL of the expansion chamber with tube discontinuity.

The data reported in Table 6 highlight the good fit between experimental data and the TL calculated with the numerical model considering the tube discontinuity: the industrial guidelines [34] are satisfied, as the differences in the $f_{min}$ value and amplitude are less than 1% and 1 dB, respectively. Moreover, the difference in terms of $f_{max}$ is reduced compared to the delta between the analytical and experimental TL (Table 3). Indeed, the analytical model (Equation (5)) simulates an ideal geometry without geometrical variations along the pipes and the chamber.

**Table 6.** Comparison between experimental and FEM with discontinuity data of simple expansion chamber.

| $n$ | FEM Discontinuty $f_{min}$ (Hz) | Delta $f_{min}$ (%) | FEM Discontinuty $f_{max}$ (Hz) | Delta $f_{max}$ (%) | FEM Discontinuty Amplitudes (dB) | Delta Amplitudes (dB) |
|---|---|---|---|---|---|---|
| 0 | 0 | 0.0 | 220 | 0.0 | 16.3 | −0.3 |
| 1 | 348 | 0.0 | 510 | +0.4 | 19.3 | +0.3 |
| 2 | 680 | −0.7 | 860 | −5.7 | 12.0 | −0.8 |
| 3 | 1040 | −0.9 | 1220 | −0.5 | 19.5 | −0.7 |
| 4 | 1370 | −0.9 | 1510 | −0.7 | 16.4 | +0.8 |

### 3.2. Model-Scale Scrubber

The simple FEM model was extended to simulate the TL of the model-scale scrubber. Considering the model of the scrubber with the tube discontinuity, a perusal of Figure 13 shows a good fit between the FEM-simulated and experimental TL curves in the frequency range up to 1500 Hz, where the discrepancy between $f_{min}$ is less than 3%, and where between amplitudes, it is less than 5 dB (Tables 7 and 8). At higher frequencies, the discrepancy between the TL amplitudes becomes higher. This inaccuracy is caused by a non-plane wave propagation of the sound at higher frequencies in the model-scale scrubber due to a change of the wave propagation direction caused by the angle between the inlet pipe and scrubber body [35]. In this regard, it should be remembered that the hypothesis of plane wave propagation was adopted both in the experimental measurements and in the numerical model.

However, the FEM model shows appreciable accuracy (Tables 7 and 8) [34] in the frequency range 20–1000 Hz, the range in which human hearing is the most sensitive [36] and the range that is taken as a reference by industry [34]. Therefore, the assessed FEM model can be employed to analyse the effects of scrubber modifications (e.g., water presence or geometry modifications) on the TL.

For this purpose, a model-scale scrubber without tube discontinuity was first modelled in order to avoid the influence of such discontinuity on the TL. Indeed, the scrubber is usually connected to the exhaust line without tube discontinuity. A perusal of Figure 14 shows that the discontinuity influences the TL amplitudes in the frequency range of interest (20–1000 Hz) and the $f_{min}$ and $f_{max}$ at higher frequencies.

**Table 7.** Comparison between experimental and FEM $f_{min}$ and $f_{max}$ of TL for the model-scale scrubber with discontinuity.

| $n$ | Experimental $f_{min}$ (Hz) | FEM $f_{min}$ (Hz) | Delta $f_{min}$ (%) | Experimental $f_{max}$ (Hz) | FEM $f_{max}$ (Hz) | Delta $f_{max}$ (%) |
|---|---|---|---|---|---|---|
| 0 | 0 | 0 | 0.0 | 119 | 120 | +0.8 |
| 1 | 195 | 190 | −2.6 | 288 | 290 | +0.7 |
| 2 | 351 | 350 | −0.3 | 448 | 460 | +2.7 |
| 3 | 511 | 510 | −0.2 | 584 | 570 | −2.4 |
| 4 | 668 | 675 | +1.0 | 731 | 750 | +2.6 |
| 5 | 838 | 840 | +0.2 | 911 | 930 | +2.1 |
| 6 | 1008 | 1020 | +1.2 | 1110 | 1130 | +1.8 |
| 7 | 1206 | 1220 | +1.2 | 1304 | 1310 | +0.5 |
| 8 | 1376 | 1375 | −0.1 | 1467 | 1480 | +0.9 |
| 9 | 1527 | 1545 | +1.2 | 1632 | 1660 | +1.7 |
| 10 | 1674 | 1680 | +0.4 | 1754 | 1710 | −2.5 |
| 11 | 1848 | 1860 | +0.6 | 1918 | 1930 | +0.6 |
| 12 | 2015 | 2020 | +0.2 | 2098 | 2100 | +0.1 |
| 13 | 2182 | 2160 | −1.0 | 2248 | 2250 | +0.1 |

**Table 8.** Comparison between experimental and FEM amplitudes of model-scale scrubber with discontinuity.

| *n* | Experimental Amplitudes (dB) | FEM Amplitudes (dB) | Delta Amplitudes (dB) |
|---|---|---|---|
| 0 | 9.4 | 6.3 | −3.1 |
| 1 | 14.8 | 12.2 | −2.6 |
| 2 | 23.0 | 21.2 | −1.8 |
| 3 | 67.3 | 62.5 | −4.8 |
| 4 | 18.0 | 16.2 | −1.8 |
| 5 | 9.0 | 7.4 | +1.6 |
| 6 | 5.2 | 5.3 | +0.1 |
| 7 | 8.6 | 7.2 | −1.4 |
| 8 | 9.4 | 13.4 | +4.0 |
| 9 | 13.7 | 23.1 | +9.4 |
| 10 | 49.9 | 48.8 | −1.1 |
| 11 | 15.0 | 10.2 | −4.8 |
| 12 | 10.2 | 7.3 | −2.9 |
| 13 | 6.1 | 9.2 | +3.1 |

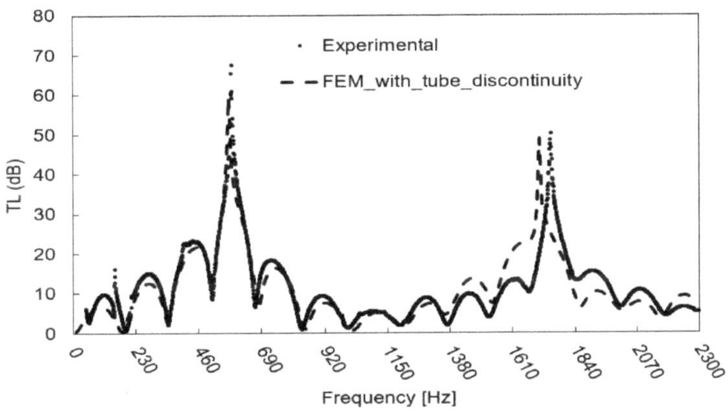


**Figure 13.** Comparison between experimental and FEM-simulated TL of the model-scale scrubber with discontinuity.

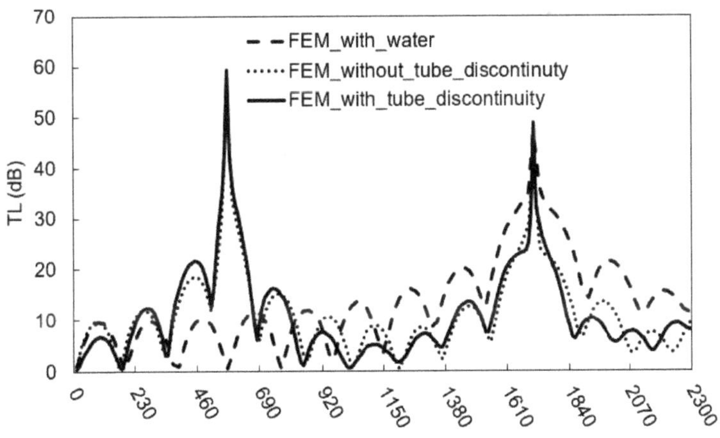


**Figure 14.** Comparison between experimental and FEM-simulated TL of the model-scale scrubber with and without discontinuity and with water.

Using the model of the scrubber without discontinuity, the influence on the TL of the presence of water in the model-scale scrubber bottom (depicted in Figure 6b) has been considered: the TL peak at around 500 Hz is completely dumped, while the TL above 920 Hz is increased (Figure 14). This effect should be kept present when interactions with exciting frequencies are considered.

### 3.3. Effects of Scrubber Design on Transmission Loss

The basic design of the scrubber (i.e., characteristics of the perforated plates and scrubber diameter and height) should comply with the requirements of the chemical processing of exhaust gases and cannot be modified to improve the acoustic performance of the component. As such, the influence of modifications on the TL that do not influence the chemical process, such as the inlet/outlet pipes diameter and length, were investigated using the assessed FEM model without tube discontinuity. Moreover, the influence on the acoustic performance of the insertion of elements as perforated pipes or filler was studied.

The first modification that was performed was the increase and the reduction of the inlet pipe diameter by 20 mm in respect to the original one (35 mm) while keeping the other dimensions constant. The results reported in Figure 15a clearly show that reducing the inlet diameter leads to an increment of the TL along the whole range of considered frequencies, whereas the opposite occurs upon increasing the inlet diameter.

Figure 15b reports that the results for the outlet diameter change by 20 mm, while the other dimensions were maintained as the originals, reducing the diameter, allowing the TL amplitudes to increase but decreasing the TL minima. The increase in the diameter also leads to a worsening of the TL in this case.

Considering the effects obtained by alternately varying the inlet and outlet diameters, it was decided the influence of the simultaneous reduction of the diameters by 20 mm should be evaluated.

Considering the effects obtained by alternately varying the diameters of the inlet and outlet pipes, the influence of the simultaneous reduction of the diameters of both the inlet and outlet pipes by 20 mm was evaluated. Figure 15c shows a clear increment of the TL due to this change. This result is consistent with the literature [22]: increasing the expansion ratio between the cross-sectional area of the expansion chamber (body of the component) and the inlet/outlet pipe, the TL amplitudes increase. In the presented study, the expansion ratio of the original model-scale scrubber and the modified one with inlet and outlet pipe diameters of 15 mm were, 5.9 and 32.0, respectively.

The lengths of the inlet and outlet pipes were also increased and decreased by 50 mm without any appreciable effects on the TL (data not reported for brevity), which is consistent with the plane wave propagations in the tube [37].

Finally, the insertion of perforated pipes and filler into the model-scale scrubber, as illustrated in Figure 16, was investigated. Two perforated pipes with the following characteristics were considered: a pipe with a diameter of 17 mm, a thickness of 1.5 mm, holes featuring a diameter of 4 mm, and holes spacing of 8 mm. The filler had a flow resistivity of 3000 $Pa{\cdot}s/m^2$, which is typical of a rigid metal foam, and was 30 mm thick. Moreover, the water at the bottom of the scrubber was removed, as a water drain can be added to the scrubber in order to increment the TL and without influencing the chemical properties.

A perusal of Figure 17 shows that the insertion of the perforated pipes not only increases the TL amplitudes but also changes the fundamental frequencies of the system; this aspect has to be taken into account when considering coupling with the engine, as the transparent frequencies must not match with the exciting ones. The insertion of the filler clearly improves the TL and changes its features due to the viscous dissipation generated by the passage of the sound waves through its porous structure. The simultaneous addition of filler and perforated pipes further increased the TL, coupling the dissipation effects of the porous structure and the holes.

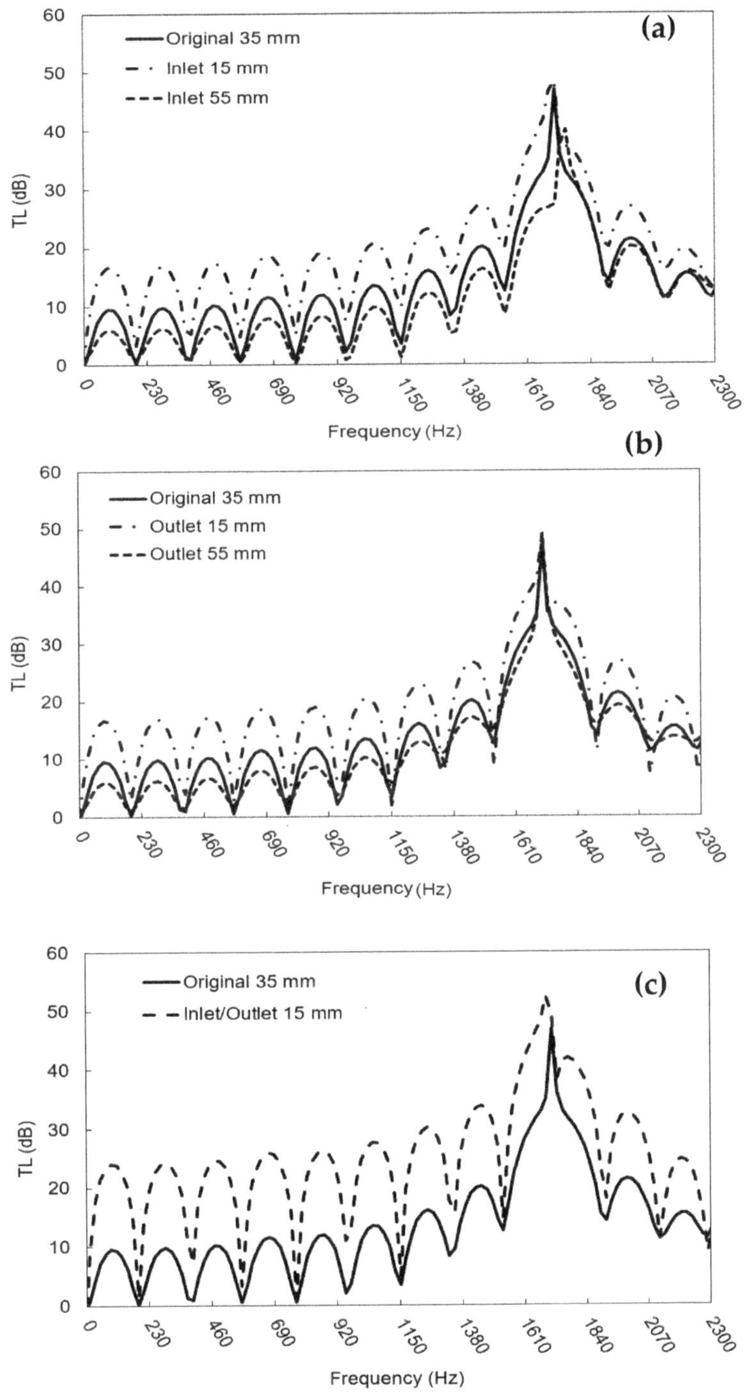


**Figure 15.** Influence of the pipe diameters on the model-scale scrubber TL: (**a**) inlet pipe, (**b**) outlet pipe, (**c**) combined inlet/outlet pipes.

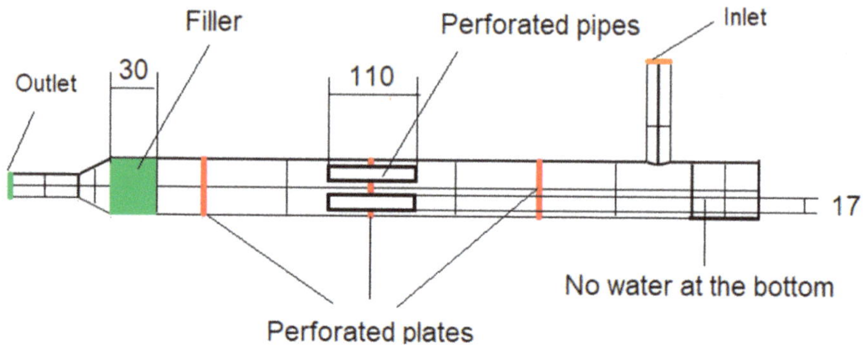


**Figure 16.** Model-scale scrubber geometry modifications.

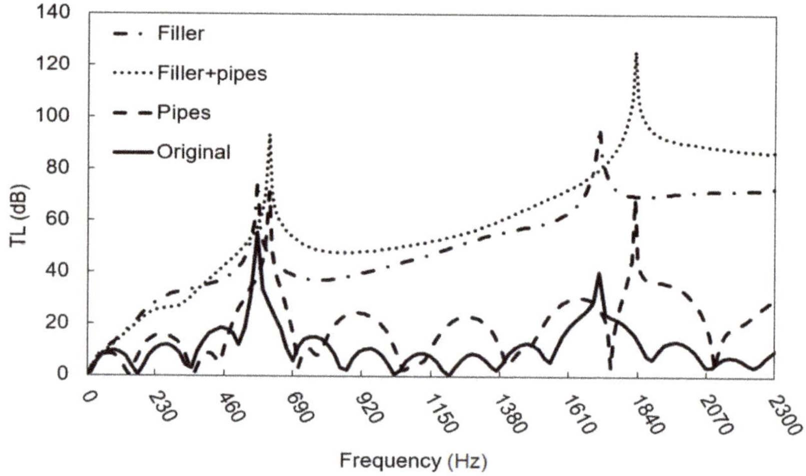


**Figure 17.** Influence of the insertion of a perforated pipes and filler on the model-scale scrubber TL.

## 4. Conclusions

The results of the present investigation clearly indicate the feasibility of an a priori approach aimed at designing a multifunctional scrubber that includes a muffler functionality within the product. As a matter of fact, the assessment of the proper FEM model on the expansion chamber creates an effective tool that can be used to forecast the TL as a function of the small and subtle modification of the existing scrubber, the construction of which is dictated by the chemical requirements for exhaust abatement. In particular, this research highlighted the following points:

1. The importance of the geometry of the experimental set-up (i.e., connection between prototype and impedance tube) on the measured TL (Sections 3.1 and 3.2);
2. The removal of the water at the bottom of the scrubber (e.g., letting the water to directly flow out), allowing the TL increase at low frequencies (Section 3.2);
3. Decreasing the inlet and outlet pipe diameter of the scrubber, allowing its TL to increase along the entire frequency range (Section 3.3);
4. The addition of perforated pipes and/or a filler inside of the scrubber is possible to increase the TL and change the fundamental frequencies of the system (Section 3.3).

Clearly, further work is necessary as to assess the effect of the potential coupling of the transparent frequencies with those exciting frequencies emitted from engines that can be measured, for example, with the method proposed by [38], as an example to consider the influence of the viscous flow inside of the component. More studies on the number and type of perforated pipes and fillers considered inside the scrubber should be performed. The evidence reported here shows that even small geometry modifications allow for the acoustic properties of the scrubber (i.e., increase the TL) to be optimized and possibly allow the transparent frequency to be tuned (i.e., be able to shift the $f_{min}$ away from the engine frequencies). These aspects lead to the possible reduction of the silencer dimension. Increasing the TL of the scrubber, the sound pressure exiting the system is minor, so the silencer has to ensure an inferior abatement in terms of dB that results in a lower volume. Notice that the volume of the scrubber is dictated by the chemistry/chemical engineering of the emission abatement process.

**Author Contributions:** Methodology, G.K.O.D.; software, G.K.O.D.; validation, G.K.O.D.; data curation, G.K.O.D.; writing—original draft preparation, G.K.O.D.; writing—review and editing, M.B., J.K. and F.M.; supervision, J.K. and F.M; funding acquisition, M.B. and J.K. All authors have read and agreed to the published version of the manuscript.

**Funding:** This work was supported by the "ABE-Abbattimento delle emissioni vibroacustiche e chimiche in ambito navale" project funded by Regione Autonoma Friuli Venezia Giulia, POR-FESR 2014–2020, attività 1.3b Incentivi per progetti di R&S da realizzare attraverso partenariati pubbico privati nelle aree di specializzazione Tecnologie Marittime e Smart Health—DGR 849/16.

**Institutional Review Board Statement:** Not applicable.

**Informed Consent Statement:** Not applicable.

**Acknowledgments:** The authors wish to gratefully thank Carlo Pestelli and Marko Gantar from Wärtsilä Italy S.p.A. for their helpful discussions and their valuable assistance in the experimental tests.

**Conflicts of Interest:** The authors declare no conflict of interest.

## References


1. Chen, Y.; Lv, L. Design and Evaluation of an Integrated SCR and Exhaust Muffler from Marine Diesels. *J. Mar. Sci. Technol.* **2015**, *20*, 505–519. [CrossRef]
2. MARPOL 73/78. *Annex VI-Regulations for the Prevention of Air Pollution from Ships, Chapter 3-Requirements for Control of Emissions from Ships, Regulation 13—Nitrogen Oxides (NOx)*; International Maritime Organization (IMO): London, UK, 2008.
3. MARPOL 73/78. *Annex VI-Regulations for the Prevention of Air Pollution from Ships, Chapter 3-Requirements for Control of Emissions from Ships, Regulation 14- Sulphure Oxides (SOx)*; International Maritime Organization (IMO): London, UK, 2008.
4. Ushakov, S.; Stenersen, D.; Einang, P.M.; Ask, T.Ø. Meeting Future Emission Regulation at Sea by Combining Low-Pressure EGR and Seawater Scrubbing. *J. Mar. Sci. Technol.* **2020**, *25*, 482–497. [CrossRef]
5. Zhao, Y.; Fan, Y.; Fagerholt, K.; Zhou, J. Reducing Sulfur and Nitrogen Emissions in Shipping Economically. *Transp. Res. Part D: Transp. Environ.* **2021**, *90*, 102641. [CrossRef]
6. Johnsen, K.; Kock, F.; Strøm, A.; Chryssakis, C. Global Sulphur Cap 2020 Update. 2019, p. 52. Available online: https://www.dnvgl.com/maritime/publications/global-sulphur-cap-2020.html (accessed on 22 December 2020).
7. Oil Price Information. Available online: https://www.petrolbunkering.com/price-information/ (accessed on 22 February 2021).
8. Lindstad, H.E.; Rehn, C.F.; Eskeland, G.S. Sulphur Abatement Globally in Maritime Shipping. *Transp. Res. Part D Transp. Environ.* **2017**, *57*, 303–313. [CrossRef]
9. Scrubbers Shown to Have Lower Climate Impact than Low-Sulphur Fuel, Wärtsilä Corporation. Available online: https://Www.Wartsila.Com/Media/News/22-09-2020-Scrubbers-Shown-to-Have-Lower-Climate-Impact-than-Low-Sulphur-Fuel-2786812 (accessed on 29 December 2020).
10. Boscarato, I.; Hickey, N.; Kašpar, J.; Prati, M.V.; Mariani, A. Green Shipping: Marine Engine Pollution Abatement Using a Combined Catalyst/Seawater Scrubber System. 1. Effect of Catalyst. *J. Catal.* **2015**, *328*, 248–257. [CrossRef]
11. International Maritime Organization. *Resolution MSC.337(91), Code on Noise Levels on Board Schips*; IMO: London, UK, 2012.
12. Lloyd's Register, ShipRight procedures, Additional Design & Construction Procedure (ADP), Procedure for the Determination of Airborne Noise Emissions from Marine Vessels, January 2019. Available online: https://www.lr.org/en/shipright-procedures/#accordion-additionaldesign&constructionprocedure(adp) (accessed on 28 December 2020).
13. NAM KyoungHun. Scrubber with Noise Reduction on Ship. Korean Patent KR102039134 (B1), 5 November 2019.
14. Sridhara, B.S.; Crocker, M.J. Review of Theoretical and Experimental Aspects of Acoustical Modeling of Engine Exhaust Systems. *J. Acoust. Soc. Am.* **1994**, *95*, 2363–2370. [CrossRef]
15. Tao, Z.; Seybert, A.F. A Review of Current Techniques for Measuring Muffler Transmission Loss. *J. Passeng. Car Mech. Syst. J.* **2003**, *112*, 2096–2100.
16. Beranek, L.L.; Ver, I.L. *Noise and Vibration Control Engineering*, 2nd ed.; John Wiley & Sons, Inc.: Hoboken, NJ, USA, 2005; ISBN 10-0471449423.

17. Dokumaci, E. *Duct Acoustics: Fundamentals and Applications to Mufflers and Silencers*; Cambridge University Press: Cambridge, UK, 2021; ISBN 9781108887656. [CrossRef]
18. Bodén, H.; Glav, R. Exhaust and intake noise and acoustical design of mufflers and silencers. In *Handbook of Noise and Vibration Control*; Crocker, M.J., Ed.; John Wiley & Sons, Inc.: Hoboken, NJ, USA, 2007; Chapter 85. [CrossRef]
19. Chung, J.Y.; Blaser, D.A. Transfer Function Method of Measuring In-duct Acoustic Properties. II. Experiment. *J. Acoust. Soc. Am.* **1980**, *68*, 914–921. [CrossRef]
20. Chung, J.Y.; Blaser, D.A. Transfer Function Method of Measuring In-duct Acoustic Properties. I. Theory. *J. Acoust. Soc. Am.* **1980**, *68*, 907–913. [CrossRef]
21. *ISO 10534: Acoustics—Determination of Sound Absorption Coefficient and Impedance in Impedances Tubes—Part 2: Transfer Function Method*; International Organization for Standardization: Geneva, Switzerland, 2001.
22. Munjal, M.L. *Acoustics of Ducts and Mufflers*, 2nd ed.; John Wiley & Sons, Inc.: Hoboken, NJ, USA, 2014; ISBN 10 1118443128.
23. Munjal, M.L.; Doige, A.G. Theory of a Two Source-Location Method for Direct Experimental Evaluation of the Four-Pole Parametersof an Acoustic Element. *J. Sound Vib.* **1990**, *141*, 323–333. [CrossRef]
24. Zhang, H.; Fan, W.; Guo, L.-X. A CFD Results-Based Approach to Investigating Acoustic Attenuation Performance and Pressure Loss of Car Perforated Tube Silencers. *Appl. Sci.* **2018**, *8*, 545. [CrossRef]
25. Zhu, D.D.; Ji, Z.L. Transmission Loss Prediction of Reactive Silencers Using 3-D Time-Domain CFD Approach and Plane Wave Decomposition Technique. *Appl. Acoust.* **2016**, *112*, 25–31. [CrossRef]
26. Bing, W.; Yongjuan, W.; Cheng, X. Study of Transmission Loss on Muffler. *RJASET* **2013**, *5*, 5556–5560. [CrossRef]
27. Tan, W.-H.; Ripin, Z.M. Analysis of Exhaust Muffler with Micro-Perforated Panel. *J. Vibroeng.* **2013**, *15*, 558–573.
28. Katz, B.F.G. Method to Resolve Microphone and Sample Location Errors in the Two-Microphone Duct Measurement Method. *J. Acoust. Soc. Am.* **2000**, *108*, 2231–2237. [CrossRef] [PubMed]
29. MSC Software-Hexagon. *Actran 2020 User's Guide, Volume, Installation, Operations, Theory and Utilities*; Free Field Technologies: Mont-Saint-Guibert, Belgium, 2020.
30. Marburg, S.; Nolte, B. *Computational Acoustics of Noise Propagation in Fluids: Finite and Boundary Element Methods*; Springer: Berlin, Germany, 2008; ISBN 978-3-540-77447-1.
31. Marburg, S. Six boundary elements per wavelength: Is that enough? *J. Comput. Acoust.* **2002**, *10*, 25–51. [CrossRef]
32. Kurowski, P.M. *Finite Element Analysis for Design Engineers*, 2nd ed.; SAE International: Warrendale, PA, USA, 2017.
33. Maa, D.-Y. Potential of Microperforated Panel Absorber. *J. Acoust. Soc. Am.* **1998**, *104*, 2861–2866. [CrossRef]
34. Wartsila Italia S.p.A.; Trieste, Italy. Personal communication, 2021.
35. Yang, Y.; Jia, H.; Lu, W.; Sun, Z.; Yang, J. Impedance-Matching Acoustic Bend Composed of Perforated Plates and Side Pipes. *J. Appl. Phys.* **2017**, *122*, 054502. [CrossRef]
36. Dubno, J.R.; Eckert, M.A.; Lee, F.-S.; Matthews, L.J.; Schmiedt, R.A. Classifying Human Audiometric Phenotypes of Age-Related Hearing Loss from Animal Models. *J. Assoc. Res. Otolaryngol.* **2013**, *14*, 687–701. [CrossRef] [PubMed]
37. Weston, D.E. The Theory of the Propagation of Plane Sound Waves in Tubes. *Proc. Phys. Soc. B* **1953**, *66*, 695–709. [CrossRef]
38. Gao, Z.; Saine, K.; Hynninen, A.; Tanttari, J. Large engine exhaust noise in real life and measurement techniques in practice. In Proceedings of the 26th International Congress on Sound and Vibration, Montreal, QC, Canada, 7–11 July 2019.

applied sciences

MDPI

Article

# A New Understanding and Modelling of TSP and BP Indices Compared to Safety IMO Ship Requirements

José M. Pérez-Canosa, José A. Orosa * and Eliseo A. Pacheco

Department of Navigation Science and Marine Engineering, University of A Coruña, ETSNyM, Paseo de Ronda 51, 15011 A Coruña, Spain; jose.pcanosa@udc.es (J.M.P.-C.); jaorosa@udc.es (E.A.P.)
* Correspondence: jose.antonio.orosa@udc.es

**Abstract:** Due to the lack of information about the concept of Tons of Steering Pull (TSP) of many escort tugs, and the lack of research works relating the TSP demanded by a tethered vessel with respect to the TSP provided by tugs, the present paper shows an original study with mathematical models on how to solve these problems. What is more, an important percentage of the towing sector always employs Bollard Pull (BP), which is considered the only parameter capable of defining performance, so this paper aims to relate BP with TSP. The present research was carried out based on more than 25 escort tugs of different towing companies. Furthermore, a real case study of different tanker vessels was used for modelling purposes of tethered vessels' TSP. Finally, once the proposed models were obtained, they were compared with International Maritime Organization (IMO) guidelines. The results showed charts with the main independent variables of tugs and vessels in order to be as useful and practical as possible to the shipping industry, mainly to ship owners and tug operators, from a safety point of view.

**Keywords:** tug; bollard pull; tons of steering pull; escort towing; ship-handling towing; mathematical model

**Citation:** Pérez-Canosa, J.M.; Orosa, J.A.; Pacheco, E.A. A New Understanding and Modelling of TSP and BP Indices Compared to Safety IMO Ship Requirements. *Appl. Sci.* **2021**, *11*, 7142. https://doi.org/10.3390/app11157142

Academic Editor: Inwon Lee

Received: 29 June 2021
Accepted: 30 July 2021
Published: 2 August 2021



## 1. Introduction

For several decades, towing operations have been lucrative commercial activities and a relevant key of safety in the shipping industry [1,2]. Furthermore, the growing significance of the size of ships in the last years, which does not match with the increase of working areas of ports, makes it evident that the margin to safety has been reduced [3]. Therefore, considering that nowadays towing operations (ship assistance and escort towing) are vital from a safety point of view, it is even more important to known that in the event of an emergency situation, employed tugs are capable of providing the assistance necessary in order to avoid an accident [4–6].

Throughout the 1970s and 1980s, the state-of-the-art began to describe the tugs' capabilities basing on polar diagrams of thrust vectors [7]. However, polar diagrams have two important handicaps to make an accurate comparison between tugs: they only represent tug capabilities at zero speed, and thrust capabilities are represented in a coordinate system referenced to the tugs themselves instead of being referenced to assisted vessels [8]. For this reason, although it is valid to analyse the abilities of tugs at very low speeds, they are not useful when assistances are carried out with an important forward speed. To overcome this lack of objective information from polar diagrams, in 2001, for the first time, researchers represented the forces generated by a purpose-built escort tug referred to as the Cartesian coordinate system of the assisted vessels [9]. This new diagram is known as a "butterfly" diagram, and through introducing the speed and the "appearance" of the tug, it is possible to read magnitudes of transverse (steering) and longitudinal (braking) forces generated on the assisted vessel at 360° and at different speeds. However, "butterfly" diagrams do not specify, for example, the displacement and the manoeuvrability of the escorted vessel. Therefore, forces generated by an escort tug can be sufficient to assist a particular vessel, but not to another, which would seriously jeopardise the overall safety of the operation.

On the other hand, BP only can be used to differentiate the capabilities at zero speed, because it is defined as the maximum force (thrust) exerted by the tug on a fixed towline in static condition [10–12]. Nowadays, this is the main parameter used to assess the capabilities of ship-handling tugs working at lower speeds, but if tugs have a certain speed over the water, BP will be different with respect to the static condition, mainly because at each speed, the hydrodynamic resistance consumes different energy.

This circumstance was more evident with the implementation of escorting assistance after the accident of *m/v* "Exxon Valdez" in 1989 and the consequent promulgation of the Oil Pollution Act in 1990 (OPA 90) [13–15]. It became clear that BP was not a good tool to assess the tugs' capabilities during escorting assistance sailing at high speeds (above 6 knots) over long distances [16]. In these operations, the hydrodynamic lift from the tug's hull will be used to develop a large force for transmitting through the rope into the tow [17], which is proportional to the tug's hull underwater area and the square of transit speed [18,19]. Therefore, the performance of escort tugs must be defined, at least, by the capability to generate steering forces using indirect methods up to 10 knots [20,21]. This force, known as Tons of Steering Pull (TSP), is more important and higher than BP, and it corresponds to a steering force used to alter the vessel's course in order to head it to a safe area; to help it to reduce the forward speed; or to counteract the effects of its rudder locked into a side, sailing at 10 knots [22–24].

Despite this, for many past decades and even nowadays, an important portion of the towing industry consider BP as the most important and only parameter that defines the characteristics of all types of tugs [4,25]. Professional works [26] confirm that tug owners publish little information about steering pull of tugs, so usually this information is estimated by multiplying BP by a figure that depends on the circumstance. In fact, even many tugs classified as Escort by a classification society do not have the TSP available in its ship's particulars. Then, the lack of reliable information about the performance of escort tugs is evident from the genesis of this assistance mode. This can be observed in the simulation study carried out by Merrick (2002) [27] in order to assess, through dynamic models, the scheme escorting in Prince William Sound (Alaska), precisely the location where *m/v* Exxon Valdez suffered the accident. This work confirms the difficulty of reaching an agreement about the purpose of escort tugs after more than 8 years of discussions with all locally involved interests. More precisely, it was proposed that in some areas where two escort tugs were being employed, afterwards, only one escort tug was needed, but no performance characteristics of the involved tugs were specified. In other research works about models and simulations with tugs [6], the efficiency of escort tugs assisting the biggest vessels is connected only with BP and the size of tugs.

Following Molyneux et al. [28], and taking into account their length, escort tugs operate in off-design hydrodynamic conditions. For that, there is much research with model experiments focused on predicting the total forces generated and the limits of safe working. Furthermore, in the literature review, the reader can find many relevant studies concerning numerical simulations and proposed mathematical models of manoeuvring ships and tugs' motion [29–31].

Regarding the tethered vessel, the manoeuvring performance, including the zigzag manoeuvres with different rudder angles, was also simulated in previous works research [26,32] with the objective of obtaining the yaw angle and lateral hull force, among others.

Regarding the influence of forces between the tug and the tethered ship, recent studies [3] use mathematical models to analyse the problems of forces generated in dynamic operations between both vessels considering the effects of drift and wind force, but in the port environment, i.e., under very low-speed conditions.

Although the research with mathematical models of forces affecting ships covers areas that are very different (hydrodynamic, hydrostatic, propulsion, and external forces), recent papers propose to develop synthetic and simple models of forces for its application to a ship in real-time mode [33]. This intention is in connection with the present work, that it is to find the simplest mathematical model able to relate the performance characteristics of tugs with the assisted vessels.

Tethered vessels must be always operated within the performance capabilities of escort tugs [26], and, therefore, the capabilities and requirements of both should be studied as a whole. However, as it can be observed in the literature revision, there are few research works that analyse, from a safety point of view, the requirements of assisted vessels together with capabilities and performance of tugs working as a unique element at high speeds. Furthermore, there is a lack of information about TSP in tugs, and BP is generally employed. Then, as the problem of relating ships' characteristics and tugs' performance has not been approached, the present paper aims to connect the tugs' efficiency at different speeds (TSP of tugs) in relation to their BP, starting from the ships' requirement (TSP of ships) and using different mathematical models. With this objective, this study proposes providing the shipping industry with new information to be used from two points of view. On one side, ship owners benefit from considering the BP tugs (always available) to know the TSP that a tug can exert to their vessels; as a consequence, they can select the type and the appropriate number of tugs to be employed. On the other side, tug operators benefit from knowing the BP needed of one or more tugs in order to satisfy the required TSP of escorted vessels as a function of their deadweight.

For this research, the next sections are composed of Materials and Methods, which describe the database used; the Results section, which makes use of ANOVA modelling, showing the graphs and the equations obtained; the Discussion section, which compares the results with the IMO ship requirements and; finally, the Conclusion section, which suggests further research that builds on the present work.

## 2. Materials and Methods

In the present paper, in order to relate the BP with TSP of the tugs, the main characteristics of more than 25 escort tugs of different towing companies were analysed, classified, and then modelled. Although this is not a number that is excessively high, mainly due to the limited number of escort tugs available in the world with their TSP published, it is a number that allows carrying out solid statistics analysis. The selected escort towing companies are located all over the world, and tugs are rendering assistance to vessels that are very different in size and characteristics. Furthermore, these tugs are employed in areas and under weather conditions very different between them, so they can be considered as representative of the escort towing fleet in the world.

Moreover, as it is also necessary to connect the TSP of tugs with the required TSP of escorted vessels, real cases data of different tanker vessels were selected and modelled, taking into consideration that each vessel has particular steering characteristics, which are a function of the speed and the steering angle [25,26,31]. Finally, after analysing the models' errors, we show the relation between variables of tugs and assisted vessels.

In order to model the multiple variables involved, Minitab version 18 was selected [34,35]. The selected variables of tugs and units for the present research are included in Table 1.

**Table 1.** Identification of tugs' variables and used units.

| Variable | Code | Units |
|---|---|---|
| Type of Tug | TT | - |
| Length | L | m |
| Beam | B | m |
| Depth | De | m |
| Draft | Dr | m |
| Bollard Pull | BP | Tons |
| Tons of Steering Pull at 10 knots | TSP | Tons |
| Maximum Tug Speed | TS | Knots |
| Horsepower | BHP | HP |
| Year of Built | YB | - |

The Type of Tug variable was codified with numerical values, as it can be seen in Table 2.

**Table 2.** Codification of Type of Tug variable.

| Type of Tug | Code |
|---|---|
| Tractor Voith | 1 |
| Azimuth Stern Drive (ASD) | 2 |
| Tractor-Z | 3 |

It was not necessary to codify the rest of the variables, so they were used with their own values, which were numerical values.

Finally, in order to compare our results with widely employed guidelines, we used the graphs of TSP published by the IMO, which were traced for oil tankers and LNG carriers sailing at 10 knots and considering the guidelines "Standards for Ship Manoeuvrability" [36,37]. These guidelines are recommendations only, and some authors [26] consider that in real cases, the needed TSP are higher than the theoretical values obtained in the graphs. However, other authors consider that the figures obtained from the graphs are excessive [38]. In any case, the scope of these works was not to analyse the reason for these discrepancies.

We used 10 knots as a reference value because it is considered by the literature as the reference escort speed. However, it is known that this speed only can be reached in certain areas of escort operations and sailing several nautical miles, because most seaports have the maximum speed established at 6–7 knots, corresponding to ship-handling assistance.

## 3. Results

### *Modelling*

As it was mentioned in previous sections, many people with interest involved in the towing industry, even in the escort sector, associate the performance and capabilities of tugs directly with BP. After a previous analysis of normality and homoscedasticity of the more common variables employed to define the BP, some of them were identified by an ANOVA study to model BP. These variables were Type of Tug; Length (m); Beam (m); Depth (m); Draft (m); Speed (knots); Horsepower (HP); and Year Built.

Therefore, in the first stage of this research work, we developed a response surface to obtain a mathematical model able to calculate the estimated BP with the correlation between each of the previous variables. The model (1) obtained is shown in Equation (1).

$$\begin{aligned} BP = -1937 + 12.73{\cdot}TT + 2.036{\cdot}L - 0.155{\cdot}B - 0.17{\cdot}De- \\ -7.03{\cdot}Dr + 0.03{\cdot}TS + 0.00681{\cdot}BHP + 0.958{\cdot}YB. \end{aligned} \tag{1}$$

Units of BP obtained are in tons and, as it can be concluded from Figure 1, which represents the real BP of samples with BP obtained with this Model 1, a high precision is obtained: (R-square, 93.40%).

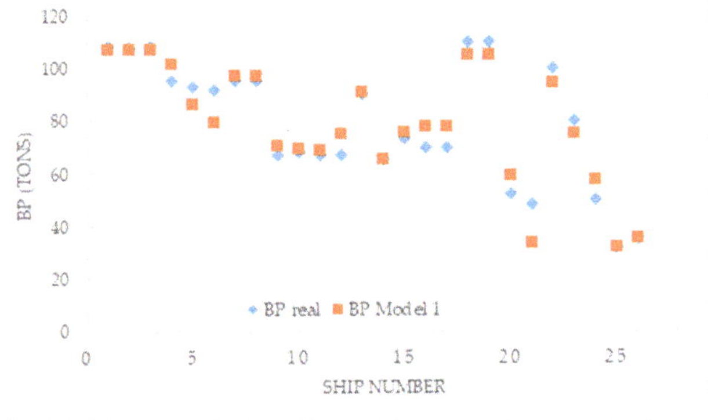


**Figure 1.** Representation of tugs' BP obtained with Model 1.

Considering that in the literature, we can find many different types of BP (as sustained BP, maximum static BP, marketing BP, or Brazilian BP) [11], the proposed model will be useful in order to calculate, with a very simple equation, the real BP. These data can be used in the next stage of the present research in the strange case that this information is not included in the tug's particulars.

Therefore, although BP is usually published by tug owners, as it was commented previously, the same does not happen with TSP data. Therefore, here, we propose another response surface modelling in order to calculate the TSP of tugs at 10 knots, and considering the same variables of the previous model, which are usually known, i.e.: Type of Tug; Length (m); Beam (m); Draft (m); Depth (m); Tug Speed (Knots); Horsepower (BHP); and Year Built.

The regression equation of mathematical Model 2 obtained is included in Equation (2):

$$\begin{aligned} \mathrm{TSP}(10\mathrm{knots}) = & -7388570 - 5828.47\cdot\mathrm{TT} - 253.689\cdot\mathrm{L} + 64.9967\cdot\mathrm{B} - 366.268\cdot\mathrm{De} - \\ & -1143.01\cdot\mathrm{Dr} - 29.2185\cdot\mathrm{TS} + 0.161301\cdot\mathrm{BHP} + 7375.24\cdot\mathrm{YB} + 1002.89\cdot\mathrm{TT}^2 + 3.21910\cdot\mathrm{L}^2 - \\ & -5.98118\cdot\mathrm{B}^2 + 24.8924\cdot\mathrm{De}^2 + 73.5723\cdot\mathrm{Dr}^2 + 1.20324\cdot\mathrm{TS}^2 - 0.000010\cdot\mathrm{BHP}^2 - 1.83738\cdot\mathrm{YB}^2 + \\ & +12.2277\cdot\mathrm{TT}\cdot\mathrm{L} + 57.7064\cdot\mathrm{TT}\cdot\mathrm{B} + 75.4063\cdot\mathrm{TT}\cdot\mathrm{De} + 181.199\cdot\mathrm{TT}\cdot\mathrm{Dr} \end{aligned} \quad (2)$$

where the units TSP are tons, and a precision of 100% was obtained in R square. Figure 2 shows the good relation between real and estimated data of tugs' TSP.

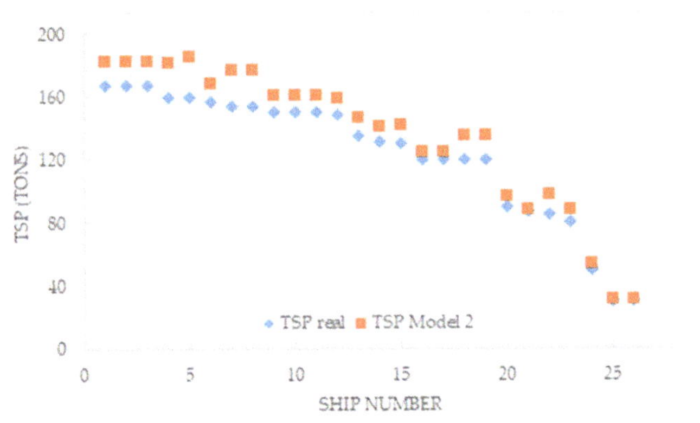


**Figure 2.** Representation of tugs' TSP real and obtained with Model 2.

Once the equations of BP and TSP are modelled with the tug's usual data, a new regression equation was obtained including BP as a variable. This regression equation of Model 3 allows calculating the TSP developed by a tug at 10 knots as a function of only two independent variables: Tug Speed (knots) and BP (tons). The result of this model is Equation (3).

$$\text{TSP}(10\text{knots}) = 563 + 4.93{\cdot}\text{BP} - 113.3{\cdot}\text{TS} - 0.0192{\cdot}\text{BP}^2 + 4.93{\cdot}\text{TS}^2 - 0.081{\cdot}\text{BP}{\cdot}\text{TS}. \quad (3)$$

A determination factor of 80.01% for R-square indicates good precision. It can be also concluded from Figure 3, which plots the real TSP and the estimated TSP of tugs as per Model 3.

**Figure 3.** Representation of tugs' TSP real and obtained with Model 3.

Finally, it is interesting to comment on the validity range of the previous models and their variables, as it shown in Table 3.

**Table 3.** Validity range of Models 1, 2, and 3.

| Variable | Min | Max | Unit |
|---|---|---|---|
| Length | 24.4 | 47.0 | m |
| Beam | 9.42 | 16.00 | m |
| Depth | 4.0 | 7.5 | m |
| Draft | 4.60 | 8.25 | m |
| Speed | 10 | 16 | Knots |
| BP | 32 | 110 | tons |
| TSP (10 kt) | 30 | 110 | tons |
| BHP | 1624 | 10,400 | HP |
| Built | 1981 | 2017 | |
| Model 1 (BP) | 35.23 | 106.80 | tons |
| Model 2 (BP) | 31.50 | 181.90 | tons |
| Model 3 (BP) | 150.01 | 35.40 | tons |

In escort towing operations, as two ships are involved (the escort tug and assisted vessel), it is necessary to know the required TSP of the assisted vessel sailing at different speeds to be counteracted in the event of an emergency situation. Therefore, in this case, it is necessary to know the TSP needed by the escorted ship and the TSP developed effectively by the tug, in order to be sure that escort operations are being carried out with a sufficient margin of safety. Then, taking into consideration the rudder forces of different tankers registered in previous studies with the rudder blocked at various angles [28], we developed a mathematical model that allows us to calculate the TSP required by a ship relating the following independent variables: Deadweight; Ship Speed; and Rudder Angle. This data collection is included in Table 4.

**Table 4.** TSP data in tons of different tanker vessels.

| Ship Speed | Deadweight 100,000 tons | | | | Deadweight 200,000 tons | | | | Deadweight 300,000 tons | | | |
|---|---|---|---|---|---|---|---|---|---|---|---|---|
| | Rudder Angle | | | | | | | | | | | |
| | 10° | 15° | 25° | 35° | 10° | 15° | 25° | 35° | 10° | 15° | 25° | 35° |
| 6 knots | 25 | 30 | 45 | 30 | 30 | 50 | 60 | 50 | 40 | 55 | 80 | 60 |
| 8 knots | 35 | 55 | 75 | 60 | 55 | 85 | 115 | 90 | 70 | 100 | 140 | 105 |
| 10 knots | 60 | 85 | 120 | 90 | 90 | 130 | 185 | 145 | 110 | 155 | 220 | 165 |
| 12 knots | 65 | 120 | 175 | 135 | 130 | 190 | 260 | 205 | 160 | 230 | 320 | 245 |

Equation (4) includes the result of Model 4 obtained:

$$\begin{aligned} TSP = &-37.8 - 23.74 \cdot SS - 0.000074 \cdot DWT + 11.80 \cdot RA + 1.406 \cdot SS^2 - \\ &-0.3196 \cdot RA^2 + 0.000069 \cdot SS \cdot DWT + 0.3847 \cdot SS \cdot RA + 0.000005 \cdot DWT \cdot RA \end{aligned} \quad (4)$$

where SS is ship speed in knots; DWT is deadweight in tons; and RA is rudder angle in degrees, being all variables referred to as tanker vessels. With this mathematical model, it we calculate the TSP (tons) required by a tanker with a precision of R-square of 96.69%, as can be seen from Figure 4.

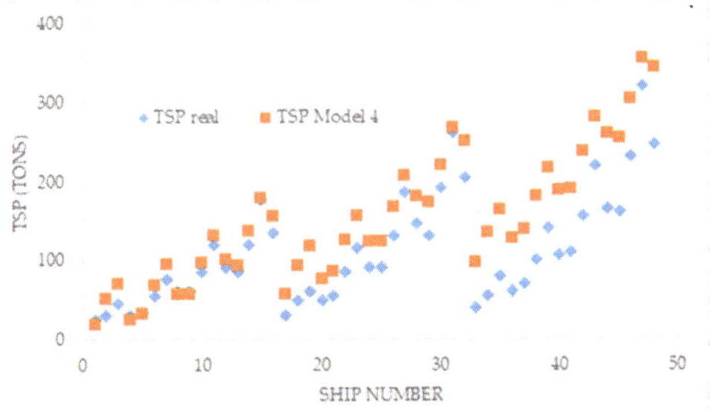


**Figure 4.** Representation of tanker vessels' real TSP and the corresponding TSP obtained with Model 4.

Once again, the validity range of Model 4 and its variables are shown in Table 5.

**Table 5.** Validity range of Model 4.

| Variable | Minimum | Maximum | Unit |
|---|---|---|---|
| Knots | 0 | 10 | Knots |
| Deadweight | 10,000 | 300,000 | tons |
| Rudder angle | 15 | 25 | degrees |
| Model 4 (TSP) | 18.5 | 341.5 | tons |

## 4. Discussion

*Models Analysis*

From a safety point of view of towing operations, when vessels are being escorted, ship operators need to know if the tugs' capabilities employed during the assistance match their needs of steering force. As it was previously mentioned, and unlike BP, the TSP of tugs are not always available, so the present section shows the relationship between both variables. Thus, anyone and at any time, without the need to carry out full-scale trials or complex simulations, can assess the tug TSP starting from their BP and vice versa.

To achieve this objective, Equation (3) of Model 3 was used to calculate TSP with a tug's characteristics, and Equation (4) of Model 4 was used to calculate TSP with a ship's particulars; then, these were matched with the objective of defining the tug BP variable as follows:

$$
\begin{aligned}
&563 + 4.93{\cdot}BP - 113.3{\cdot}TS - 0.0192{\cdot}BP^2 + 4.93{\cdot}TS^2 - 0.081{\cdot}BP{\cdot}TS = \\
&= -37.8 - 23.4{\cdot}SS - 0.000074{\cdot}DWT + 11.80{\cdot}RA + 1.406{\cdot}SS^2 - 0.3196{\cdot}RA^2 + \\
&+ 0.000069{\cdot}SS{\cdot}DWT + 0.3847{\cdot}SS{\cdot}RA + 0.000005{\cdot}DWT{\cdot}RA.
\end{aligned} \tag{5}
$$

Therefore, Equation (5) allows us to define tug BP as a function of the required TSP of a tethered vessel using the deadweight, ship speed, and rudder angle as independent variables.

Considering that the literature review indicates that maximum steering forces (TSP) are required from a rudder angle of 25° [25], BP was calculated for series of assisted ship's rudder angles of 15° and 25° according to Equation (5). In this simulation, the tug speed of 10 knots was considered as a constant. In each series, an individualised BP was calculated for each ship's speed, from 10 knots to zero knots; once the escort tug begins to exert force, the ship's speed will decrease progressively until it is under control. Therefore, more than 800 runs were tabulated, as can be seen in Figures 5 and 6.

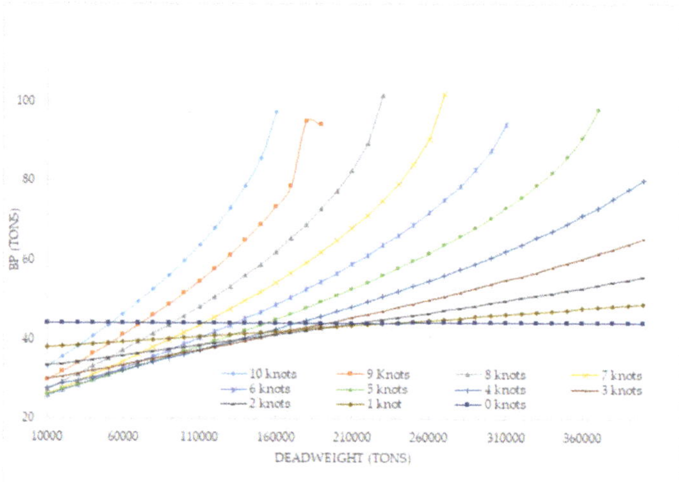


**Figure 5.** Tug BP needed as a function of ship's characteristics (constant rudder angle of 15°).

Figure 5 shows the BP needed from tugs to assist tethered tanker vessels sailing throughout the range of escort speeds with a rudder angle of 15°. Figure 6 represents the same but with a rudder angle of 25°.

If we now compare Figures 5 and 6, it can be concluded that when the assisted ship is in static condition (zero speed), the BP hardly varies as a function of deadweight. This means that unlike the safety policy of many port authorities, which require excessive BP regardless of the type of traffic, it would not be necessary to employ tugs in ship-handling towing with BP as high as those currently used in many cases. In this circumstance, at zero speed, BP and TSP concur.

Furthermore, comparing both figures and at higher speeds, it is observed that the BP required is higher with a higher rudder angle (25°). This aspect would be in line with the literature stating that the maximum TSP required by an assisted vessel is produced with a rudder angle of 25°, and as a consequence, the corresponding BP needed from tugs.

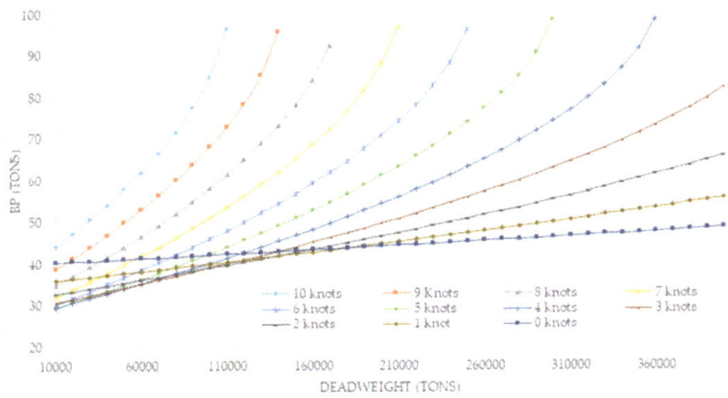


**Figure 6.** Tug BP needed as a function of ship's characteristics (constant rudder angle of 25°).

On the other hand, taking into consideration the tendency of curves at higher speeds, especially with a rudder angle of 25°, it could be concluded that at present, there are no escort tugs available in the market with that BP required. Nevertheless, this circumstance is solved in real life using two tugs working in tandem [39], so the BP of each escort tug is added as it is regulated in Los Angeles/Long Beach port [40,41].

Another particular case of interest observed in both Figures 5 and 6 is that except for 10 knots, in the range of lower deadweight, the BP required is lower than the BP required in static condition. Although this result could be considered illogical, it should be noted that these vessels are not sailing straight ahead, so the greater the rudder angle, the greater the yaw angle, and as a consequence, the greater the lateral surface of the underwater hull working against the flow of water. This hydrodynamic force will reduce the shipping speed and, therefore, tugs can control the TSP demanded with a lower BP.

In order to know if our models satisfy the tools used by the towing industry at present, Figure 7 represents the IMO recommendations about upper and lower TSP limits required by a tanker vessel of different deadweight sailing at 10 knots.

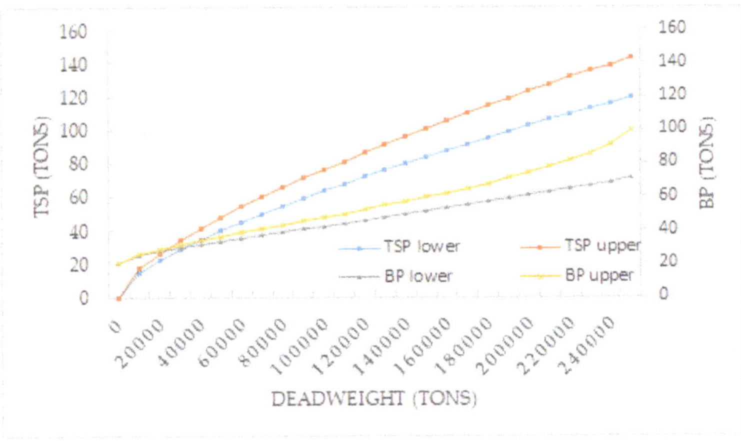


**Figure 7.** IMO TSP limits and the corresponding BP limits as per Model 3.

As there are no published equations that define the IMO TSP limits, once we obtained several values of IMO graphs, they were used in our Model 3 in order to calculate the BP limits corresponding to IMO TSP limits. The results are also represented in Figure 7, which shows a good relationship between both variables.

In this sense, considering that the TSP provided by escort tugs are not always available, it is interesting to understand that starting from the TSP demanded by assisted vessels according to IMO, it is possible to assess the minimum BP (which are always available) that tugs would have to have to satisfy the TSP required.

At the same time, with the objective of analysing the relation between IMO TSP and TSP demanded by a tanker vessel according to our Model 4, we developed a simulation of 160 runs for rudder angles of 10°, 15°, 25°, and 35°, where the ship speed variable was considered as constant (10 knots). From Figure 8, it can be concluded that IMO TSP limits correspond very well with our Model 4 when the tanker vessel is working with a rudder angle of 10°. Although in IMO TSP no information is published about rudder angles, this conclusion can be considered as acceptable taking into account that, calculating the correct wheel-over point, merchant vessels rarely steer with large angles of the rudder.

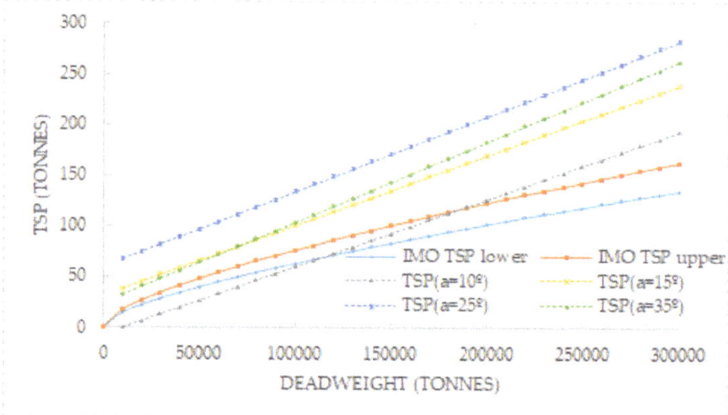


**Figure 8.** IMO TSP limits and calculated TSP as per Model 4 for different rudder angles.

From Figure 8, it is also interesting to observe that at higher rudder angles, the TSP needed according to Model 4 is considerably higher than the upper limits of TSP required by IMO. This conclusion could be related to previous simulations and manoeuvring studies [26], where it is stated that the needed steering pull is 50% higher than upper IMO guidelines. Therefore, in the absence of more objective information that considers the navigation and manoeuvring characteristics of assisted vessels, the proposed models and the relationship between independent variables of tug and vessel could be useful for the towing sector.

Finally, as it was previously mentioned, the parameter usually employed in order to classify a tug is BP. Therefore, this section intends to show and relate the main three variables playing in a tug–vessel system in a 3D map: deadweight of the tethered ship; tug BP and IMO TSP limits. For this mission, the upper and lower IMO TSP limits are deduced from its graphs, and tug BP considering the previous TSP is calculated following our Model 3. Figures 9 and 10 represent the surface graphs where the appearance is very similar due to the difference between the upper and lower TSP limits being barely 20%.

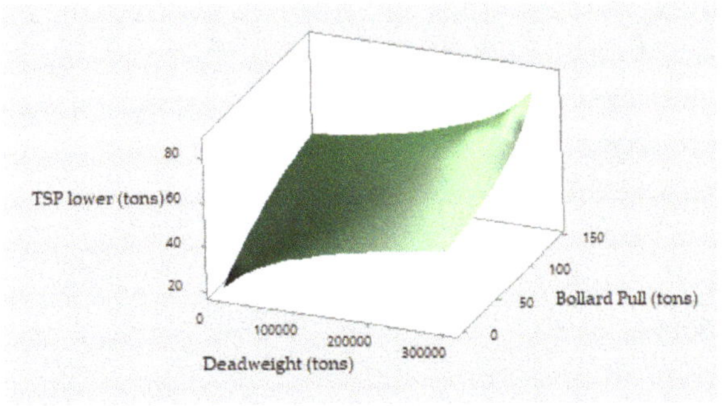


**Figure 9.** Surface graph for IMO TSP lower limit.

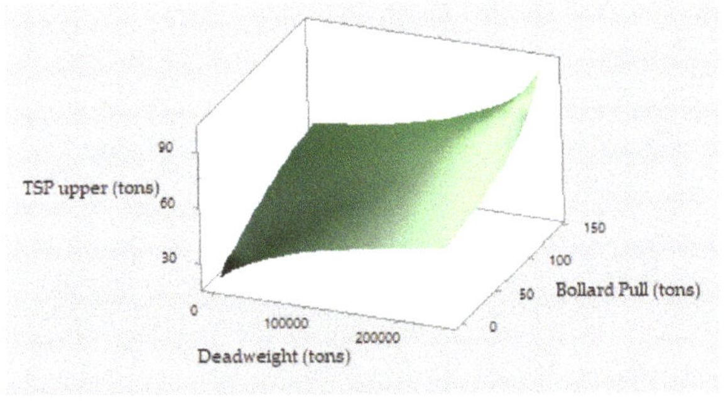


**Figure 10.** Surface graph for IMO TSP upper limit.

## 5. Conclusions

With the genesis of escorting operations, it was soon shown that BP was not the only parameter that allows us to analyse the full performance of tugs, nor can it be used for comparative purposes between different tugs. Furthermore, as it is vital to know the TSP developed by an escort tug at different speeds (unknown many times), and given the lack of research works in the TSP subject, a case study with a number representative of an escort tug fleet was conducted. With this information, one mathematical model was proposed to ascertain BP; in the strange event that this parameter is not available or in case of doubts, a mathematical model was proposed to calculate TSP exerted by tugs based on their main characteristics, and another model was developed to ascertain the TSP of tugs using only two variables. One of these variables is BP, so this will allow us to relate TSP and BP. The last model proposed, employing a real case study, was used to obtain the TSP demanded by assisted vessels.

Considering that all the proposed mathematical models had a very good precision, afterwards, we matched the TSP equation demanded by tanker vessels with the TSP equation provided by tugs. In this way, the requirements of tugs expressed in BP can be obtained as a function of rudder angle and speed. Subsequently, using the third proposed model, IMO TSP is expressed in BP, i.e., the best-known parameter of a tug fleet. Finally, we carried out a comparison between the TSP IMO guidelines and the results of the proposed models, concluding that IMO guidelines should be over-reviewed in excess and/or considering the rudder angle maximum to be applied by the assisted ship.

**Author Contributions:** Conceptualisation, J.M.P.-C., J.A.O. and E.A.P.; methodology, J.M.P.-C., J.A.O. and E.A.P.; validation, J.M.P.-C., J.A.O. and E.A.P.; formal analysis, J.M.P.-C., J.A.O. and E.A.P.; investigation, J.M.P.-C., J.A.O. and E.A.P.; data curation, J.M.P.-C., J.A.O. and E.A.P.; writing—original draft preparation, J.M.P.-C., J.A.O. and E.A.P.; writing—review and editing, J.M.P.-C., J.A.O. and E.A.P. All authors have read and agreed to the published version of the manuscript.

**Funding:** This research was funded by the Spanish Ministry of Education and the University of A Coruña (Spain) in their research project to reduce energy consumption in ships (Grant No. 64900).

**Institutional Review Board Statement:** Not applicable.

**Informed Consent Statement:** Not applicable.

**Data Availability Statement:** Not applicable.

**Acknowledgments:** The authors wish to express their deepest gratitude to the Sustainability Specialisation Campus of the University of A Coruña for the administrative and technical support.

**Conflicts of Interest:** The authors declare no conflict of interest.

## References

1. Pulido Begines, J.L. Cuestiones Generales. In *Los Contratos de Remolque Marítimo*, 1st ed.; J. M. Bosch: Barcelona, Spain, 1996; pp. 27–129.
2. Jayarathne, N.; Ranmuthugala, D.; Leong, Z. Safe tug operations during ship assist manoeuvres. *J. Navig.* **2019**, *72*, 813–831. [CrossRef]
3. Wu, G.; Zhao, X.; Sun, Y.; Wang, L. Cooperative maneuvering mathematical modeling for multi-tugs towing a ship in the port environment. *J. Mar. Sci. Eng.* **2021**, *9*, 384. [CrossRef]
4. Paulauskas, V.; Martynas Simutis, M.; Placiene, B.; Barzdžiukas, R.; Jonkus, M.; Paulauskas, D. The influence of Port Tugs on improving the navigational safety of the port. *J. Mar. Sci. Eng.* **2021**, *9*, 342. [CrossRef]
5. Dongxing, X.; Xiufeng, Z.; Yong, Y. The computer simulation study of tug pushing operation system. In Proceedings of the 35th Chinese Control Conference (CCC), Chengdu, China, 27–29 July 2016; pp. 2074–2080.
6. Figari, M.; Martinelli, L.; Martelli, M.; Enoizi, L.; Benedetto Piaggio, B. A new escort tug family designed to anticipate new safety requirements and operational needs. In *Technology and Science for the Ships of the Future*, 1st ed.; Marino, A., Bucci, V., Eds.; IOS Press: Amsterdam, The Netherlands, 2018; Volume 1, pp. 986–993. [CrossRef]
7. Baer, W. Assessment of tug performance. In *Proceedings of the First International Tug Conference*; Troup, K.D., Ed.; Thomas Reed and Company Ltd.: London, UK, 1970; p. 4.
8. Scalzo, S.; Hogue, D. Escort tug performance results. In *Proceedings of the 14th International Tug and Salvage Convention*; Troup, K.D., Ed.; Thomas Reed Publications: Surrey, UK, 1996; pp. 57–74.
9. Gray, D.L. Development of tanker escort regulations in the United States with emphasis on the U.S. Pacific Coast. In *Work Boat World, Asia 2001, Proceeding of the Singapore International Convention and Exhibition Centre, Singapore, 20–22 February 2001*; Baird Publications: Melbourne, Australia, 2001.
10. Brownlie, K. *Controllable Pitch Propellers*, 1st ed.; The Institute of Marine Engineers: London, UK, 1998; pp. 3–8.
11. Jukola, H.; Skogman, A. Bollard Pull. In *Proceedings of the 17th International Tug and Salvage Convention*; Smith, A., Ed.; The ABR Company Limited: Wiltshire, UK, 2002; pp. 189–197.
12. El Zaalik, M.A.A.; Kotb, M.A.; Sharara, A.I. Theoretical and experimental measurements of bollard pull with emphasis on propeller dimensions. *Int. J. Multidiscip. Res. Dev.* **2015**, *3*, 777–783. [CrossRef]
13. Cohen, M. In the Wake of the Exxon Valdez. The Oil Pollution Act of 1990. An Analysis. In *Proceedings of the 11th International Tug Convention and Marine Salvage Symposium*; Troup, K.D., Ed.; Thomas Reed Publications: Surrey, UK, 1990; pp. 5–28.
14. Jansen, M. Paradigm Shifts in Tugnology. In *Proceedings of the 23rd International Tug, Salvage & OSV Convention and Exhibition*; Gorman, D., Ed.; The ABR Company Ltd.: Wiltshire, UK, 2014; pp. 245–249.
15. Allan, R.G. The Evolution of Tug Design through ITS Eyes. In *Proceedings of the 25th International Tug and Salvage Convention and Exhibition*; Wraight, C., Bury, J., Eds.; The ABR Company Ltd.: Wiltshire, UK, 2018; pp. 31–50.
16. Allan, R. The evolution of escort tug technology: Fulfilling a promise. *SNAME Trans.* **2000**, *108*, 99–122.
17. Baniela, S.I.; Melón, G.E. The Voith turbo fin (VTF) a new system to improve the performance of escort tractor Voith Tug. *J. Navig.* **2006**, *59*, 505–519. [CrossRef]
18. Artyszuk, J. Types and power of harbour tugs. The latest trends. *Pr. Nauk. Politech. Warsz. Transp.* **2013**, *98*, 9–20.
19. Noble, J.M. Bollard Pull–Fact or Fiction. How the experts determine tow Requirements. In *Proceedings of the 18th International Tug and Salvage Convention*; Smith, A., Ed.; The ABR Company Ltd.: Wiltshire, UK, 2004; pp. 75–80.
20. Piaggio, B.; Martelli, M.; Viviani, M.; Figari, F. Manoeuvring model and simulation of the non-linear dynamic interaction between tethered ship and tug during escort. In Proceedings of the Maritime Transportation and Harvesting of Sea Resources, Lisbon, Portugal, 9–11 October 2017.
21. Iglesias-Baniela, S.; Vinagre-Ríos, J.; Pérez-Canosa, J.M. Ship handling in unprotected waters: A review of new technologies in escort tugs to improve safety. *Appl. Mech.* **2021**, *2*, 46–62. [CrossRef]
22. Allan, R.; Molyneux, D. Escort tug design alternatives and a comparison of their hydrodynamic performance. *Trans. Nav. Soc. Nav. Archit.* **2004**, *112*, 191–205.
23. DNV·GL. Full Scale Testing of Escort Vessels. Class Guideline DNVGL-CG-0155. February 2016. Available online: http://www.dnvgl.com (accessed on 10 June 2021).

24. Van Kasteren, J. The ART of tugnology: The ultimate balance between safe ship assist and safe escort duties. *Rotor Tug* **2012**, *1*, 1–11.
25. Hensen, H. Tug Use in Port. A Practical Guide. Including Port. In *Port Approaches and Offshore Terminals*, 3rd ed.; The ABR Company Ltd.: Wiltshire, UK, 2018; pp. 235–263.
26. Bay, J.; Schultz, S. Manoeuvring Study of Escorted Tankers to and from Kitimat. In *Real-Time Simulations of Escorted Tankers Bound for a Terminal at Kitimat*, 1st ed.; Force Technology: Lyngby, Denmark, 2010.
27. Merrick, J.R.W. Evaluation of tug escort schemes using simulation of drifting tankers. *Simulation* **2002**, *78*, 380–388. [CrossRef]
28. Molyneux, D.; Xu, J.; Bose, N. Flow vectors around an escort tug at a large yaw angle. In Proceedings of the Canadian Marine Hydrodynamics and Structures Conference, St. John's, NL, Canada, 16–17 October 2017.
29. Yoshimura, Y. Mathematical model for manoeuvring ship motion (MMG Model). Proceeding of the Workshop on Mathematical Models for Operations involving Ship-Ship Interaction, Tokyo, Japan, 4–5 August 2005.
30. Sutulo, S.; Guedes-Soares, C. Mathematical models for simulation of manoeuvring performance of ships. In *Marine Technology and Engineering*, 1st ed.; Guedes-Soares, C., Garbatov, Y., Fonseca, N., Texeira, A.P., Eds.; CRC Press: Boca Raton, MA, USA, 2011; pp. 661–699. [CrossRef]
31. Tzeng, C.-Y.; Chen, J.-F. Fundamental properties of linear ship steering dynamic models. *J. Mar. Sci. Technol.* **1999**, *7*, 79–88.
32. Aksu, E.; Köse, E. Evaluation of mathematical models for tankers' maneuvering motions. *J. ETA Marit. Sci.* **2017**, *5*, 95–109. [CrossRef]
33. Sen, D.O.; Vinh, T.C. Developing mathematical model for calculating forces affecting to Ship motions. *Fluid Mech. Res. Int.* **2018**, *2*, 66–71. [CrossRef]
34. Minitab v.18. Response Surfaces. Available online: https://support.minitab.com/en-us/minitab/18/help-and-how-to/modeling-statistics/doe/supporting-topics/response-surface-designs/response-surface-central-composite-and-box-behnken-designs (accessed on 29 July 2021).
35. Orosa, J.A.; Costa, A.M.; Vergara, D.; Fraguela, F. The influence of climate parameters on maintenance of wind farms—A Galician case study. *Sensors* **2021**, *21*, 40. [CrossRef] [PubMed]
36. International Maritime Organization, IMO 2002 Standards for Ship Manoeuvrability. Maritime Safety Committee (IMO MSC), 76th Session, Agenda Item 23. Report on the Maritime Safety Committee on its Seventy-Sixth Session, MSC 76/23/Add.1. Annex 6, Resolution MSC.137 (76) (Adopted on 4 December 2002), 1–6. London, UK, 18 December 2002. Available online: https://wwwcdn.imo.org/localresources/en/KnowledgeCentre/IndexofIMOResolutions/MSCResolutions/MSC.137(76).pdf (accessed on 2 August 2021).
37. Davis, H.; Polawski, M. *Classification Society Tug Review for Prince William Sound Regional Citizens' Advisory Council*, 1st ed.; Det Norske Veritas: Sunrise, FL, USA, 2011.
38. Allan, R.G. *A Review of Best Available Technology in Tanker Escort Tugs. Project 212-090*; Revision 3; Robert Allan Ltd.: Vancouver, BC, Canada, 2013.
39. Baniela, S.I.; Díaz, A.P. The first escort tractor Voith tug with a bulbous bow: Analysis and consequences. *J. Navig.* **2008**, *61*, 143–163. [CrossRef]
40. Schisler, V.J.; Brooks, G. Team Towing: Using Relatively Small Tractors on Heavy Ships. Professional Mariner 2001. Available online: http://www.towingsolutionsinc.com (accessed on 9 June 2021).
41. California Code of Regulations. Title 14, Division 1. Subdivision 4, Office of Spill Prevention and Response Chapter 4. Vessel Requirements. Subchapter 2. Tank Vessel Escort Program for the Los Angeles/Long Beach Harbor. Sections 851.20–851.32. August 2012. Available online: https://wildlife.ca.gov/OSPR/Legal/OSPR-Regulations-Index (accessed on 2 August 2021).

applied sciences

MDPI

Article

# Resilience Assessment for the Northern Sea Route Based on a Fuzzy Bayesian Network

**Weiliang Qiao [1], Xiaoxue Ma [2,*], Yang Liu [3] and He Lan [3]**

1 Marine Engineering College, Dalian Maritime University, Dalian 116026, China; xiaoqiao_fang@dlmu.edu.cn
2 Public Administration and Humanities College, Dalian Maritime University, Dalian 116026, China
3 School of Maritime Economics and Management, Dalian Maritime University, Dalian 116026, China; liuyang0120@dlmu.edu.cn (Y.L.); lh909@dlmu.edu.cn (H.L.)
* Correspondence: maxx1020@dlmu.edu.cn

**Abstract:** The safety level of the northern sea route (NSR) is a common concern for the related stakeholders. To address the risks triggered by disruptions initiating from the harsh environment and human factors, a comprehensive framework is proposed based on the perspective of resilience. Notably, the resilience of NSR is decomposed into three capacities, namely, the absorptive capacity, adaptive capacity, and restorative capacity. Moreover, the disruptions to the resilience are identified. Then, a Bayesian network (BN) model is established to quantify resilience, and the prior probabilities of parent nodes and conditional probability table for the network are obtained by fuzzy theory and expert elicitation. Finally, the developed Bayesian networkBN model is simulated to analyze the resilience level of the NSR by back propagation, sensitivity analysis, and information entropy analysis. The general interpretation of these analyses indicates that the emergency response, ice-breaking capacity, and rescue and anti-pollution facilities are the critical factors that contribute to the resilience of the NSR. Good knowledge of the absorptive capacity is the most effective way to reduce the uncertainty of NSR resilience. The present study provides a resilience perspective to understand the safety issues associated with the NSR, which can be seen as the main innovation of this work.

**Keywords:** northern sea route; resilience assessment; Bayesian network; fuzzy theory

**Citation:** Qiao, W.; Ma, X.; Liu, Y.; Lan, H. Resilience Assessment for the Northern Sea Route Based on a Fuzzy Bayesian Network. *Appl. Sci.* **2021**, *11*, 3619. https://doi.org/10.3390/app11083619

Academic Editor: José A. Orosa

Received: 22 March 2021
Accepted: 13 April 2021
Published: 16 April 2021



## 1. Introduction

Under the influence of global warming, a rapid decline in sea ice coverage and an increase in the depth of Arctic waters have been observed since 2000 Gascard et al., 2017 [1]. The potential commercial and political importance of arctic shipping routes has been the focus of various countries, shipping companies, and international organizations. Among the three shipping routes in Arctic waters, the arctic northeast route, also known as the northern sea route (NSR), is the most critical route because it greatly reduces the sailing distance between Asia, Europe, and North America compared with rational routes via the Suez [2]. According to [3], the reduced distance can be as much as 40%. As a result, shipping companies from Germany, China, Korea, Russia, Norway, and other countries have expressed interest in investing this route to obtain a competitive edge over competitors [4]. Although the opening and commercial operation of the NSR provide vast potential opportunities, Arctic waters still present major challenges for the shipping industry [5]. Harsh natural environments and fragile ecosystems make it difficult to maintain a satisfactory safety level for shipping operations along the NSR, which greatly increases the concerns of national and international parties. For example, the International Maritime Organization (IMO) issued the Polar Code in 2010 to ensure the safety level of shipping operations in Arctic waters [6]. In addition, the International Hydrographic Organization (IHO) has initiated a program to provide an accurate navigation chart for ships traveling along the NSR. Overall, the safety level of shipping operations within Arctic waters, particularly along the NSR, has garnered attention from various levels of companies at the national and international scales. However, the safety level of the NSR is still far from satisfactory, and accidents associated with shipping operations still occur at times. Therefore, a great deal of work has been done to better understand how to improve the security capacity and mitigate the various risks along the NSR. Therefore, in the present study, the concept of system resilience is introduced to attempt to obtain knowledge about the security capacity, and an investigation is conducted to improve the system resilience of the NSR.

### 1.1. Literature Review

The risk management of arctic shipping operations seeks to minimize the risk level. One of the most important issues is to identify the causal factors that contribute to risk or accidents. Therefore, many studies have been conducted to investigate the risk factors based on accidents and the unique natural environment in Arctic waters; such methods involve applying various theories, models, and techniques. Fault tree analysis was utilized by [7] to investigate the risk factors involved in chemical tanker accidents in Arctic waters. Later, an integrated methodology including fault tree analysis and the Human Factor Analysis and Classification System (HFACS) model was developed by [8] to study the collision risk during icebreaker assistance. Records of navigational shipping accidents that occurred in the Northern Baltic Sea during 2007 and 2013 were collected and investigated by [9], who found that ridged ice of a certain thickness cannot be considered the main cause of accident occurrence. Ref. [10] investigated the risk factors associated with Arctic maritime accidents from 1993 to 2011 by performing root cause analysis. Ref. [10] summarized various navigational risk contributors when sailing along an Arctic route, and the corresponding numerical weights were also given by using fuzzy analytic hierarchy process (AHP) and expert elicitation. The risk factors can also be identified by the application of Bayesian networks. Later, a novel Bayesian network (BN) model was proposed by [11] to identify the shipping operation risk in Arctic waters. Some studies based on the perspective of navigators have been performed to evaluate the shipping operation risk in Arctic waters, and such studies include those of [12,13], who studied the navigation risk in icy waters in the Baltic Sea. Ref. [6] investigated the key factors that influence the operation of arctic shipping routes.

Resilience is widely regarded as a specified capacity factor pertaining to an organization or a system that reflects how well the organization or system can recover from a disrupted or destroyed state. The concept of resilience was first introduced by [14], who argued that resilience can enable an ecosystem to continue to operate after being disturbed. Later, many studies tried to define resilience [15–17]; however, there is currently no universal consensus on the definition of resilience. Based on its various definitions, resilience is widely used in different fields [18–21] due to its contribution to the development of the relationship between capacity and a disruption or disaster. Many important applications of system resilience are subject to infrastructure systems, such as urban infrastructure systems [22], electrical infrastructure systems [23], inland waterway ports [24], and full-service deep-water ports [25]. In addition, organizational resilience was defined by [17] as the ability of enterprises and other organizations to improve performance. Similarly, in the social domain, the resilience capacity is regarded as "the ability of communities to cope with external stresses and disturbances as a result of social, political, and environmental change" [26]. In past decades, substantial differences existed among various descriptions of system resilience; however, the quantification of system resilience has always been a common concern. Generally, system resilience can be regarded as a time-dependent variable, as shown in Figure 1. Initially, the system is stable in the original state; then, with the appearance of a disruptive event, the performance of the system deteriorates gradually, which can trigger system resilience to recover the system performance to the normal level. According to [27], system resilience can be decomposed into absorptive capacity, adaptive capacity, and restorative capacity, which can absorb the shocks caused by a disruption, adapt the system to the new disrupted conditions, and restore the system to normal conditions as much as possible, respectively.

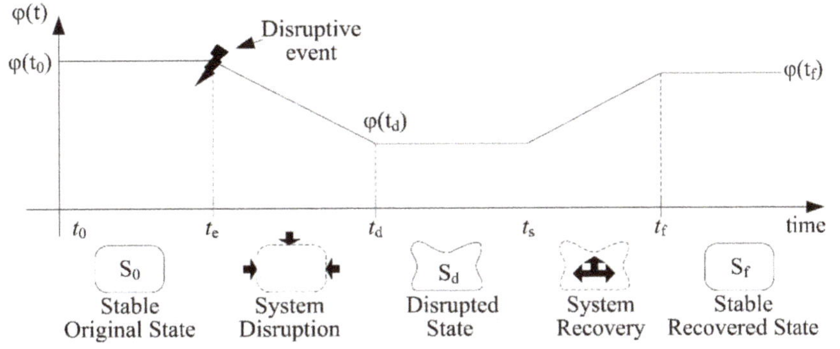


**Figure 1.** Resilience description associated with system performance and state transition [28].

Bayesian networks (BNs) are widely accepted as effective tools for conducting risk assessments and making decisions due to their notable advantage in easily updating information. Ref. [29] reviewed the applications of Bayesian inference for probabilistic risk assessment through 2007. Ref. [30] discussed different aspects of Bayesian network-based risk assessment. Actually, in engineering practice, BNs are usually integrated with other models, such as event trees [31], fault trees [32], bow-tie diagrams [33], HFACS [34], and so on. In addition to their successful application in risk assessment, BNs have been further utilized to evaluate the system performance and resilience capacity of specific systems. In the supplier selection BN model developed by [35], the resilience capacity was quantified in aspects of withstanding, adapting to, and recovering from a disruption. Moreover, [36] established a supply chain BN to forecast how risk factors affect the performance of the biomass supply chain network. Recently, in many studies, researchers have gradually explored the application of BNs in measuring system resilience. For instance, [37] established a comprehensive model for the resilience measurement of Systems of Systems (SoS) based on a BN.As a case study, the Littoral Combat Ship (LCS) was taken as an example, and conditional resilience metrics were proposed to assess the importance of individual subsystems. Later, [38] proposed a BN model to quantify the system resilience of infrastructure and took an inland waterway port as an example to demonstrate the application of the model. Then, a resilience measurement model based on a BN was further extended to explore aninterdependent electrical infrastructure system [23] and a full-service deep-water port [25].

### *1.2. Discussion of Existing Studies*

Based on the literature review, extensive efforts have been made by international organizations and the relevant authorities to maintain the safety level of the NSR. Many studies on the NSR have been conducted, and safety issues for shipping operations that move through Arctic waters are still the main concerns globally due to the unavailability and uncertainty of information associated with navigation along the NSR. Improving the capacity to safeguard various ships transiting Arctic waters, especially the NSR, is a great challenge for the commercial operation of the NSR. The lack of safeguard capacity evaluations considering the safety of shipping operations along the NSR is a current gap in the literature. To address this gap, the present study proposes a comprehensive assessment model involving a fuzzy BN and information entropy to quantitatively measure the resilience of the NSR.

The present study mainly aims to pioneer the establishment of an integrated evaluation model based on fuzzy BN information entropy theory to measure the resilience of the NSR safeguard system and investigate the influence of contributing factors on system resilience. Under the framework of resilience, the disruptions to the NSR safeguard system were first identified by a literature review and expert elicitation. Then, resilience decomposition was performed to identify the contributing factors for system resilience, which were mapped into the BN structure together with the factors related to disruptions as the parent nodes in the BN. Experts with navigation experience or academic experience associated with the NSR were employed as consultants to quantify variables by using fuzzy theory. Finally, we ran the developed BN with GeNIe 2.3 software and generated conclusions. The main features of the present study are summarized as follows:

- A system resilience analysis framework is explored to improve the safeguard capacity of the NSR in mitigatingvarious disruptions that affect safe transit along the NSR;
- A methodology is established to conduct a resilience assessment of the NSR based on the integration of fuzzy theory, a BN, and information entropy theory;
- The proposed fuzzy BN-based model can effectively cope with the uncertainty and unavailability of information associated with Arctic waters;
- The conduction of different types of analyses, such as forward and backward propagation, sensitivity, and information entropy analyses, helps us obtain a complete understanding of NSR resilience.

### *1.3. Organization*

The remainder of the paper is structured as follows. Section 2 illustrates an overview of the proposed resilience measurement methodology, the employed techniques, and the relevant theory. In Section 3, a model aimed at measuring the system resilience of the NSR is established based on the fuzzy Bayesian network. The results and discussion of the present study are given in Section 4, and finally, the conclusion is presented in Section 5.

## 2. Proposed Resilience Measurement Methodology

The resilience level of a specified system is widely regarded as the measurement of the interaction between system resilience and disruptions exerted on the system. Thereinafter, disruption identification and resilience capacity decomposition are essential for resilience measurement, based on which some quantification technology can be applied to calculate the resilience level and investigate the resilience characteristic. In the present study, a methodology that includes five steps based on a fuzzy Bayesian network is proposed to evaluate system resilience, as illustrated in Figure 2.

Step 1—Identify the disruptions exerted on the system. The first step is to identify the potential disruptions that impacts the performance of the NSR resilience. A potential disruption may be caused by a severe nautical environment, the failure of equipment or facilities, and navigational service outages; these issues will be analyzed in detail based on expert elicitation and a literature review.

Step 2—System resilience decomposition. This step will focus on the design of the resilience capacity structure that fully relates to the different stages of disruption identified in Step 1. In this step, the system resilience capacities are primarily decomposed into three capacities: absorptive capacity, adaptive capacity, and restorative capacity.

Step 3—Quantification of resilient variables. It is critical for the resilience measurement to quantify the identified disruption and relevant capacities. Therefore, in this step, a methodology associated with the combination of fuzzy theory and expert elicitation is applied based on the present data concerning the NSR resilience. Additionally, the noisy-OR function is used, and it can be regarded as the input for the subsequent Bayesian network.

Step 4—System resilience measurement. The resilience level of the Arctic NSR is calculated through the establishment of a Bayesian network in which the prior probabilities of the nodes without parent nodes and the conditional probability table (CPT) for the child nodes with multiple parent nodes are obtained according to Step 3.

Step 5—Result analysis. Different techniques are employed in this step to gain insights based on the developed Bayesian network in Step 4. The techniques include forward propagation, backward propagation, sensitivity analysis, and information uncertainty analysis.

Step 6—Proposed recommendation. Some potential recommendations are proposed according to the abovementioned steps to improve the resilience level of the Arctic NSR to defend against disruptions from the environment.

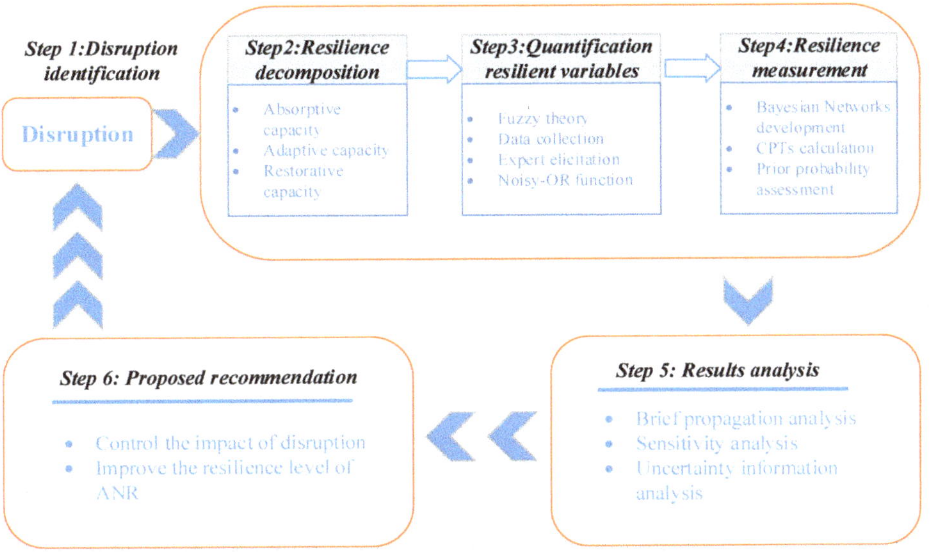


**Note:** conditional probability table (CPT); Arctic northern sea route (ANR)

**Figure 2.** Proposed resilience measurement model for the northern sea route (NSR).

## 2.1. Fuzzy Theory

### 2.1.1. Fuzzy Number Selection to Design a Questionnaire

Since first proposed by [39], fuzzy set theory has been widely introduced in the analytic hierarchy process (AHP) because experts prefer to express fuzzy judgements rather than crisp comparisons. According to fuzzy set theory, we can define a fuzzy number, $M$, in the set of real numbers, $R$, where $F(R)$ represents the fuzzy sets; then, $M \in F(R)$. The concept of a fuzzy number can be expressed as [40]:

$$\left.\begin{array}{ll} \mu_M(x_0) = 1 & x_0 \in R \\ A_\alpha = [x, \mu_{A_\alpha}(x) \geq a] & \alpha \in [0,1] \end{array}\right\} \tag{1}$$

where $\mu$ is the membership function and $\alpha$ is a random real number. Generally, triangular and trapezoidal fuzzy numbers have been widely applied for the fuzzy judgements of experts in varied case studies. According to the definition of a fuzzy number, the membership function of the triangular fuzzy number can be expressed as:

$$\mu_M(x) = \left\{\begin{array}{ll} \frac{x}{a_2-a_1} - \frac{a_1}{a-2a_1}, & x \in [a_1, a_2] \\ \frac{a_3}{a_3-a_2} - \frac{x}{a-2a_1}, & x \in [a_2, a_3] \\ 0 & otherwise \end{array}\right\} \tag{2}$$

where the triangular fuzzy number is defined as $M(a_1, a_2, a_3)$ for $a_1 \leq a_2 \leq a_3$; $a_1$ and $a_3$ are the lower and upper values of fuzzy number, $M$, respectively; and $a_2$ is the modal value.

For the limit case of $a_1 = a_2 = a_3$, $M$ is a nonfuzzy number. Similarly, for a trapezoidal fuzzy number, $N(a_1, a_2, a_3, a_4)$, the membership function is expressed as:

$$\mu_N(x) = \left\{ \begin{array}{ll} \frac{x}{a_2 - a_1} - \frac{a_1}{a - 2a_1}, & x \in [a_1, a_2] \\ 1 & x \in [a_2, a_3] \\ \frac{a_4}{a_4 - a_3} - \frac{x}{a - 4a_3}, & x \in [a_3, a_4] \\ 0 & otherwise \end{array} \right\} \tag{3}$$

To illustrate the operational rules of fuzzy numbers, two different triangular fuzzy numbers are taken as examples, $M_1(a_1, a_2, a_3)$ and $M_2(b_1, b_2, b_3)$, and the algebraic operation rules are:

$$M_1 \oplus M_2 = (a_1, a_2, a_3) \oplus (b_1, b_2, b_3) = (a_1 + b_1, a_2 + b_2, a_3 + b_3) \tag{4}$$

$$M_1 \otimes M_2 = (a_1, a_2, a_3) \otimes (b_1, b_2, b_3) = (a_1 b_1, a_2 b_2, a_3 b_3) \tag{5}$$

$$\lambda \odot M_1 = \lambda \odot (a_1, a_2, a_3) = (\lambda a_1, \lambda a_2, \lambda a_3) \quad \lambda > 0,\ \lambda \in R \tag{6}$$

$$M_1^{-1} = (a_1, a_2, a_3)^{-1} = (\frac{1}{a_1}, \frac{1}{a_2}, \frac{1}{a_3}) \tag{7}$$

Generally, the linguistic expressions of experts are meaningful for handling complex circumstances and eliciting meaningful advice. However, it is difficult to directly quantify the qualitative expressions of experts; therefore, the application of fuzzy numbers can solve this problem. Several attempts have been made to transfer qualitative linguistic expressions into corresponding fuzzy numbers [41–43], and the technique proposed by [43] is widely accepted; this approach includes eight scales associated with verbal expressions. Later, [44] suggested that the estimation of human memory aptitude should include seven terms, plus or minus two patches, indicating that the number of verbal terms should be between five and nine for good expert elicitation.

In the present study, transformation scale six, which includes five verbal expressions [45], was selected to design the expert questionnaire, and fuzzy trapezoidal numbers were selected to express expert judgements. The reason for selecting scale six was mainly that the experts preferred to give their judgements through five verbal expressions. In addition, the reliability of scale six has been verified by its successful application in various fields, such as the oil and gas [46], medicine [47] and marine [48] fields. Table 1 illustrates the corresponding relationship between linguistic expressions and fuzzy numbers based on scale six. According to Equation (3), the expression of the membership function of the trapezoidal fuzzy is illustrated in Figure 3.

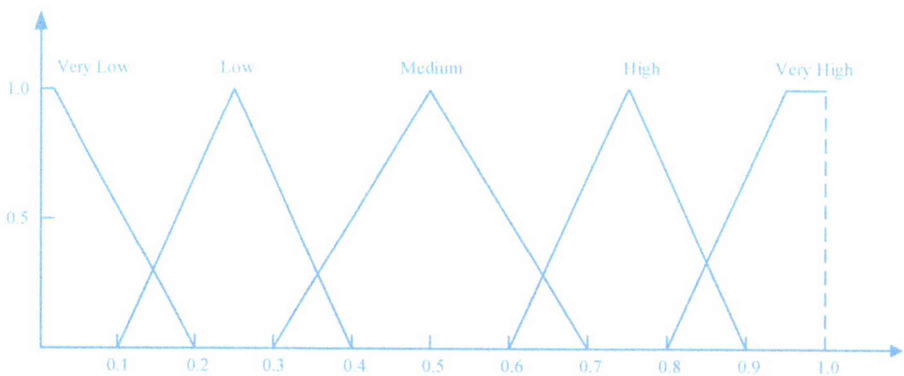


**Figure 3.** Scale six transformation [45].

**Table 1.** Linguistic expressions and their corresponding fuzzy numbers for scale six.

| Linguistic Expressions | Fuzzy Trapezoidal Numbers |
|---|---|
| Very Low (VL) | (0,0,0.1,0.2) |
| Low (L) | (0.1,0.25,0.25,0.4) |
| Medium (M) | (0.3,0.5,0.5,0.7) |
| High (H) | (0.6,0.75,0.75,0.9) |
| Very High (VH) | (0.8,0.9,1,1) |

#### 2.1.2. Weight Determination for the Expert Capacity

With different professional positions, experience levels, education levels, and so on, the experts employed gave various judgements regarding the issues in the present study; therefore, it is necessary to weigh individual expert capacities to value the judgements made by high-level experts more than those from low-level experts. Before evaluating the expert capacity, indicators representing the expert capacities should be designed. Then, a score rating for each indicator should be developed to quantify various capacities. Next, pairwise comparison matrices can be established based on expert information. These matrices can be further processed as follows until the weight of an individual expert is calculated [49].

- With respect to fuzzy numbers in pairwise comparison matrices, the geometric mean technique is applied to obtain the synthetic pairwise comparison matrix, $\widetilde{B} = \left[\widetilde{b}_{ij}\right]$, as follows:

$$\widetilde{b}_{ij} = \left(\widetilde{a^{(1)}}_{ij} \otimes \widetilde{a^{(2)}}_{ij} \otimes \cdots \otimes \widetilde{a^{(k)}}_{ij}\right)^{\frac{1}{4}} \quad (8)$$

  where $\widetilde{A^{(k)}} = \left[\widetilde{a^{(k)}}_{ij}\right]$ is the pairwise comparison matrix of the $K$thindicator for the expert capability evaluation.
- The fuzzy weights of the criteria for each expert can be calculated by the following equation:

$$\widetilde{r}_i = \left(\widetilde{b}_{i1} \otimes \widetilde{b}_{i2} \otimes \cdots \otimes \widetilde{b}_{in}\right)^{\frac{1}{n}} \quad (9)$$

  where the fuzzy weight of the $i$thexpert is indicated by $\widetilde{r}_i$.
- The fuzzy weights for each criterion are defined as follows:

$$\widetilde{w}_i = \widetilde{r}_i \otimes (\widetilde{r}_1 \oplus \widetilde{r}_2 \oplus \cdots \oplus \widetilde{r}_n)^{-1} \quad (10)$$

  where $\widetilde{w}_i(lw_i, mw_i, uw_i)$ denotes the fuzzy weight of the $i$thcriterion, for which $lw_i$, $mw_i$, $uw_i$ indicate the lower, middle, and upper values of the fuzzy weights of the $i$thcriterion, respectively.
- The weight of each expert is computed by employing the center of area technique, as follows:

$$P(E_i) = \left(\frac{1}{3}\right)[(uw_i - lw_i) + (mw_i - lw_i)] + lw_i \quad (11)$$

#### 2.1.3. Expert Viewpoint Aggregation

To eliminate any cognitive biases, it is necessary to aggregate expert opinions. Suppose that each expert, $E_i(i = 1, 2, 3, \ldots, n)$, states his or her personal viewpoints about a certain issue by utilizing a predefined set of linguistic expressions. These linguistic variables can then be transferred into corresponding triangular or trapezoidal fuzzy numbers, which can be processed further until defuzzification is achieved.

(1) Calculation of the degree of similarity. $S_{uv}(\widetilde{E}_u, \widetilde{E}_v)$ is defined as the degree of agreement for different opinions among experts. Suppose $\widetilde{E}_u(a_1, a_2, a_3)$ and $\widetilde{E}_v(b_1, b_2, b_3)$

represent two standard triangular fuzzy numbers ($u \neq v$); then, the degree of agreement between $\widetilde{E}_u$ and $\widetilde{E}_v$ can be obtained by:

$$S_{uv}(\widetilde{E}_u, \widetilde{E}_v) = 1 - \frac{1}{J}\sum_{i=1}^{J} |a_i - b_i| \quad i = 1,2,3 \tag{12}$$

where $J$ is the number of fuzzy set members; e.g., $J = 3$ is for standard triangular fuzzy numbers and $J = 4$ is for standard trapezoidal fuzzy numbers. Additionally, the greater the values of $S_{uv}(\widetilde{E}_u, \widetilde{E}_v)$ are, the greater the similarity between experts $E_u$ and $E_v$.

(2) Calculation of the average agreement (AA) degree for each expert viewpoint.

$$AA(E_u) = \frac{1}{U-1}\sum_{u \neq v, v=1}^{U} S_{uv}(\widetilde{E}_u, \widetilde{E}_v) \tag{13}$$

where $U$ is the total number of experts.

(3) Calculation of the relative agreement (RA) degree between two kinds of experts. The value of $RA(E_u)$ for the $u$thexpert can be obtained by:

$$RA(E_u) = \frac{AA(E_u)}{\sum_{u=1}^{U} AA(E_u)} \tag{14}$$

(4) Estimation of the consensus coefficient (CC) for each expert. The value of $CC(E_u)$ for the $u$thexpert can be obtained by:

$$CC(E_u) = \beta \times P(E_u) + (1-\beta) \times RA(E_u) \tag{15}$$

where the coefficient $\beta (0 \leq \beta \leq 1)$ is introduced to represent the importance of $P(E_u)$ over $RA(E_u)$, namely, the greater $\beta$ is, the more important $P(E_u)$ is. Actually, when $\beta = 0$, no weight is distributed to $P(E_u)$, indicating that a homogenous group of experts is employed; for another limit case, $\beta = 1$, the consensus degree among the various expert viewpoints is adequately high.

(5) Calculation of the aggregated results of the expert viewpoints. The aggregated results denoted by $\widetilde{R}_A$ can be computed by:

$$\widetilde{R}_A = CC(E_1) \otimes \widetilde{E}_1 \oplus CC(E_2) \otimes \widetilde{E}_2 \oplus \cdots \oplus CC(E_U) \otimes \widetilde{E}_U \tag{16}$$

(6) Defuzzification of the aggregated results. The defuzzification of fuzzy numbers is critically important for the application of fuzzy set theory. The center of area (CoA) method is widely used for the defuzzification operation, and it is expressed as:

$$X = \frac{\int \mu_M(x) x dx}{\int \mu_M(x) dx} \tag{17}$$

where $X$ represents the defuzzification result and $\mu_M(x)$ indicates the aggregated membership functions defined in (1) and (2) for fuzzy triangular and trapezoidal numbers, respectively. Therefore, the fuzzy numbers of the aggregated results, denoted as $\widetilde{R}_A(c_1, c_2, c_3)$ for fuzzy triangular numbers and $\widetilde{R}_A(c_1, c_2, c_3, c_4)$ for fuzzy trapezoidal numbers, can be defined by (18) and (19), respectively.

$$R_A = \frac{\int_{c_1}^{c_2} \frac{x-c_2}{c_2-c_1} x dx + \int_{c_2}^{c_3} \frac{c_3-x}{c_3-c_2} x dx}{\int_{c_1}^{c_2} \frac{x-c_2}{c_2-c_1} dx + \int_{c_2}^{c_3} \frac{c_3-x}{c_3-c_2} dx} = \frac{c_1+c_2+c_3}{3} \tag{18}$$

$$R_A = \frac{\int_{c_1}^{c_2} \frac{x-c_1}{c_2-c_1} x dx + \int_{c_2}^{c_3} x dx + \int_{c_3}^{c_4} \frac{c_4-x}{c_4-c_3} x dx}{\int_{c_1}^{c_2} \frac{x-c_1}{c_2-c_1} dx + \int_{c_2}^{c_3} dx + \int_{c_3}^{c_4} \frac{c_4-x}{c_4-c_3} dx} = \frac{1}{3} \frac{(c_4+c_3)^2 - (c_2+c_1)^2 - c_4c_3 + c_1c_2}{c_4+c_3-c_2-c_1} \quad (19)$$

### 2.2. Bayesian Network Theory

A Bayesian network is a multielement graphical model involving a set of variables and their conditional dependencies via a directed acyclic graph [50]. With the advantages of a flexible structure and probabilistic reasoning function, BNs are widely applied in the field of safety evaluation [32,51]. Series of nodes and edges are necessary for a basic BN to represent system variables and association rules, respectively. Each node in a BN is assigned a probability distribution, and each edge is affiliated with a direction, which is directed from parent node to child node. The nodes with multiple parent nodes should be assigned conditional probabilities that are included in the conditional probability tables (CPTs). The CPTs are based on the type and strength of the causal relationships between parent and child nodes.

Considering the conditional dependencies of the variables, a BN represents the joint probability distribution of a set of variables, $V = \{Z_1, \cdots, Z_n\}$, as in [52]:

$$P(V) = \prod_{i=1}^{n} P(Z_i|P_a(Z_i)) \quad (20)$$

where $P_a(Z_i)$ is the parent set of variable $Z_i$. The probability of variable $Z_i$ can be obtained by:

$$P(Z_i) = \sum_{Z_j, j \neq i} P(V) \quad (21)$$

Based on Bayes' theorem, a BN can update the prior probability of events under newly given observations, called evidence, to yield the posterior probability. The posterior probability can be calculated by:

$$P(V \backslash E) = \frac{P(V,E)}{P(E)} = \frac{P(V,E)}{\sum_V P(V,E)} \quad (22)$$

where $P(V \backslash E)$ is the desired posterior probability and $E$ represents the evidence.

#### 2.2.1. Prior Probability Calculation for Nodes without Parents

The prior probability for nodes without parents in the Bayesian network can be quantified by the integration of fuzzy theory and the aforementioned expert elicitation. The calculation can be conducted following the steps described below.

Step 1—The questionnaire is designed to obtain expert viewpoints according to the approach described in Section 2.1.1.

Step 2—The capacities of experts employed in the present study are weighted and scored to compensate for any cognitive biases and aggregate expert opinion based on the calculation process mentioned in Section 2.1.2.

Step 3—The fuzzy aggregation of expert opinions obtained from Step 1 is implemented by applying the calculation methods expressed in Section 2.1.3.

#### 2.2.2. Conditional Probability Table Calculation with the Noisy-OR Function

The calculation of the CPT is crucial for reliable model inference. However, the CPT is generally regarded as complicated, especially for the case of a large BN with many nodes. Therefore, many studies have focused on CPT calculations, of which expert elicitation associated with fuzzy theory has been widely employed [34,53,54]. Based on expert elicitation, the Noisy-OR model [55] is utilized to calculate the CPT with the following assumptions (suppose a child node $Y$ has $n$ parents, $X_1$, $X_2$, ..., $X_i$, ..., $X_n$, as shown in Figure 4a):

(1) All the nodes in the proposed Bayesian network can be regarded as Boolean variables; that is, the nodes have binary states, true or false, representing positive or negative outcomes, respectively;
(2) The causes (parent nodes) of $X_1$, $X_2$, $X_3$, ... $X_n$ are mutually independent;
(3) The probability that $Y$ is true when only one causal factor, $X_i$, is true while all other factors except $X_i$ are false can be expressed as:

$$P_i = P(Y|\overline{X_1}, \overline{X_2}, \ldots, X_i, \ldots, \overline{X_n}) \tag{23}$$

where $P_i$ represents the connection probability between the parent node, $X_i$, and the corresponding child node, $Y$. The connection probability can be derived by:

$$p_i = \frac{p(Y|X_i) - p(Y|\overline{X_i})}{1 - p(Y|\overline{X_i})} \tag{24}$$

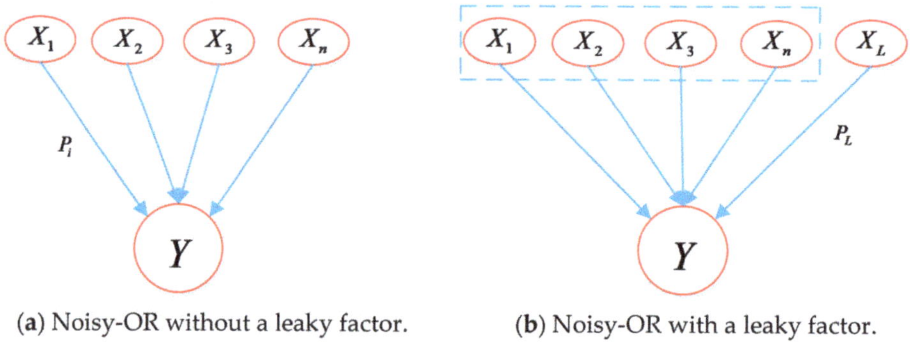


**Figure 4.** Typical Noisy-OR model.

Suppose every causal factor, $X_i$, in Figure 4a is a member of the set $\{X_p\}$ and all the causal factors with a "true" state label belong to $\{X_T\}$. Additionally, the causal factors with a "false" label are members of $\{X_F\}$. Then, the CPT of $Y$ can be obtained by

$$p(Y|X_p) = 1 - \prod_{X_i \in X_T} (1 - p_i) \tag{25}$$

For the limit case in which $\{X_T\}$ is a null set, that is, the states of all the causal factors are "false", then $p(Y|X_p) = 0$. However, it is almost impossible for the probability, $p(Y|X_p)$, to equal zero. In practice, the subsequent event, $Y$, may occur even if all the causal factors are "false" because some unknown causal factors may still exist beyond the identified causal factors. Therefore, the leaky Noisy-OR model (illustrated in Figure 4b) is introduced to handle the unknown causal factors; in this case, the conditional probability of Y can be calculated by:

$$p(Y|X_p) = 1 - \prod_{X_i \in X_T} (1 - p_i)(1 - p_L) \tag{26}$$

where $P_L$ represents the connection probability between $Y$ and the leaky causal factors $X_L$.

### 2.3. *Information Entropy Theory*

Currently, the Arctic NSR is not a well-developed commercial waterway; as a result, some information associated with the NSR may be absent, which makes it difficult to quantify the uncertainty of the unknown information. To improve the reliability of the underlying information, information entropy theory is applied in the present study. In the framework of information entropy, the uncertainty of a variable can be measured by a probability distribution. Suppose $X$ is a discrete random variable with the probability distribution $P$, where $P_i$ represents the probability of $x_i \in X$. Then, the information pertaining to a specific value, $x_i$, for this random variable can be defined as:

$$I(x_i) = -\log(P_i) \tag{27}$$

where the base in the logarithm is usually taken as 2 ore; in the present study, the base is 2.

The information function theoretically controls the volume of information conveyed by the specific state, such as $x_i$. The information entropy function can be defined as the expectation of the information, which is expressed as:

$$H(X) = \mathrm{E}[I(X)] \tag{28}$$

The information entropy function can be further processed mathematically as:

$$H(X) = \sum_{i=1} P_i \cdot I(x_i) = -\sum_{i=1} P_i \cdot \log_2(P_i) \tag{29}$$

For the discrete random variable, X, the Function (29) can be expressed as:

$$H(X) = -\sum_{x \in X} P(x) \log_2 P(x) \tag{30}$$

where $0 \le P(x) \le 1$ and $\sum_{x \in X} P(x) = 1$. The higher the value of the information entropy is, the more uncertainty the variable contains. The entropy can reach the maximum when the probability distribution is uniform, i.e., $P(x)$ is equal for all $x$.

Suppose $X$ is a target variable and $Y$ is a contributing factor, similar to the relationship between the child node and parent node in the Bayesian network. If the target variable $X$ is conditionally dependent on the dependent variable $Y$, then we have:

$$H(Y|X) = \sum_i P(Y_i) \cdot H(Y_i|X_i) \tag{31}$$

where $i$ denotes the number of states, $i = 2$ for a Boolean variable, and $H(Y|X)$ represents the conditional entropy of target $X$ given the contributing factor $Y$. Then, the mutual information between the target variable $X$ and the contributing factor $Y$ can be defined as:

$$I(X,Y) = H(X) - H(Y|X) \tag{32}$$

## 3. Model Established for NSR Resilience Measurement

### *3.1. Scenario Development*

The Arctic NSR is characterized by a special nautical environment and unique location compared to other national and international waterways. Traditionally, the NSR depicted in Figure 5 is made up of five legs: the Barents Sea, the Kara Sea, the Laptev Sea, the East Silerian Sea, and the Chukchi Sea. The presence of sea ice, including drifting icebergs, has made this route almost inaccessible for marine transportation [56]. The area of ice-covered waters within the NSR is shown in Table 2. It is widely accepted that the navigating period for the NSR can be continuous from September to March of the following year.

**Figure 5.** The Arctic Northeast Route [11].

**Table 2.** Average ice-covered waters in the marginal seas of the NSR region [57].

| Sea | March (Million km$^2$) | September (Million km$^2$) | Seasonal Changes |
|---|---|---|---|
| Barents Sea | 0.855 | 0.128 | 85% |
| Kara Sea | 0.830 | 0.266 | 68% |
| Laptev Sea | 0.536 | 0.196 | 63% |
| East Silerian Sea | 0.770 | 0.516 | 33% |
| Chukchi Sea | 0.595 | 0.196 | 67% |

With the difficulties associated with communication and data acquisition in the NSR region, the safety issues and environmental risks related to NSR transport are traditionally managed based on empirically determined rules and regulations, which can be useful for shipping activities characterized by low traffic density with small ships [58]. However, the ships sailing along the NSR are becoming increasingly large in size, and there is a lack of relevant empirical data on which mitigation of the hazards with respect to the NSR can be based [59]. Therefore, safety issues concerning the NSR are an urgent problem that must be solved.

### 3.2. Disruption Identification

A disruption of the system resilience, also known as a potential threat, can be beneficial for the improvement of the resilience level of the system. Only if a disruption is identified effectively can the resilience be quantitatively measured; however, the disruptions are different from risks and causal factors contributing to accidents. Actually, only part of a disruption is able to ultimately cause accidents, with evolution to risk and causal factors. Therefore, it is difficult to identify all disruptions practically. In the present study, the disruptions identified from safety meetings, accident reports, interviews, and literature reviews are listed in Table 3.

**Table 3.** Descriptions of the disruptions to the NSR resilience.

| No. | Disruption | Description |
|---|---|---|
| 1 | Weather forecast inaccuracy | The weather along the Arctic Northeast Route is complex and variable, making it difficult to predict; as a result, the NSR security system suffers from information uncertainty associated with weather forecasts [60] |
| 2 | Sea chart incomplete | The navigation chart for the NSR is still incomplete due to complex factors such as the geological conditions, lack of hydrographic ship information, and political intervention [61] |
| 3 | Communication/positioning unavailable | According to the database for maritime accidents (DAMA of Det Norske Veritas), missing safety instructions and defective communication can impact the safety level of navigation in Arctic waters [62] |
| 4 | Malfunction of power plant | Based on the comments from experts of the Canadian Transport Agency, the malfunction of power plants, such as engine failure and power and back-up power failures, are the primary causes of ship collisions, foundering, and grounding along the NSR [11] |
| 5 | Damaged propeller/steering gear | The propeller and steering gears of ships that use the NSR can be severely damaged by icebergs, which cause considerable disruptions to the NSR safety [60] |
| 6 | Malfunction of deck machinery | In the case of extremely low temperatures, some deck machinery may malfunction, thus impacting normal ship operation along the NSR [61] |
| 7 | Restricted function of nav. instruments | The navigation instruments on ships may not work properly due to the influence of high latitudes, which can put the ship at risk [63] |
| 8 | Low temperature | Low temperatures can easily affect the performance of security-related equipment such as the hull, windlass, and mooring winch [64] |
| 9 | Iceberg/floating ice | The existence of sea ice, such as icebergs, floating ice, and old ice, is the main feature that distinguishes the NSR from other sea lanes globally; the impact of ice on the safety of ships is continuous and inevitable [65] |
| 10 | Poor visibility | Poor visibility caused by steam fog, ice fog, blowing snow, and other processes along the NSR is frequently encountered and limits the watchkeeping of navigation officers [64] |
| 11 | Rough sea | Most of the currents in the NSR are along the coast of a shallow-water continental shelf, and the currents in the narrow straits between the islands are strong, thus creating challenges for polar navigation [66] |
| 12 | Magnetic storm | Magnetic storms can greatly disturb the NSR security system in the field of communication and influence navigation instruments [35] |
| 13 | Obstacles other than ice | Underwater obstacles such as reefs, beaches, and unknown explosives are potential disruptions that threaten the effectiveness of the NSR precautions [67] |
| 14 | Seafarer competency | Seafarers on ships sailing along the NSR are faced with considerable ship handling and emergency challenges, which can potentially affect the function of the NSR precautions [61] |
| 15 | Geopolitics | The ships sailing along the NSR have to consider different laws and regulations, including those at the local, national, and international level; additionally, sometimes political considerations are involved, making the decision-making process very complex. [68] |

As shown in Table 3, 15 kinds of disruptions were identified in the present study. Subsequently, a safety meeting was conducted. After consulting experts, such as experienced mariners, senior scholars, and safety managers associated with the Arctic NSR, all disruptions were finally divided into four groups: navigation services out of order ($N = 3$), malfunction/failure of machinery ($N = 4$), harsh nautical environment ($N = 5$), and other disruptions ($N = 3$). The classification results are presented in Figure 6. Furthermore, classifying all the disruption factors into different groups can facilitate the calculations for the CPT of the Bayesian network in Section 3.4.

### 3.3. *Resilience Capacity Decomposition for the Arctic Northeast Route*

The resilience capacity can be defined as the capacity of a certain system to absorb, adapt to, and recover from any shock due to a disruption [23]. It is widely accepted that resilience capacities are classified as the absorptive capacity, adaptive capacity, and restorative capacity for a specified system [25,27]. In the present study, the underlying factors pertaining to these three kinds of capacities for the Arctic NSR resilience are identified, as illustrated in Figure 7. All the factors presented in Figure 7 will be included in the developed BN.

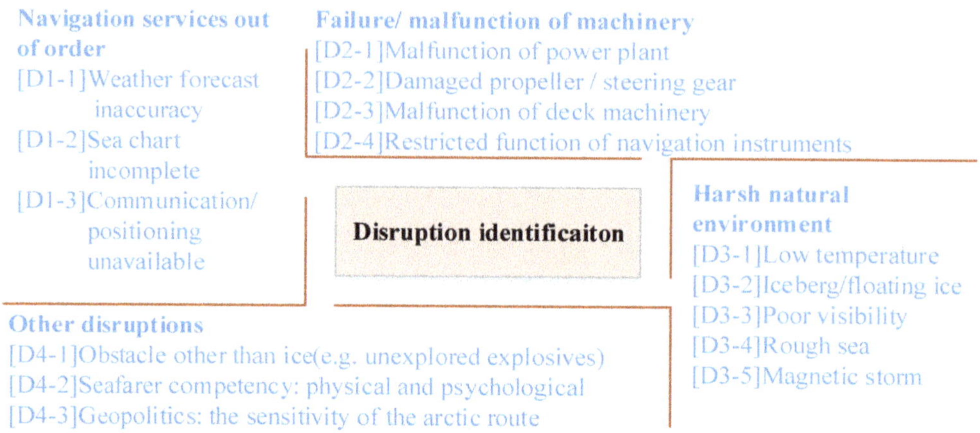


**Figure 6.** Disruptions to safe navigation along the NSR.

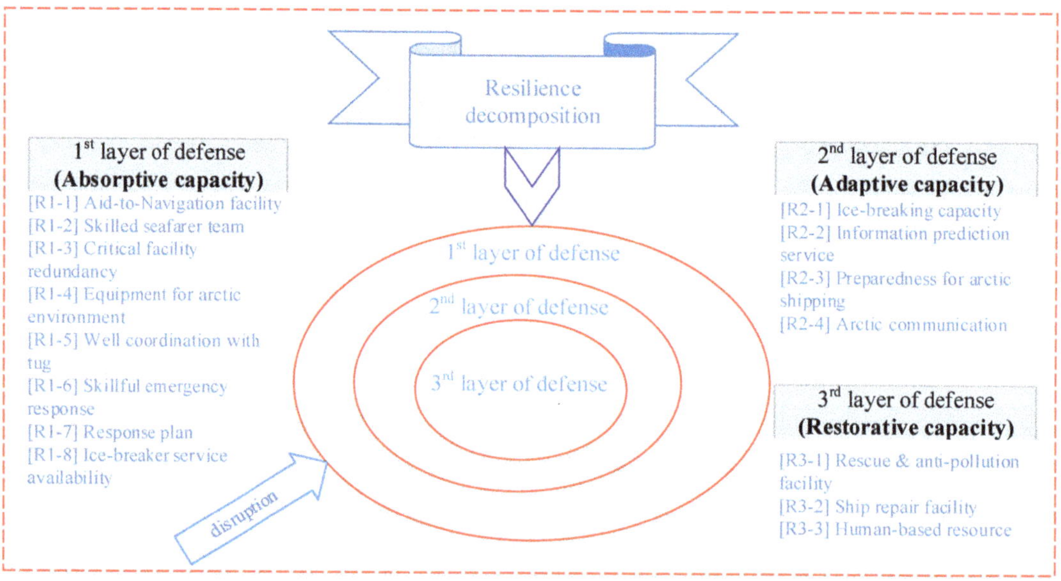


**Figure 7.** Resilience decomposition designed for the NSR.

### 3.3.1. Absorptive Capacity

The absorptive capacity can be defined as the extent to which the Arctic Northeast Route security system is able to automatically defend against shocks due to disruptions. As shown in Figure 6, the absorptive capacity is considered the first line of defense to resist the adverse impact of the disruption. Inspired by the "inherent resilience" theory proposed by [69], the absorptive capacity can be further divided into the defense capacity under ordinary circumstances and the emergency response capacity under crisis conditions. Finally, a list of seven contributing factors for the absorptive capacity of the Arctic Northeast Route security system was obtained, and it is presented in Table 4.

**Table 4.** Absorptive capacity of NSR resilience.

| Features | Description |
|---|---|
| Aid-to-navigation facility | The aid-to-navigation (A-to-N) facilities, including visual and audible A-to-N facilities, racons, radar marks, and shore-based Automatic Identification System (AIS) stations, are critical for navigating safely along the NSR. There are approximately 1240 coastal visual signs and 300 floating markers associated with the NSR [68]. |
| Skilled seafarer team | Compared with conventional shipping route, the NSR is characterized by lots of special risks. The skilled seafarer team can swiftly evaluate the situation and take effective countermeasures with teamwork to deal with potential risks. In addition, a skilled seafarer team can develop a harmonious atmosphere in which everyone can limit the defense risk. |
| Critical facility redundancy | The redundant critical facilities, such as those that provide emergency power backup, high power conservation, extra positioning techniques, and backup navigation instruments, are able to strengthen the robustness of the security system and restrict the consequences of disruptions. |
| Equipment for arctic environment | The equipment designed for the arctic environment can effectively defend against or absorb the risks caused by weather condition, such as low temperatures, frost, and moisture, especially the equipment allocated on deck, including cranes, mooring winches, and windlasses, which are exposed to the external environment. |
| Coordination with icebreaker | Icebreaker assistance operations play an essential role in ice-covered waters to reduce the risk of accidents, such as ice collisions and propeller or rudder damage. In addition, the case of trapped vessels by ice can also be avoided, and the disruption introduced by large amounts of floating ice can be absorbed [8]. |
| Ice pilots/navigators | The assistance provided by ice pilots or navigators can defend against or attenuate various risks caused by disruptions; this approach is suitable for the case in which a shipmaster has little experience navigating in ice along the NSR. The organizations providing ice pilot services can be obtained from the NSRA [68]. |
| Skilful emergency response | Under unpredictable situations related to heavy fog, floating ice, and strong wind in the NSR region, a proficient emergency response group can establish counteractive measures to response to the threats. Besides, a skilled seafarer can fully utilize the available resources, which are essential for mitigating various risks [27]. |
| Response plan | A detailed response plan corresponding to various predictable disruptions and risks encountered along the NSR is useful for guiding seafarers or operators to take appropriate actions during disruptions, thus keeping the disruption or risk controllable. |
| Navigational publications | Navigational publications refer to charts and other navigational publications, such as those used for guidance in arctic navigation, lists of radio signals, and notices to mariners; ice charts, which are useful for monitoring the risk related to sea ice, are particularly important [67,70] |

#### 3.3.2. Adaptive Capacity

The adaptive capacity can be defined as the level to which a system self-organizes itself and takes dynamic measures to restrict and control the severe impact of a disruption. Compared with the absorptive capacity, the adaptive capacity requires flexible and dynamic efforts [71] and mitigates a disruption as a "second line of defense", as presented in Figure 6. It is also regarded as part of the "post-accident strategy". The features contributing to the adaptive capacity of the NSR resilience are described in Table 5.

**Table 5.** Adaptive capacity of NSR resilience.

| Features | Description |
|---|---|
| Ice-breaking capacity | The ice-breaking capacity allows the vessels to adapt to a risk or disruption caused by floating ice in ice-covered waters along the NSR. The ice-breaking capacity can be improved by strengthening the hull to bear ice loads for safe navigation in ice fields [70]. |
| Information prediction services | Information prediction services include those for sea ice coverage, thickness, and motion, as well as weather conditions [1]. Under the conditions of predictable information, in the case of a disruption, seafarers and operators can take action in advance, for example, by anchoring to avoid a disruption or risk, changing the designed route, or requesting assistance from an icebreaker. |

Table 5. Cont.

| Features | Description |
| --- | --- |
| Preparedness for arctic shipping | Before navigating into ice-covered waters, vessels must be fully prepared, including obtaining extra fuel reserves, psychological preparation, understanding all the regulations that must be observed, preparing for low temperatures, etc.; in case of a disruption, the vessel can make some changes based on their preparations to adapt to the new situation caused by the disruption [70]. |
| Arctic communication | Arctic communication, including radar, radio, and International Maritime Satellite Organization (INMARSAT) communication, can coordinate operations from ship to ship and ship to icebreaker to adapt to the harsh environment and sailing conditions along the NSR [67]; additionally, Arctic communication helps vessels in the NSR region get assistance and guidance during disruptions, thus making risks controllable. |

#### 3.3.3. Restorative Capacity

The restorative capacity can be defined as the extent to which a specified system is able to recover from or repair a destroyed or failed function due to the impact of a disruption. Based on Figure 6, the restorative capacity acts as the last line of defense for the system to resist a disruption; that is, in the case that the absorptive and adaptive capacities fail to tolerate the impact caused by a disruption, the restorative capacity remains. Generally, it takes a relatively long time for the restorative capacity to recover the normal function of the system, and various human resources, services, tools, and so on are essential. Three aspects of the restorative capacity pertaining to the Northeast Route security system are described in Table 6.

Table 6. Restorative capacity of NSR resilience.

| Features | Description |
| --- | --- |
| Rescue and anti-pollution facility | Rescue and anti-pollution facilities are utilized to restore the shipping capacity of the NSR after natural disasters and accidents. Currently, there are three rescue and research centers established temporally along the NSR from July to October. In addition, the rescue and anti-pollution capacities of the ports (20 approximately) along the NSR need to be improved for the quick restoration of the damaged route. |
| Ship repair facility | Ship repair facilities, such as yards, docks, gate operations, cranes, and warehouses, are essential for vessels that experience hull damage, machinery malfunctions, and propeller damage. Notably, the restorative capacity aimed at damaged ship repair will be limited if the stakeholders have little interest in investment. |
| Human-based resources | The restorative capacity associated with human-based resources includes service restoration and technology restoration, which are substantial parts of the post-disaster strategy. Restoration may include communication, navigation, pilot service, ice-breaker assistance, and weather and hydrology information systems. |

### 3.4. *Resilience Measurement Using the Fuzzy Bayesian Network*

In this section, a Bayesian network is established to quantify the resilience of the Arctic Northeast Route security system. Based on the identified disruptions and the resilience capacity decomposition in Sections 3.2 and 3.3, the visual structure of the proposed Bayesian Network described in Figure 8 is developed using GeNIe 2.3 software.

#### 3.4.1. Reliability Quantification for the Employed Experts

Expert elicitation is useful and valuable in most situations when available resources are lacking or limited by physical circumstances [72], and the Arctic NSR is one of these cases. However, the competence of the employed experts is essential for scientific conclusions. According to the study of [34], the selection criteria used for capable experts in this research can be described as follows:

- A heterogeneous group of experts is usually preferred to a homogenous group. In a heterogeneous group, the individual experience of each expert receives considerable attention;
- With respect to the education and experience of the experts in a field, the longer they have focused on a subject (academic or practical subject), the more accurate their intuitionistic judgement is;
- With respect to expert familiarity with a subject, especially through practical experience, an experienced specialist can theoretically master every detail of the subject.

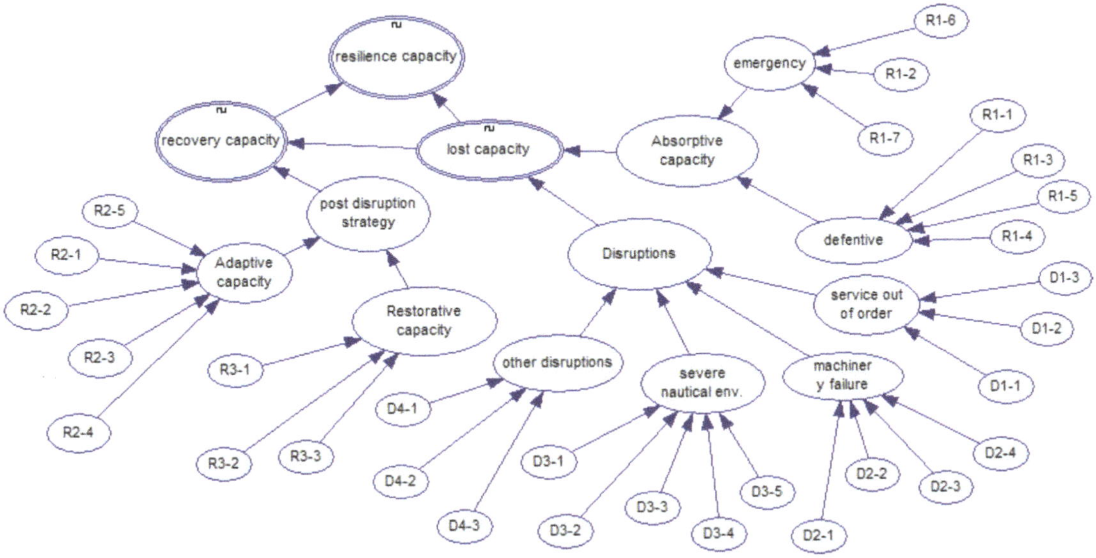


**Figure 8.** Bayesian network (BN) established for measuring the resilience of the NSR.

Based on the abovementioned criteria, a heterogeneous group of five experts (their information is listed in Table 7) was developed to comment on the various disruption and resilience capacities associated with the Arctic NSR.

**Table 7.** Details of the experts.

| Item | Age | Occupation | Educational Level | Certificate Rank | Job Tenure |
|---|---|---|---|---|---|
| Expert 1 (E1) | 53 | Senior seafarer | Bachelors of navigation | Senior Captain | He has been working on board a ship for nearly 25 years; as a senior captain, he sailed the Arctic Northeast Route recently. |
| Expert 2 (E2) | 50 | Senior seafarer | Bachelors of navigation | 2nd Officer | He has been working on board a ship for nearly 15 years; currently, he is certified as a Chief Officer working on a ship capable of navigating the NSR. |
| Expert 3 (E3) | 48 | Professor | Ph.D. of navigation | Senior Captain | He has been working on board ships since 1991 and obtained the certificate of senior captain; currently, he is an associate professor focusing on Arctic sea transport. |

**Table 7.** *Cont.*

| Item | Age | Occupation | Educational Level | Certificate Rank | Job Tenure |
|---|---|---|---|---|---|
| Expert 4 (E4) | 41 | Associate professor | Masters of marine engineering | Chief engineer | He has been working on ships since 2001, beginning as a cadet and eventually becoming a chief engineer; currently, he is an associate professor focusing on risk assessments of Arctic transport. |
| Expert 5 (E5) | 43 | Safety manager | Masters of navigation technology | Chief officer | He has been working on board ships since 2001, eventually becoming a chief officer; he is familiar with marine operation and equipment management for ships transiting Arctic waters. |

The weight for each expert was assessed based on their education level, expertise, professional position, work experience, and age [46,73–75]. In addition, the competency certificate level of the selected experts was also regarded as an important indicator. Finally, the rating criteria were developed, as presented in Table 8.

**Table 8.** Scores of indicators for expert evaluation.

| Indicator | Classification | Score | Indicator | Classification | Score |
|---|---|---|---|---|---|
| Professional position | Senior academic<br>Junior academic<br>Engineer<br>Technician<br>Worker | 5<br>4<br>3<br>2<br>1 | Experience (2) | 6–9<br>≤5 | 2<br>1 |
| | | | Education level | Ph.D.<br>Masters<br>B.S. or B.E.<br>Junior college<br>School level | 5<br>4<br>3<br>2<br>1 |
| Age | ≥50<br>40–49<br>30–39<br>≤30 | 4<br>3<br>2<br>1 | Certificate rank | Senior Cap. or C/E<br>Cap. or C/E<br>C/O or 2/E<br>Operational Officer/engineer<br>ratings | 5<br>4<br>3<br>2<br>1 |
| Experience (1) | ≥30 years<br>20–29<br>10–19 | 5<br>4<br>3 | | | |

Based on the information from the employed experts shown in Table 7, the capacity was first scored by every individual expert according to the rating criteria presented in Table 8; then, pairwise comparison matrices were derived. Finally, the weight for each expert was obtained by applying Formulas (8)–(11), as summarized in Table 9.

**Table 9.** Evaluation results of expert capabilities.

| Expert | Position | Experience | Education | Age | Certificate | Weight |
|---|---|---|---|---|---|---|
| Expert 1 | Senior seafarer | 30 | Bachelors | 53 | Senior Captain | 0.189 |
| Expert 2 | Senior seafarer | 27 | Bachelors | 50 | 2nd Officer | 0.121 |
| Expert 3 | Professor | 18 | Ph.D. | 48 | Senior Captain | 0.300 |
| Expert 4 | Associate professor | 22 | Masters | 41 | Chief engineer | 0.206 |
| Expert 5 | Safety manager | 8 | Masters | 43 | Chief officer | 0.184 |

### 3.4.2. Calculation of the Prior Probabilities of the Nodes without Parent Nodes

Based on the qualitative analysis discussed in Sections 3.2 and 3.3, there are a total of 30 parent nodes included in the established Bayesian network illustrated in Figure 5. The prior probabilities of these parent nodes can be calculated according to the calculation procedures mentioned in Section 2.2.1. The experts involved in this process are considered independently during decision making. The results of the aggregation and defuzzification of expert judgements are presented in Table 10.

**Table 10.** Aggregation results for each Boolean variable based on expert elicitation.

| Item | Aggregated Value | Item | Aggregated Value | Item | Aggregated Value |
|---|---|---|---|---|---|
| D1-1 | 0.6528 | D3-4 | 0.5325 | R1-6 | 0.67 |
| D1-2 | 0.4615 | D3-5 | 0.5445 | R1-7 | 0.6198 |
| D1-3 | 0.7225 | D4-1 | 0.4355 | R2-1 | 0.5445 |
| D2-1 | 0.7084 | D4-2 | 0.7807 | R2-2 | 0.7085 |
| D2-2 | 0.6036 | D4-3 | 0.6438 | R2-3 | 0.5021 |
| D2-3 | 0.7338 | R1-1 | 0.5887 | R2-4 | 0.6635 |
| D2-4 | 0.7138 | R1-2 | 0.8142 | R2-5 | 0.8594 |
| D3-1 | 0.6468 | R1-3 | 0.4994 | R3-1 | 0.7897 |
| D3-2 | 0.7818 | R1-4 | 0.5079 | R3-2 | 0.6083 |
| D3-3 | 0.6641 | R1-5 | 0.7892 | R3-3 | 0.6097 |

3.4.3. Calculation of the Condition Probability Table for the Network

To obtain the conditional probability table for the nodes with multiple parents, the leaky Noisy-OR function is applied considering that the states of child nodes may be false even if all the parent nodes are true. Based on the calculated connection probability, $P_i$, for each parent node, the Leaky Noisy-OR function can be expressed as:

$$NoisyOR(P_1,\ X_1, P_2,\ X_2, \ldots, P_n,\ X_n, l) \tag{33}$$

where $l$ represents the leaky factor, indicating the degree to which the missing causal factors contribute to a consequence being true. Generally, the leaky factor is a nonzero probability, even if all the causal factors are false. Therefore, the leaky factor represents a probability that implies that the child node will be true if the parent nodes are false, as expressed below.

$$l = P(Y = true|X_1 = false, X_2 = false, \ldots, X_n = false) \neq 0 \tag{34}$$

In the present study, five experts with attributes described in Section 3.4.1 were employed to give their judgements regarding the conditional probabilities of nodes with multiple parent nodes in the form of linguistic expressions. The expert insights were further processed based on the fuzzy methodology proposed in Section 2.1, after which the conditional probabilities for child nodes can be expressed by the Noisy-OR function as follows:

$$P(Nav.\ service\ out\ of\ order) = Noisy-OR([D1-1], 0.6277, [D1-2], 0.5870, [D1-3], 0.3626, 0.05)$$

$$P(equipment\ failure) = Noisy-OR([D2-1], 0.49, [D2-2], 0.26, [D2-3], 0.80, [D2-4], 0.58, 0.032)$$

$$P(severe\ nautical\ env.) = NoisyOR([D3-1], 0.43, [D3-2], 0.38, [D3-3], 0.87, [D3-4], 0.61, [D3-5], 0.39, 0.011)$$
$$P(other\ disruptions) = NoisyOR([D4-1], 0.24, [D4-2], 0.57, [D4-3], 0.33, 0.218)$$
$$P(restorative\ capacity) = NoisyOR([R3-1], 0.84, [R3-2], 0.68, [R3-3], 0.69, 0.016)$$
$$P(adaptive\ capacity) = NoisyOR([R2-1], 0.94, [R2-2], 0.88, [R2-3], 0.17, [R2-4], 0.71, [R2-5], 0.92, 0.02)$$
$$P(defentive\ capacity) = NoisyOR([R1-1], 0.41, [R1-3], 0.57, [R1-5], 0.21, [R1-6], 0.96, 0.009)$$
$$P(emergency\ capacity) = NoisyOR([R1-2], 0.90, [R1-7], 0.89, [R1-8], 0.67, 0.04)$$

The conditional probability table for the proposed Bayesian network can be induced by combining the Noisy-OR function of child nodes with Equation (26). To concisely represent the computing process for the CPT, the node "service out of order" ($Y$) with three parent nodes ($X_1$, $X_2$, $X_3$) was taken as an example, and the exhaustive information is given in Table 11.

**Table 11.** Calculation process for the conditional probabilities of the node "nav. service out of order".

| Expert Judgement | | | | | |
|---|---|---|---|---|---|
| $Y \leftarrow X_1$: | M | H | H | M | H |
| $Y \leftarrow X_2$: | M | L | L | M | VH |
| $Y \leftarrow X_3$: | VH | H | M | H | H |
| Aggregation of expert judgement (the same method as adopted for the aggregated value in Table 10) | | | | | |
| $P(Y \leftarrow X_1) = 0.6528$ $P(Y \leftarrow X_2) = 0.4615$ $P(Y \leftarrow X_3) = 0.7225$ | | | | | |
| Conditional probability calculation by the Noisy-OR model | | | | | |
| $P(Y \leftarrow X_1, X_2) = 1 - (1 - P(Y \leftarrow X_1)) * (1 - P(Y \leftarrow X_2)) = 0.8130$ | | | | | |
| $P(Y \leftarrow X_1, X_3) = 1 - (1 - P(Y \leftarrow X_1)) * (1 - P(Y \leftarrow X_3)) = 0.9037$ | | | | | |
| $P(Y \leftarrow X_2, X_3) = 1 - (1 - P(Y \leftarrow X_2)) * (1 - P(Y \leftarrow X_3)) = 0.8506$ | | | | | |
| …… | | | | | |
| Conditional probability table for the node "nav. Service out of order" ($Y$) | | | | | |
| $P(Y \leftarrow X_1, \overline{X_2}, \overline{X_3}) = 0.6528$ $P(Y \leftarrow X_1, X_2, \overline{X_3}) = 0.8130$ | | | | | |
| $P(Y \leftarrow X_2, \overline{X_1}, \overline{X_3}) = 0.4615$ $P(Y \leftarrow X_1, \overline{X_2}, X_3) = 0.9037$ | | | | | |
| $P(Y \leftarrow X_3, \overline{X_1}, \overline{X_2}) = 0.7225$ $P(Y \leftarrow \overline{X_1}, X_2, X_3) = 0.8506$ | | | | | |
| $P(Y \leftarrow \overline{X_1}, \overline{X_2}, \overline{X_3}) = 0.0519$ $P(Y \leftarrow X_1, X_2, X_3) = 0.9481$ | | | | | |

### 3.5. Resilience Quantification

Many studies have been performed to evaluate the resilience of a specified system, and different models have been proposed to quantify system resilience. In the present study, the model developed by [28] is utilized to measure the resilience of Arctic Northeast Route security. Specifically, resilience is defined as the ratio of the recovery capacity to the loss capacity of the system, which can be expressed by:

$$R = \frac{recovery}{loss} \tag{35}$$

In the present study, the loss capacity variable of the system is conditioned based on three variables, namely, the probability of disruption occurrence (PDO), the absorptive capacity, and the actual security capacity (ASC). In practice, the loss capacity is greatly dependent on the degree to which the absorptive capacity is capable of withstanding the system shocks caused by disruptions; that is, the probability of the absorptive capacity being in a true state. If the state of the absorptive capacity is true (the probability is 100%), then the loss capacity will be zero. The actual security capacity is set as a constant in the present study because both the absorptive capacity and recovery capacity are associated with the actual security capacity in the expression of the resilience measurement. Therefore, the loss capacity can be obtained according to Table 12.

**Table 12.** Conditional calculation of the loss capacity.

| State of the Absorptive Capacity | True | False |
|---|---|---|
| Expression | PDO × ASC | 0 |

The recovery capacity of the system is dependent on the effectiveness of the post-disruption strategy (the combination of the adaptive capacity and restorative capacity) and the degree of the Loss Capacity (LC). If the post-disruption strategy is fully effective, that is, the state of the post-disruption strategy is true, the security system will recover 85% of the loss capacity; otherwise, the value will be zero. Therefore, the recovery capacity can be calculated according to Table 13.

**Table 13.** Conditional calculation of the recovery capacity.

| State of the Post-Disruption Capacity | True | False |
|---|---|---|
| Expression | LC × 0.80 | 0 |

## 4. Results and Discussion

The resilience level of the developed Arctic Northeast Route security system can be evaluated quantitatively based on the fuzzy Bayesian Network, and the expected resilience is 78.5%, as shown in Figure 9. In this section, we will conduct various types of analyses, such as belief propagation, sensitivity, and uncertainty analyses, based on the Bayesian network structure, as depicted in Figure 9.

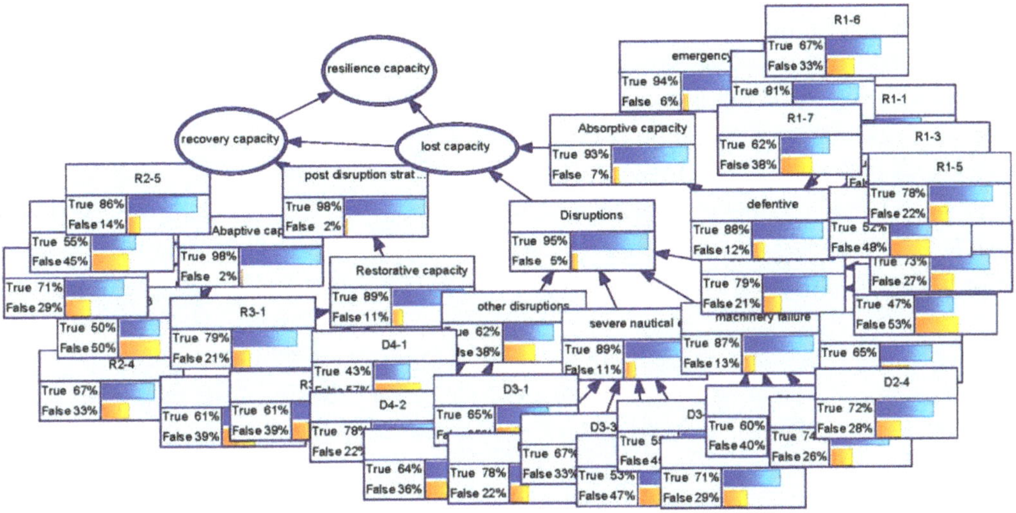


**Figure 9.** The base model of BN used to measure the resilience of the NSR.

### 4.1. Belief Propagation Analysis

Bidirectional reasoning and probabilistic inference can be performed in the Bayesian network by updating evidence formation, which is called "propagation analysis" [76]. Generally, propagation analysis in Bayesian networks includes forward-propagation analysis and backward-propagation analysis. The former is basically a cause-to-effect analysis in which the parent nodes measure their impact on the child nodes, and the latter is widely applied to update the probability distribution of the parent nodes for certain specified states of the target nodes.

As with forward-propagation analysis, three disruptive scenarios are defined to observe the corresponding impacts on system resilience and the decomposed capacities. The scenario description and BN inference results are presented in Table 14. In Scenario 1, it is observed that there is a lack of skilled seafarer teams (the state of (R1-2) is *false*), which reduces the absorptive capacity of system resilience from 93.4% to 89.2%; eventually, the resilience of the system descends to 78.1% from 78.5% in the base model. Scenario 2 simulates the case of two contributing factors in which the skilled seafarer team and the information prediction service are unavailable (the states of (R1-2) and (R2-2) are set as *false*). The results show that both the adaptive capacity and system resilience are reduced, the former from 98% to 94% and the latter from 78.5% to 77.8%, respectively. In Scenario 3, another contributing factor is considered on the basis of Scenario 2, that is, the state of a new factor, namely, the rescue and anti-pollution facility (R3-1) is further defined as false.

After simulating the developed BN model, the results indicate that the resilience of the NSR suffers from the most notable effect, decreasing to 75.9%, while the restorative capacity is lowered to 65.7% from 88.2%. A general insight obtained from the forward-propagation analysis indicates that all individual capacities are critical for the development of the resilience of the NSR.

**Table 14.** Different scenario sets for forward-propagation analysis.

| Scenario | R1-2 | R2-2 | R3-1 | Absorptive Capacity (%) | Adaptive Capacity (%) | Restorative Capacity (%) | Resilience (%) |
|---|---|---|---|---|---|---|---|
| Base | – | – | – | 93.4 | 98.0 | 88.2 | 78.5 |
| 1 | *false* | – | – | 89.2 (↓) | 98.0 | 88.2 | 78.1 (↓) |
| 2 | *false* | *false* | – | 89.2 | 94.8 (↓) | 88.2 | 77.8 (↓) |
| 3 | *false* | *false* | *false* | 89.2 | 94.8 | 65.7 (↓) | 75.9 (↓) |

Note: ↓ denotes a reduction in a certain variable value in the defined scenario compared with the value in the base case.

### 4.2. Sensitivity Analysis

Sensitivity analysis is one of the characteristic features of the application of BNs in identifying the most important independent variable(s) based on a particular dependent variable under a given set of conditions [77]. Herein, GeNIe 2.3 software is applied to simulate the established BN model and investigate the extent to which the parent nodes differ from the target nodes.

In the present study, the fuzzy ratio of variation (RoV) associated with the prior and posterior probabilities is calculated according to Equation (33) to evaluate the desired probability level of the contributing variables.

$$RoV(X_i) = \frac{\xi(X_i) - \zeta(X_i)}{\zeta(X_i)} \tag{36}$$

where $X_i$ denotes the ith contributing factor and $\xi(X_i)$ and $\zeta(X_i)$ represent the posterior and prior probabilities of $X_i$, respectively. Actually, all the parent nodes in the Bayesian network can theoretically be analyzed.

As discussed in Section 3.3, the NSR system resilience is mainly represented by the absorptive capacity, adaptive capacity, and restorative capacity. Therefore, in this section, we will set these individual capacities as the target nodes to investigate the corresponding effects of the contributing factors. Deductive reasoning is applied to update the probabilities of the contributing factors for the resilience related to the occurrence of the individual capacity (absorptive capacity, adaptive capacity, or restorative capacity), and then the value of the ratio of variation can be obtained by Equation (33). Finally, the results of the sensitivity analysis are illustrated in Figure 10.

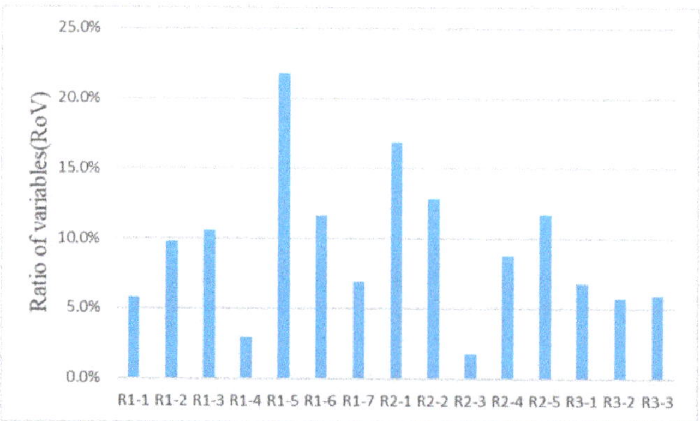


**Figure 10.** Sensitivity analysis for the absorptive capacity.

Figure 10 shows the effects of individual contributing factors on the corresponding capacity. For the absorptive capacity, when the state is set as "true", it is obviously found that the contributing factor "skilful emergency response" (R1-5) has the most important influence on the desired absorptive capacity; moreover, the "well-coordinated with tugs" (R1-4) has the lowest impact on the improvement of the absorptive capacity. When the state of the adaptive capacity is set as "true", after updating the probabilities of parent nodes, we can obtain the values of RoV for the contributing factors, as shown in Figure 7, from (R2-1) to (R2-5). Similarly, the results indicate that the "ice-breaking capacity" has the highest influence on the expected adaptive capacity, and the "information prediction service" ranks second; however, the effect of the "preparedness for arctic shipping" on the adaptive capacity is almost negligible. For the restorative capacity, there is no evident difference among the three contributing factors, and the "rescue and anti-pollution facility" (R3-1) may be characterized as the most potential influence on the restorative capacity.

### *4.3. Uncertainty Analysis Based on Information Entropy Theory*

Based on the discussion in Section 3.3, theoretically, the NSR security system is conditionally dependent on the absorptive capacity, adaptive capacity, and restorative capacity, which can be illustrated as shown in Figure 11. Actually, there is no direct and obvious causal relationship between system resilience and the decomposed capacities based on the analysis of the fuzzy Bayesian network. The system resilience and its supporting capacities are connected by the dotted line in Figure 8.

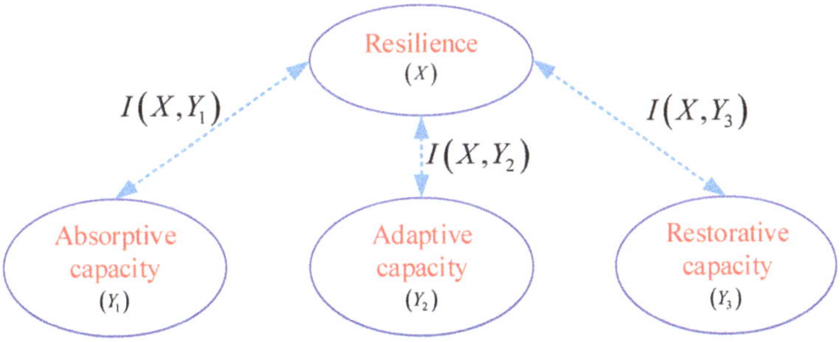


**Figure 11.** Mutual information links between resilience and the corresponding decomposed capacities.

As shown in Figure 11, the system resilience is set as the target variable $X$, and the decomposed absorptive capacity, adaptive capacity, and restorative capacity are regarded as $Y_1$, $Y_2$, and $Y_3$, respectively. In this section, $Y_1$, $Y_2$, and $Y_3$ are set as the predictive variables (predictor) associated with system resilience $X$. The predictive importance of each predictor can be determined by calculating the entropy and mutual information $I(X, Y_i)$, and the most important predictor will provide maximum information gain for system resilience.

The prior probability of system resilience can be obtained by simulating the developed Bayesian network as:

$$P(X = true) = 0.785,\ P(X = false) = 0.205 \tag{37}$$

Then, the entropy of system resilience can be calculated by:

$$H(X) = -\sum_{i=1}^{2} P_{x_i}(X)\cdot \log_2 P_{x_i}(X) = 0.785 \times \log_2(0.785) + 0.205 \times \log_2(0.205) = 0.743 \tag{38}$$

$$\begin{gathered} H(X|Y_1) = \sum_{i=1}^{2} P_{x_i}(X) H_i(X|Y_1) = \\ P(X = true)\cdot H(X = true|Y_1 = true) + P(X = false)\cdot H(X = false|Y_1 = false) \end{gathered} \tag{39}$$

where the conditional entropies $H(X = true|Y_1 = true)$ and $H(X = false|Y_1 = false)$ are unknown terms to be solved according to the developed Bayesian network. Notably, the state of the absorptive capacity in the network is set as "true" and "false", and the BN is simulated with GeNIe 2.3 software; the conditional entropy can be calculated as:

$$H(X = true|Y_1 = true) = -[0.805 \times \log_2(0.805) + 0.185 \times \log_2(0.185)] = 0.304 \quad (40)$$

$$H(X = false|Y_1 = false) = -[0.298 \times \log_2(0.298) + 0.692 \times \log_2(0.692)] = 0.89 \quad (41)$$

By substituting the results of Equations (37), (40) and (41) into Equation (39), we are able to obtain the conditional entropy of $H(X|Y_1)$ by:

$$H(X|Y_1) = 0.785 \times 0.304 + 0.205 \times 0.888 = 0.421 \quad (42)$$

Finally, the mutual information between $X$ and $Y_1$ can be evaluated based on the results of Equations (38) and (42) by:

$$I(X, Y_1) = H(X) - H(X|Y_1) = 0.743 - 0.421 = 0.322 \quad (43)$$

In the framework of information entropy, the result obtained from Equation (40) indicates that the absorptive capacity can reduce the uncertainty of system resilience by approximately 32.2%; that is, if we have good knowledge associated with the absorptive capacity, the extent of uncertainty of the system resilience can be reduced by 32.2%. Similarly, the conditional entropy and mutual information of the adaptive capacity and restorative capacity can also be calculated. Finally, the results for all three capacities obtained from information entropy theory are listed in Table 15.

**Table 15.** Summary of the information entropy analysis.

| Capacity Type ($Y_i$) | Conditional Entropy $H(X\|Y_i)$ | Mutual Information $I(X, Y_i)$ | Implication of Mutual Information |
|---|---|---|---|
| Absorptive capacity ($Y_1$) | 0.421 | 0.322 | The uncertainty of system resilience can be reduced by 32.2% in the case of a good knowledge of absorptive capacity |
| Adaptive capacity ($Y_2$) | 0.440 | 0.303 | The uncertainty of system resilience can be reduced by 30.3% in the case of a good knowledge of adaptive capacity |
| Restorative capacity ($Y_3$) | 0.542 | 0.201 | The uncertainty of system resilience can be reduced by 20.1% in the case of a good knowledge of restorative capacity |

Based on the information presented in Table 15, by comparing the values of mutual information between system resilience and its three supporting capacities, it can be observed that the absorptive capacity of system resilience is able to reduce the extent of uncertainty regarding the resilience level of the NSR, and the restorative capacity is the least valuable in reducing the uncertainty of the resilience level. Generally, the results of the present study agree with the general belief that the absorptive capacity is complex and expandable and that the restorative capacity is not as difficult to identify. The result of $I(X, Y_1) > I(X, Y_2) > I(X, Y_3)$ can be used by the related authorities and organizations to focus on theabsorptive capacity with regard to obtaining a better understanding of NSR resilience because the uncertainty of resilience is mainly related to the absorptive capacity.

## 5. Conclusions

The present work attempts to propose a comprehensive framework by which the safety level of the NSR can be improved. The safety level of the NSR is a common concern for mariners, authorities, and related organizations. Although huge endeavors have been made, the safety level of the NSR is still far from being satisfied. In the present study, system resilience is introduced to analyze the NSR safeguard capacity. Based on the existing studies about system resilience, the resilience capacity of the NSR is decomposed into three capacities: absorptive capacity, adaptive capacity, and restorative capacity. Then, we identify the contributing factors pertaining to these capacities. Moreover, the disruptions to the NSR are also identified. Next, all the identified disruptions, contributing factors, and capacities are regarded as nodes in the BN and, as a result, a BN model is developed. The prior probabilities for the nodes without parents and the CPT for the network inference are obtained by the fuzzy methodology proposed by [34]. Finally, the established BN model is simulated with GeNIe 2.3 software, and a sensitivity analysis and an information entropy analysis are conducted with regard to the results obtained from the simulation. Overall, the present study provides a resilience perspective to understand and evaluate the NSR safety issue, which can be seen as the main innovation of this work. The analysis framework proposed in the present study can improve our understanding of the system resilience of the NSR and aid in mitigating disruptions to the NSR.

Although the framework proposed in the present study is demonstrated to effectively quantify and interpret the resilience of the NSR, the structure of the proposed BN can be optimized based on additional information from experts and stakeholders. In addition, the resilience level of the system is generally regarded as a variable that changes over time; therefore, in the future, potential work can be devoted to improving the performance of resilience modelling for the NSR based on the framework given in the present study. The work associated with the abovementioned issues remains in progress.

**Author Contributions:** Conceptualization and project administration, X.M.; Methodology, W.Q.; investigation, Y.L.; Writing, H.L. All authors have read and agreed to the published version of the manuscript.

**Funding:** This research was funded by National Social Science Foundation of China, grant numbers are 19BZZ104 and 20VHQ012. The APC was funded by Fundamental Research Funds for the Central Universities (grant no. 3132019190).

**Institutional Review Board Statement:** Not applicable.

**Informed Consent Statement:** Not applicable.

**Conflicts of Interest:** The authors declare no conflict of interest.

## References

1. Gascard, J.-C.; Riemann-Campe, K.; Gerdes, R.; Schyberg, H.; Randriamampianina, R.; Karcher, M.; Zhang, J.; Rafizadeh, M. Future sea ice conditions and weather forecasts in the Arctic: Implications for Arctic shipping. *Ambio* **2017**, *46*, 355–367. [CrossRef]
2. Wang, H.; Zhang, Y.; Meng, Q. How will the opening of the northeast Sea Route influence the Suez Canal Route? An empirical analysis with discrete choice models. *Transp. Res. A Policy Pract.* **2018**, *107*, 75–89. [CrossRef]
3. Schøyen, H.; Bråthen, S. The Northern Sea Route versus the Suez Canal: Cases from bulk shipping. *J. Transp. Geogr.* **2011**, *19*, 977–983. [CrossRef]
4. Sun, K.; Zhang, J.J. Corporate engagement in the age of Arctic development and its implications for Chinese Corporations. *J. Ocean Univ.* **2017**, *2*, 71–77.
5. Jiang, X.; Fan, H.; Zhang, Y.; Yuan, Z. Using interpretive structural modeling and fuzzy analytic network process to identify and allocate risks in Arctic shipping strategic alliance. *Polar Sci.* **2018**, *17*, 83–93. [CrossRef]
6. Shyu, W.-H.; Ding, J.-F. Key Factors Influencing the Building of Arctic Shipping Routes. *J. Navig.* **2016**, *69*, 1261–1277. [CrossRef]
7. Senol, Y.E.; Aydogdu, Y.V.; Sahin, B.; Kilic, I. Fault Tree Analysis of chemical cargo contamination by using fuzzy approach. *Expert Syst. Appl.* **2015**, *42*, 5232–5244. [CrossRef]
8. Zhang, M.; Zhang, D.; Goerlandt, F.; Yan, X.; Kujala, P. Use of HFACS and fault tree model for collision risk factors analysis of icebreaker assistance in ice-covered waters. *Saf. Sci.* **2019**, *111*, 128–143. [CrossRef]
9. Goerlandt, F.; Goite, H.; Banda, O.A.V.; Höglund, A.; Ahonen-Rainio, P.; Lensu, M. An analysis of wintertime navigational accidents in the Northern Baltic Sea. *Saf. Sci.* **2017**, *92*, 66–84. [CrossRef]
10. Kum, S.; Sahin, B. A root cause analysis for Arctic Marine accidents from 1993 to 2011. *Saf. Sci.* **2015**, *74*, 206–220. [CrossRef]
11. Baksh, A.A.; Abbassi, R.; Garaniya, V.; Khan, F. Marine transportation risk assessment using Bayesian Network: Application to Arctic waters. *Ocean Eng.* **2018**, *159*, 422–436. [CrossRef]

12. Jalonen, R.; Riska, K.; Hanninen, S. *A Preliminary Risk Analysis of Winter Navigation in the Baltic Sea*; Finnish Maritime Administration: Helsinki, Finland, 2005.
13. Kujala, P.; Hänninen, M.; Arola, T.; Ylitalo, J. Analysis of the marine traffic safety in the Gulf of Finland. *Reliab. Eng. Syst. Saf.* **2009**, *94*, 1349–1357. [CrossRef]
14. Holling, C.S. Resilience and Stability of Ecological Systems. *Annu. Rev. Ecol. Syst.* **1973**, *4*, 1–23. [CrossRef]
15. Fiksel, J. Designing Resilient, Sustainable Systems. *Environ. Sci. Technol.* **2003**, *37*, 5330–5339. [CrossRef] [PubMed]
16. Hollnagel, E.; Woods, D.D.; Leveson, N. *Resilience Engineering: Concepts and Precepts*; Ashgate Publishing Limited: Farnham, UK, 2006.
17. Vogus, T.J.; Sutcliffe, K.M. Organizational resilience: Towards a theory and research agenda. In Proceedings of the 2007 IEEE International Conference on Systems, Man and Cybernetics, Montreal, QC, Canada, 7–10 October 2007; pp. 3418–3422.
18. Gilberto, C.G. Linkages between vulnerability, resilience, and adaptive capacity. *Glob. Environ. Chang.* **2006**, *16*, 293–303.
19. Smale, S. Assessing Resilience and Vulnerability: Principle, Strategies and Actions. *Ann. Bot.* **2008**, *101*, 403–419.
20. Engel, K.E.; Warner, J.F. Resilience in Talcahuano, Chile: Appraising local disaster response. *Disaster Prev. Manag. Int. J.* **2019**, *28*, 585–602. [CrossRef]
21. Di Nardo, M.; Mariano, C.; Teresa, M.; Madonna, M. An adaptive resilience approach for a high capacity railway. *Int. Rev. Civ. Eng.* **2020**, *11*, 98–105. [CrossRef]
22. Attoh-Okine, N.O.; Cooper, A.T.; Mensah, S.A. Formulation of Resilience Index of Urban Infrastructure Using Belief Functions. *IEEE Syst. J.* **2009**, *3*, 147–153. [CrossRef]
23. Hossain, N.U.I.; Jaradat, R.; Hosseini, S.; Marufuzzaman, M.; Buchanan, R. A framework for modeling and assessing system resilience using a Bayesian network: A case study of an interdependent electrical infrastructure system. *Int. J. Crit. Infrastruct. Prot.* **2019**, *25*, 62–83. [CrossRef]
24. Hossain, N.U.I.; Amrani, S.E.; Jaradat, R.; Marufuzzaman, M.; Buchanan, R.; Rinaudo, C.; Hamilton, M. Modeling and assessing interdependencies between critical infrastructures using bayesian network: A case study of inland waterway port and surrounding supply chain network. *Reliab. Eng. Syst. Saf.* **2020**, *198*, 106898. [CrossRef]
25. Hossain, N.U.I.; Nur, F.; Hosseini, S.; Jaradat, R.; Marufuzzaman, M.; Puryear, S.M. A Bayesian network based approach for modeling and assessing resilience: A case study of a full service deep water port. *Reliab. Eng. Syst. Saf.* **2019**, *189*, 378–396. [CrossRef]
26. Adger, W.N. Social ecological resilience: Are they related? *Prog. Hum. Geogr.* **2019**, *24*, 347–364. [CrossRef]
27. Vugrin, E.D.; Warren, D.E.; Ehlen, M.A. A resilience assessment framework for infrastructure and economic systems: Quantitative and qualitative resilience analysis of petrochemical supply chain to a hurricane. *Process Saf. Progr.* **2011**, *30*, 280–290. [CrossRef]
28. Henry, D.; Ramirez-Marquez, J.E. Generic metrics and quantitative approaches for system resilience as a function of time. *Reliab. Eng. Syst. Saf.* **2012**, *99*, 114–122. [CrossRef]
29. Kelly, D.L.; Smith, C.L. Bayesian inference in probabilistic risk assessment? The current state of the art. *Reliab. Eng. Syst. Saf.* **2009**, *94*, 628–643. [CrossRef]
30. Fenton, N.; Neil, M. *Risk Assessment and Decision Analysis with Bayesian Networks*; CRC Press: Boca Raton, FL, USA, 2012.
31. Rathnayaka, S.; Khan, F.; Amyotte, P. SHIPP methodology: Predictive accident modeling approach. Part II. Validation with case study. *Process. Saf. Environ. Prot.* **2011**, *89*, 75–88. [CrossRef]
32. Khakzad, N.; Khan, F.; Amyotte, P. Safety analysis in process facilities: Comparison of fault tree and Bayesian network approaches. *Reliab. Eng. Syst. Saf.* **2011**, *96*, 925–932. [CrossRef]
33. Khakzad, N.; Khan, F.; Amyotte, P. Dynamic risk analysis using bow-tie approach. *Reliab. Eng. Syst. Saf.* **2012**, *104*, 36–44. [CrossRef]
34. Qiao, W.L.; Liu, Y.; Ma, X.X.; Liu, Y. Human Factors Analysis for Maritime Accidents Based on a Dynamic Fuzzy Bayesian Network. *Risk Anal.* **2020**, *1*, 13444. [CrossRef]
35. Hosseini, S.; Barker, K. A Bayesian network model for resilience-based supplier selection. *Int. J. Prod. Econ.* **2016**, *180*, 68–87. [CrossRef]
36. Amundson, J.; Faulkner, W.; Sukumara, S.; Seay, J.; Badurdeen, F. A Bayesian Network Based Approach for Risk Modeling to Aid in Development of Sustainable Biomass Supply Chains. *Comput. Aided Chem. Eng.* **2012**, *30*, 152–156. [CrossRef]
37. Han, S.Y.; Marais, K.; de Laurentis, D. Evaluating System of Systems resilience using interdependency analysis. In Proceedings of the 2012 IEEE International Conference on Systems, Man, and Cybernetics (SMC), Seoul, Korea, 14–17 October 2012; pp. 1251–1256.
38. Hosseini, S.; Barker, K. Modeling infrastructure resilience using Bayesian networks: A case study of inland waterway ports. *Comput. Ind. Eng.* **2016**, *93*, 252–266. [CrossRef]
39. Van Laarhoven, P.; Pedrycz, W. A fuzzy extension of Saaty's priority theory. *Fuzzy Sets Syst.* **1983**, *11*, 229–241. [CrossRef]
40. Chang, D.-Y. Applications of the extent analysis method on fuzzy AHP. *Eur. J. Oper. Res.* **1996**, *95*, 649–655. [CrossRef]
41. Miller, G.A. The magical number seven, plus or minus two: Some limits on our capacity for processing information. *Psychol. Rev.* **1956**, *101*, 343–352. [CrossRef] [PubMed]
42. Nicolis, J.S.; Tsuda, I. Chaotic dynamics of information processing: The "magic number seven plus-minus two" revisited. *Bull. Math. Biol.* **1985**, *47*, 343–365.
43. Chen, S.J.; Hwang, C.L. *Fuzzy Multiple Attribute Decision Making: Methods and Applications, Lecture Notes in Economics and Mathematical Systems, No. 375*; Springer: Berlin/Heidelberg, Germany, 1992.

44. Ramzali, N.; Lavasani, M.R.M.; Ghodousi, J. Safety barriers analysis of offshore drilling system by employing Fuzzy Event Tree Analysis. *Saf. Sci.* **2015**, *78*, 49–59. [CrossRef]
45. Gupta, S.; Bhattacharya, J. Reliability Analysis of a conveyor system using hybrid data. *Qual. Reliab. Eng. Int.* **2007**, *23*, 867–882. [CrossRef]
46. Omidvari, M.; Lavasani, S.; Mirza, S. Presenting of failure probability assessment pattern by FTA in Fuzzy logic (case study: Distillation tower unit of oil refinery process). *J. Chem. Health Saf.* **2014**, *21*, 14–22. [CrossRef]
47. Kabir, S.; Yazdi, M.; Aizpurua, J.I.; Papadopoulos, Y. Uncertainty-Aware Dynamic Reliability Analysis Framework for Complex Systems. *IEEE Access* **2018**, *6*, 29499–29515. [CrossRef]
48. Qiao, W.L.; Liu, Y.; Ma, X.X.; Liu, Y. A methodology to evaluate human factors contributed to maritime accident by mapping fuzzy FT into ANN based on HFACS. *Ocean Eng.* **2020**, *197*, 106892. [CrossRef]
49. Buckley, J.J. Fuzzy hierarchical analysis. *Fuzzy Sets Syst.* **1985**, *17*, 233–247. [CrossRef]
50. Jensen, F.V.; Nielsen, T.D. *Bayesian Networks and Decision Graphs*; Springer Science and Business Media LLC: Berlin/Heidelberg, Germany, 2007.
51. Kabir, S.; Papadopoulos, Y. Applications of Bayesian networks and Petri nets in safety, reliability, and risk assessments: A review. *Saf. Sci.* **2019**, *115*, 154–175. [CrossRef]
52. Jensen, F.V.; Nielsen, T.D. *Bayesian Networks and Decision Graphs*, 2nd ed.; Springer Science and Business Media LLC: New York, NY, USA, 2009.
53. Wang, Y.F.; Xie, M.; Chin, K.S.; Fu, X.J. Accident analysis model based on Bayesian Network and Evidential Reasoning approach. *J. Loss Prev. Process Ind.* **2013**, *26*, 10–21. [CrossRef]
54. Zarei, E.; Khakzad, N.; Cozzani, V.; Reniers, G. Safety analysis of process systems using Fuzzy Bayesian Network (FBN). *J. Loss Prev. Process. Ind.* **2019**, *57*, 7–16. [CrossRef]
55. Agnieszka, O. Learning Bayesian Network Parameters from Small Data Sets Application of Noisy-OR Gates. *Int. J. Approx. Reason.* **2001**, *27*, 165–182.
56. Ellis, B.; Brigham, L. 2009: Arctic Marine Shipping Assessment 2009 Report. Arctic Council Rep., 189p. Available online: https://oaarchive.arctic-council.org/handle/11374/54 (accessed on 5 November 2020).
57. Zakharov, V.F. *Sea Ice in the Climate System: A Russian View*; NSIDC Special Report 16; NSIDC: Boulder, CO, USA, 1997; Available online: https://nsidc.org/pubs/special/16/NSIDC-specialreport-16.pdf (accessed on 11 November 2020).
58. Royal Institution of Naval Architects. Safety Guidance for Naval Architects. Second Editions. 2010. Available online: http://www.rina.org.uk/hres/safety%20guidance%20for%20naval%20architects%20_%20second%20edition%20_%20march%202010%20_%20final%20inc%20tb%20edits.pdf (accessed on 10 January 2021).
59. Lloyd's Register. Written Evidence (ARC0048). Lloyd's Register. 2015. Available online: https://www.parliament.uk/globalassets/documents/lords-committees/arctic/Lloyds-Register-ARC0048.pdf (accessed on 11 November 2020).
60. Afenyo, M.; Khan, F.; Veitch, B.; Yang, M. Arctic shipping accident scenario analysis using Bayesian Network approach. *Ocean Eng.* **2017**, *133*, 224–230. [CrossRef]
61. Fu, S.; Zhang, D.; Montewka, J.; Zio, E.; Yan, X. A quantitative approach for risk assessment of a ship stuck in ice in Arctic waters. *Saf. Sci.* **2018**, *107*, 145–154. [CrossRef]
62. Sahin, B.; Kum, S. Risk Assessment of Arctic Navigation by Using Improved fuzzy-AHP Approach. *Int. J. Marit. Eng.* **2015**, *157*, 241.
63. Halliday, W.D.; Tetu, P.L.; Dawson, J.; Insley, S.J.; Hilliard, R.C. Tourist vessel traffic in important whale areas in the western Canadian Arctic: Risks and possible management solutions. *Mar. Policy* **2018**, *97*, 72–81. [CrossRef]
64. Olsen, D.J.; Pitman, N.D.; Basak, S.; Wunsche, B.C. Sketch-Based Building Modelling. Graphics Group Technical Report #2011-002, Department of Computer Science, University of Auckland. 2011. Available online: http://www.cs.auckland.ac.nz/burkhard/Reports/GraphicsGroupTechnicalReport2011_002.pdf (accessed on 5 November 2020).
65. Stoddard, M.A.; Etienne, L.; Fournier, M.; Pelot, R.; Beveridgle, L. Making sense of Arctic maritime traffic using. The Polar Operational Limits Assessment Risk Indexing System (POLARIS). *IOP Conf. Ser. Earth Environ. Sci.* **2016**, *34*, 012034. [CrossRef]
66. Zhang, M.; Zhang, D.; Fu, S.; Yan, X.; Goncharov, V. Safety distance modeling for ship escort operations in Arctic ice-covered waters. *Ocean Eng.* **2017**, *146*, 202–216. [CrossRef]
67. Khan, B.; Khan, F.; Veitch, B.; Yang, M. An operational risk analysis tool to analyze marine transportation in Arctic waters. *Reliab. Eng. Syst. Saf.* **2018**, *169*, 485–502. [CrossRef]
68. Milaković, A.-S.; Gunnarsson, B.; Balmasov, S.; Hong, S.; Kim, K.; Schütz, P.; Ehlers, S. Current status and future operational models for transit shipping along the Northern Sea Route. *Mar. Policy* **2018**, *94*, 53–60. [CrossRef]
69. Rose, A. Economic Resilience to Disasters, Community and Regional Resilience Initiative Research Report 8. 2009. Available online: http://www.resilientus.org/library/Research_Report_8_Rose_1258138606.pdf (accessed on 20 November 2020).
70. Maritime Safety Administration of China. *Guidance on Arctic Navigation in the Northeast Route*; China Comm Press: Beijing, China, 2014. (In Chinese)
71. Biringer, B.; Vugrin, E.; Warren, D. *Critical Infrastructure System Security and Resiliency*; CRC Press: Boca Raton, FL, USA, 2016.
72. Yazdi, M.; Daneshvar, S.; Setareh, H. An extension to fuzzy developed failure mode and effects analysis application. *Saf. Sci.* **2017**, *98*, 113–123. [CrossRef]
73. Celik, M.; Lavasani, S.M.; Wang, J. A risk-based modelling approach to enhance shipping accident investigation. *Saf. Sci.* **2010**, *48*, 18–27. [CrossRef]

74. Miri-Lavasani, M.R.; Wang, J.; Yang, Z.; Finlay, J. Application of fuzzy fault tree analysis on oil and gas offshore pipelines. *Int. J. Mar. Sci. Eng.* **2011**, *1*, 29–42.
75. Lavasani, S.M.; Zendegani, A.; Celik, M. An extension to Fuzzy Fault Tree Analysis (FFTA) application in petro-chemical process industry. *Process Saf. Environ. Protect.* **2015**, *93*, 75–88. [CrossRef]
76. Fenton, N.; Neil, M.; Lagnado, D.A. A general structure for legal arguments about evidence using Bayesian networks. *Cogn. Sci.* **2013**, *37*, 61–102. [CrossRef]
77. Salteli, A.; Chan, K.; Scott, E.M. *Sensitivity Analysis*; John Wiley and Sons: New York, NY, USA, 2000.

applied sciences

MDPI

*Communication*

# Toward More Sustainable River Transportation in Remote Regions of the Amazon, Brazil

**Jassiel Vladimir Hernández-Fontes [1], Harlysson Wheiny Silva Maia [1], Valeria Chávez [2,*] and Rodolfo Silva [2]**

1 Departamento de Engenharia Naval, Escola Superior de Tecnologia, Universidade do Estado do Amazonas, Manaus 69050-020, Brazil; jvfontes@uea.edu.br (J.V.H.-F.); hwmaia@uea.edu.br (H.W.S.M.)
2 Coordinación de Hidráulica, Instituto de Ingeniería, Universidad Nacional Autónoma de México, Mexico City 04510, Mexico; rsilvac@iingen.unam.mx
* Correspondence: vchavezc@iingen.unam.mx

**Abstract:** This paper explores means of achieving more efficient and sustainable river transport in remote regions by making relatively simple, practical modifications to boats or implementing new technologies for propulsion and energy generation. The research focuses on the case of the simple boats used to transport children to school in riverine communities of the Brazilian Amazon. A range of options to improve the efficiency of existing boats is described. Under normal operational conditions, small improvements to these boats may have long-term environmental and socioeconomic benefits. Implementing changes such as those suggested, it may also be possible to boost sources of employment in these regions and elsewhere, where industrial and technological limitations are significant.

**Keywords:** efficient transport; riverine communities; sustainable cities; renewable energy; environment

**Citation:** Hernández-Fontes, J.V.; Maia, H.W.S.; Chávez, V.; Silva, R. Toward More Sustainable River Transportation in Remote Regions of the Amazon, Brazil. *Appl. Sci.* **2021**, *11*, 2077. https://doi.org/10.3390/app11052077

Academic Editor: José A. Orosa

Received: 5 February 2021
Accepted: 23 February 2021
Published: 26 February 2021



## 1. Introduction

Navigation using small boats is one of the main means of transport for remote riverine communities in many developing countries [1]. Often, such boats are the only way to access islands and other remote localities [2], and they become multi-purpose boats, used for passengers and cargo, as well as offering a wide range of services to the community, such as transportation of students to schools [3,4]. In the United Nations Goals for Sustainable Development [5–7], the fourth and tenth objectives are to provide quality education and to reduce inequalities between all peoples. In many parts of the world, to travel to their school, students have no alternative but to use small boats, sometimes on quite long journeys. In the Brazilian Amazon, it is estimated that in around 350 riverine communities, this type of river transport is the only means their children have to travel to school [8]. In remote regions, efficient, sustainable river transport for daily activities is required [9–11]. In several Amazonian regions, most riverboats are designed and constructed locally, using traditional shipbuilding techniques inherited over generations [12], rather than modern technological methods [12–14]. Despite the vital importance of these boats, they are often inadequate vessels, which may be slow and vulnerable to rain, winds, and currents, exposing the users to risk and discomfort [15,16].

Navigation in the Amazon waterways is mainly regulated by the Brazilian Navy, which establishes the minimum requirements for vessels [17]. However, in remote regions, it is often difficult to implement statutory shipbuilding and operational procedures when regulation is scarce [16,18,19]. This may have effects on the stability of the vessels used, human safety, and fuel consumption. Making the small boats used in remote regions safer, more efficient, and sustainable could help to minimize accidents and to preserve protected areas [20,21].

To improve the efficiency of boats used for river transport, factors related to the reduction of hydrodynamic resistance and improving the propulsion system must be addressed [22]. To find the optimal configuration of a specific boat for different operational conditions, detailed analyses of the interaction of the hull, engine, shaft, and propeller are required, as described by [23,24]. This generally requires advanced study, not often found in the small shipyards of remote river communities [25], where the components of a propulsion system are often selected based on previous empirical experience [12], with outboard engines being the most common choice for small riverboats. One means of improving the performance of boats is to implement permanent modifications to the hull, perhaps by optimization methods. This could possibly lessen the hydrodynamic resistance [22,26–31] and fuel consumption for specific operational conditions. Small modifications to the shape of the hull of an existing boat could easily be carried out in any small shipyard, even in a remote river community, and bring a permanent improvement to increase the performance of the boat [32]. Another area that can be looked into is the incorporation of hybrid technologies, that is, combining clean, renewable energy with traditional fossil fuels [22,33–36].

Research on improving hull shape is not new [28,29]. In fact, several works have investigated the optimization of ship hulls and the effects this has on performance [26,37–41]. Such optimization is commonly carried out using Computational Fluid Dynamics (CFD) methods, which are often applied in algorithms that allow numerical tests to be performed to reduce the hydrodynamic resistance of a vessel in various operational conditions, as reported by [37,38,40]. Similar work has examined the optimization of the hulls of ocean renewable energy devices [42–44], large commercial ships [45,46], and high-speed vessels [47,48]. Aside from hull optimization methods, other means could increase the performance and sustainability of small transport vessels. In this study, we aim to fill this gap. An integrated study is proposed to explore different ways of improving the efficiency and sustainability of existing boats that perform river transport activities in remote regions of developing countries. In many such cases, attaining sustainability is crucial in the face of environmental and socioeconomic restrictions. This paper examines the case of transporting school students by small boats in riverine communities in the Amazon region.

The study is divided as follows: Section 2 describes the main considerations to improve the efficiency of the boats. Then, Section 3 presents some challenges of scholar transport in the Amazon region, including a brief discussion of possible environmental and socioeconomic impacts. Finally, Section 4 presents different means of improving the sustainability of the boats, and Section 5 summarizes the main conclusions.

## 2. Main Considerations to Improve the Efficiency of Boats

There is no direct method to improve the efficiency of the small boats used for riverine transportation in remote regions. In fact, reducing the fuel consumption and increasing the performance of the boat requires several optimization projects since there are different fuel-optimal working conditions for each boat draft and speed [22].

Hochkirch and Bertram [22] explained that there are several ways of decreasing the consumption of fuel by a vessel: reducing the required power for operation and propulsion; using fuel energy for propulsion and on-board equipment more efficiently; and using hybrid technologies for operation, combining fossil fuels with renewables, such as solar energy. According to [22], there are several factors to be considered to reduce the power required for propulsion, thus improving the performance of a ship. These can be classified into improving the propulsion system and reducing the boat's hydrodynamic resistance, as shown in Figure 1. To improve ship propulsion (Figure 1a), the propeller must operate optimally. Propulsion losses due to rotation of the shaft and propeller must be reduced, as must friction loads between the water and the propulsor, as well as those caused by the generation of tip and hub vortices. The propeller must operate in optimal wake conditions. In addition, the efficiency of the interactions between the components of the engine-shaft-propeller system must be maximized to increase the transmission of power delivered from the engine to the propeller.

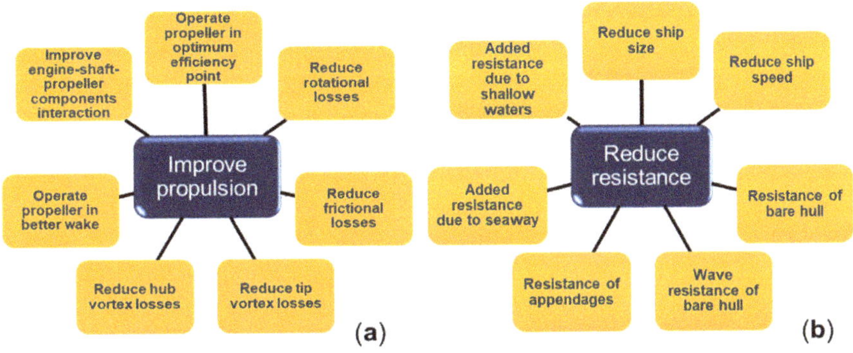


**Figure 1.** Factors that will reduce the power required for propulsion (**a**) by improving propulsion and (**b**) by reducing resistance. Adapted from [22].

On the other hand, to reduce resistance (Figure 1b), reducing the size of the vessel and its speed may help; however, this practice does not seem to be economically viable for most applications. Other factors that may reduce a vessel's total resistance are: the resistance of the bare hull; the resistance of the waves generated when the bare hull is in motion; the resistance caused by appendages of the hull; the resistance caused by the seaway; the resistance caused by wind; and unexpected environmental conditions. In a practical way, it is possible to define the total resistance of a ship as the sum of frictional resistance (due to the friction of the fluid and the hull surface) and residuary resistance (due to the combination of viscous pressure and the wave-making resistance) [24,49].

Molland et al. [24] describe how current attempts to decrease the hull resistance of vessels are related to speed, trim, added resistance in waves, bow shape, bulbous bows, stern shape, hull shape using CFD investigations, and air-bubble lubrication methods.

As shown in Figure 1, increasing a vessel's sustainability by improving its performance involves several factors. Hollenbach and Friesch [32] explained that a reduction in fuel oil consumption of a vessel depends mainly on its type, speed, and working conditions. Therefore, in systematic studies, some parameters must be kept fixed for experimental and numerical investigations [50–52].

A permanent long-term means of improving the performance of a vessel is to modify the shape of its hull to reduce resistance [26,37–41], as this could yield a constant benefit. Therefore, following Hollenbach and Friesch [32], various means of modifying the forebody, midship, and aft body of the hull shapes of vessels can be considered, as shown in Figure 2. The possible gains that each of these modifications produces in reducing ship resistance can also be analyzed. Although these values are theoretical, they provide a preliminary idea about how permanent hull modifications can be useful. For instance, according to these data, gains of ~2–5% can be produced by optimization strategies applied to the forebody hull form.

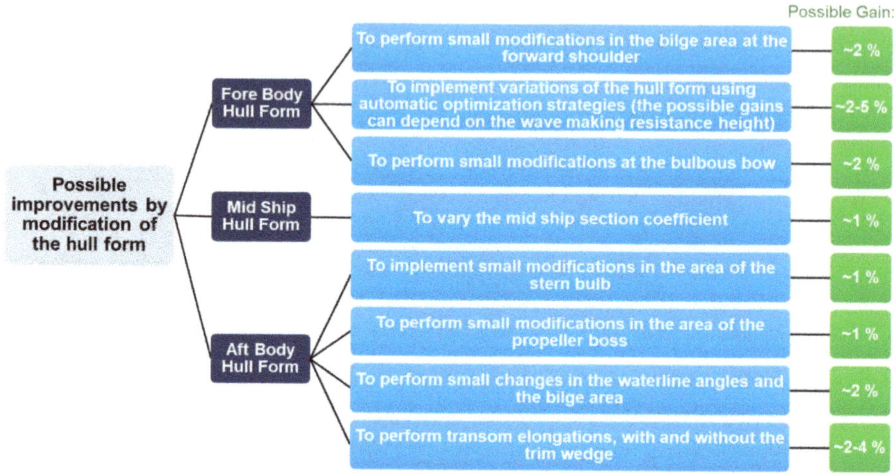


**Figure 2.** Possible gains in resistance reduction through permanent modifications of the hull, considering that the main dimensions of the vessel and the propeller diameter are unchanged. Modified from [32].

## 3. Challenges in the Amazon Region: The Case of School Transport

To explore alternatives that improve river transport using small boats in remote regions, the boats used for school transport in small riverine communities of the Brazilian Amazon were considered in this work. In these regions, river vessels are the main means of transport for passengers and cargo. However, the complexity of these waterways, due to their extension and interaction with preserved areas, makes transportation here a daily challenge. Many routes are unmarked. Figure 3 shows some of the river systems in the state of Amazonas, where many students who live in outlying areas use boats to travel to schools in the bigger communities, such as Manaus and Manaquiri.

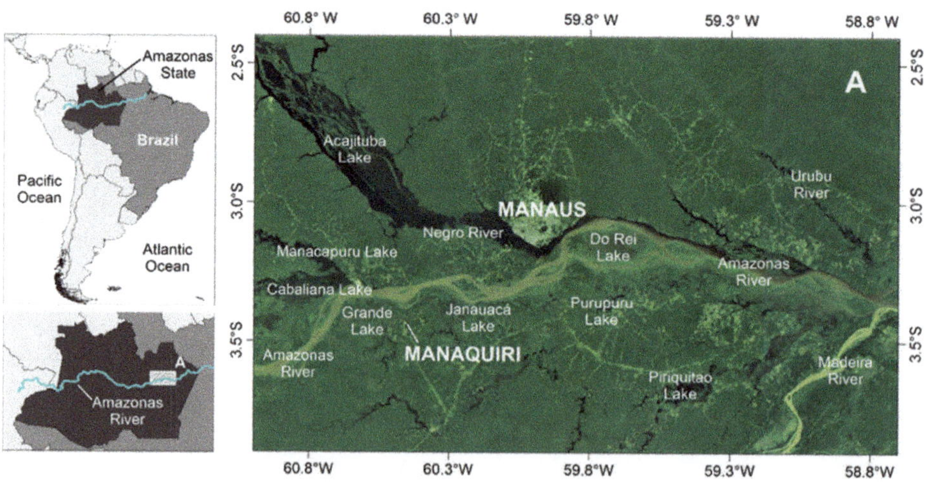


**Figure 3.** Example of the complexity of the fluvial system around the municipalities of Manaus and Manaquiri, Amazonas State, Brazil. Adapted from Google Maps.

Overall, it is estimated that in the north of Brazil, around 300,000 students use river transport to travel to school [53]. The Secretary of Education for the city of Manaus (Semed) has identified several rural schools that use river transport [54]: 20 on the river Amazon and 29 on the river Negro. Students may pass up to 3 h, twice a day, to travel to and from their school by boat [55], as in the remote communities of the municipality of Manaquiri, on the Amazon, for example, as shown in (Figure 4) [53].

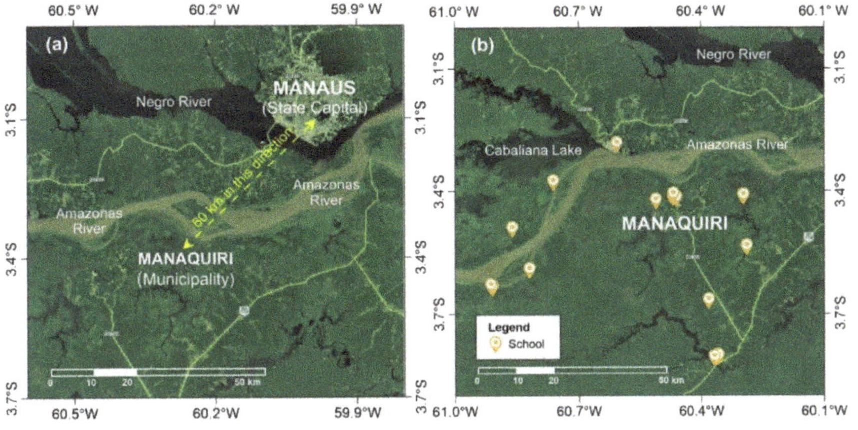


**Figure 4.** An example of a municipality in Amazonas State where river transport is necessary for education. (**a**) Location of Manaquiri municipality with respect to Manaus, the capital of Amazonas State. (**b**) Location of the main schools around Manaquiri. Adapted from Google Maps.

To overcome the challenges of school transportation, the National Fund for the Development of Education (FNDE) has provided resources for the development of school boats via the Caminhos da Escola (School Paths) program [56]. This initiative aims to offer faster, safer transportation for those using river transport to reach their classrooms. Figure 5 illustrates activities on the rivers related to school transport in the Amazon, Brazil. Figure 5a shows a type of boat widely used for school transport several years ago that is still used for school transport in some regions. Figure 5b,c show the school boats promoted by the local government, and in Figure 5d, these boats are seen at a station in the port of Manaus. Figure 5d shows a school boat performing daily activities in riverine communities. In Figure 5f, a school in Tapauá is presented, and finally, Figure 5g shows a different type of boat used for transporting people in remote regions of the state.

**Figure 5.** School transport in the state of Amazonas, Brazil. (**a**) An old boat used to transport children to school about ten years ago (Kleyphide Pereira da Silva 2020). (**b**) Front view of an Amazon school boat promoted by the government in

Manaus, Amazonas (Luiz Henrique Moreira Sousa 2020). (**c**) Profile view of the Amazon school boat in Itacoatiara, Amazonas (Fernando V. Dias Balieiro 2013). (**d**) School boat station in the port of Manaus, Amazonas (Harlysson Maia and Francisco Xavier de Carvalho Neto 2020). (**e**) A school boat performing its daily activities in a remote community on the Amazon (modified from [57]). (**f**) A typical school in the municipality of Tapauá, Amazonas (Fernando V. Dias Balieiro 2013). (**g**) A typical boat used for transporting people between remote communities in the state (Fernando V. Dias Balieiro 2013).

### *3.1. Possible Economic and Environmental Impacts*

#### 3.1.1. Economic Impact

As shown in Section 2, improving the sustainability and efficiency of the school boats depends on several factors. While a fuel reduction analysis is beyond the scope of this work, reducing the hull resistance of ships, also applicable to riverboats, may minimize the effort required by the engine and thus reduce the amount of fuel needed [22]. Small changes to the hull shape can reduce hydrodynamic resistance. Based on Figure 2, after hull optimization, it can be assumed that in specific working conditions, the fuel consumption of a typical Amazon school boat is reduced by ~5%. In the long-term, this could produce substantial economic benefits, as the boats travel long distances in areas where access to fuel may be limited. If a typical motorboat consumes an average of ~95 L per hour at cruising speed [58], and the current price of fuel in Brazil is 0.76 $USD/liter (6-26-2020, [59]), with constant activity 8 h per day, the estimated 5% reduction in fuel consumption would provide savings of ~38 L (~29 $USD) per day or ~13870 L (~10,541 $USD) per year for one boat. In a fleet of 10 boats, several thousand liters could be saved per year. This direct economic benefit also has an effect on daily activities in the Amazon region that use similar small boats, such as school transport, food commerce, fishing, health campaigns, and scientific activities.

#### 3.1.2. Environmental Impact

By optimizing the hulls of these boats, the height of the waves they produce may be lessened. As these boats often travel close to the riverbanks, waves can cause erosion through applied shear effects [60]. Lower waves will produce less applied shear, and the riverbanks will be preserved, which, in turn, avoids riverbank instability that can affect trees, shrubs, and plants growing nearby, as well as aquatic species [61].

The reduction in gas emissions due to the use of less fuel is an obvious positive effect of optimizing the riverboats. Some of the greenhouse gases produced by the combustion of fossil fuels in inland navigation [62] are shown in Table 1 [63]. The Intergovernmental Panel on Climate Change [63] proposes emission factors for various types of engines (diesel, gasoline) for marine and inland vessels. A full overview of these for lake, river, coastal, and ocean vessels is found in [62]. Table 1 shows the approximate gas emission factors for European ships and boats on inland waters [63], estimated for four-stroke gasoline engines. In this table, the possible reduction in gas emissions by the optimized school boat is shown per day and per year, assuming the 5% reduction in fuel consumption described in the previous subsection. The reduction in gas emission per day (in grams) was obtained by multiplying the mass of the saved 38 L of fuel (Section 3.1.1), assuming that a liter of fuel weighs 0.750 kg, by the corresponding emission factor (in g/kg fuel). The reduction per day was then multiplied by 365 to obtain the reduction of gas emissions per year (Table 1). The reduction of greenhouse gases by a single optimized boat is seen to be significant, particularly for carbon dioxide ($CO_2$), which could be reduced by ~33 tons per year. The reduction in gas emissions may also have long-term benefits to the environment. However, it is important to consider other possible external factors related to the adequate functioning of the propulsion system and the operational conditions of the boats, which can sometimes be subject to stochastic environmental interactions.

**Table 1.** Possible reduction in gas emissions by the optimized school boat.

| Gas | Emission Factor * (in g/kg fuel) | Reduction of Emissions per Day ** (in g) | Reduction of Emissions per Year ** (in g) |
|---|---|---|---|
| Carbon Dioxide ($CO_2$) | $3.2 \times 10^3$ | $91.20 \times 10^3$ | $33.28 \times 10^6$ |
| Methane ($CH_4$) | 1.70 | 48.50 | $17.68 \times 10^3$ |
| Nitrous Oxide ($N_2O$) | 0.08 | 2.28 | $0.83 \times 10^3$ |
| Carbon Monoxide (CO) | $1 \times 10^3$ | $28.50 \times 10^3$ | $10.40 \times 10^6$ |
| Nitrogen Oxides (NO, $NO_2$) | 9.7 | $0.28 \times 10^3$ | $10.09 \times 10^4$ |
| Non-Methane Volatile Organic Compounds (NMVOCs) | 34 | $0.97 \times 10^3$ | $35.37 \times 10^4$ |

* IPCC default emission factors for European ships and boats on inland waterways for gasoline 4-stroke engines [62,63]. ** Considering fuel density as 750 g/liter.

## 4. Alternatives to Improve the Performance of Small River Boats in the Amazon Region

Hull optimization is only one means of improving the performance of existing small boats that operate on rivers in remote regions. Other alternatives are shown in Figure 6. Practical engineering guidelines are needed to facilitate hull optimization improvements in the shipyards of the area. These guidelines should include typical specifications of any project, including the requirements of the boat owner, preliminary design, project contract, project planning and detailing, and construction details [64,65]. Considering that shipbuilding and repair is an important economic activity in Amazonas State, with over 10,000 people employed in shipyards [14], it must be possible to facilitate practical hull modifications here.

The various transport activities that take place on the rivers in the remote regions of the Amazon employ different types of small boats that use various means of propulsion [25]. Often, these methods are selected with no regard for the hydrodynamic relation between the propellers and the shape and size of the hull of the boat [66]. Although the adequate selection of the propulsor depends on the hull form and operation conditions [24], the diffusion of guidelines for a proper selection and operation of commercial propulsors in communities can bring positive impacts. Research and development (R&D) initiatives are required to improve existing propulsion technologies for these boats or to encourage a search for new ones [66,67]. The use of hydrogen-based technologies for propulsion could be an alternative in the future [68–71]. Adequate operation of the propulsor is required to reduce possible propeller vibrations since low-frequency noise due to cavitation and a non-uniform wake can have environmental impacts. The frequency band of the sound generated may affect many organisms [24].

Hydrofoil technologies would reduce the hydrodynamic resistance of the vessels. Although hydrofoils are nothing new in marine transport [72,73], their development in remote river regions, such as in the Amazon, would require R&D activities. Modern and efficient concepts could also be considered, such as those recently shown in [74,75].

Current advances in renewable energy technologies mean that boats could be engineered to operate using technologies that combine fossil fuels and renewable energies [76]. Hybrid technologies would make their functioning more sustainable, and various energy resources are available in the Amazon region. In the early stages of innovation, existing commercial technologies could be installed to harness renewable energies at a small scale in the boats, for example, to activate navigation controls or to maintain a backup battery. This battery could then provide illumination for activities performed at night, as needed in remote areas without energy.

Since the solar energy potential in the Amazon can be estimated in tens to hundreds of MWp (Mega Watt peak) [77,78], the potential for photovoltaic (PV) devices is huge. Solar panels could easily be installed on the roof of a boat. There are already some solar-powered vessels in the Amazon region, used for tourism [79] and transport [80]. It is also worth mentioning that solar challenges have been introduced in Brazil, which has encouraged universities and research centers to develop solar-powered devices in national and international competitions [81–83]. The use of solar energy to power marine vessels is currently increasing. For instance, solar energy is being considered by automobile companies for electric yachts [84].

Wind energy could be harnessed using small commercial wind turbines placed on the roof of a boat for small-scale power generation, such as those shown by [85]. Although wind currents are not constant in the Amazon region, these small devices could take advantage of the currents when they occur, thus contributing to battery charging.

With respect to hydrokinetic energy, water currents offer constant availability in the numerous rivers of the region. The flow velocity close to the water surface is small in some rivers (~0.4-0.6 m/s) but can reach ~2 m/s in other places (for details, see [86]). An array of small turbines could be deployed beneath the water surface to take advantage of river currents (see a simplified concept in Figure 6). While the boat moves forward, this may not be convenient due to the possible increase in hydrodynamic resistance. However, when the boat is at rest, these currents could be made use of. Some commercial devices that work at low current velocities (e.g., ~0.9 m/s [87]) could be adapted for this application. Some of these are practical and sufficiently portable to charge small electronic devices by harnessing wind or water currents [88]. Arrangements of these devices could be used to generate energy for electronic devices on the boat, particularly when it is at rest.

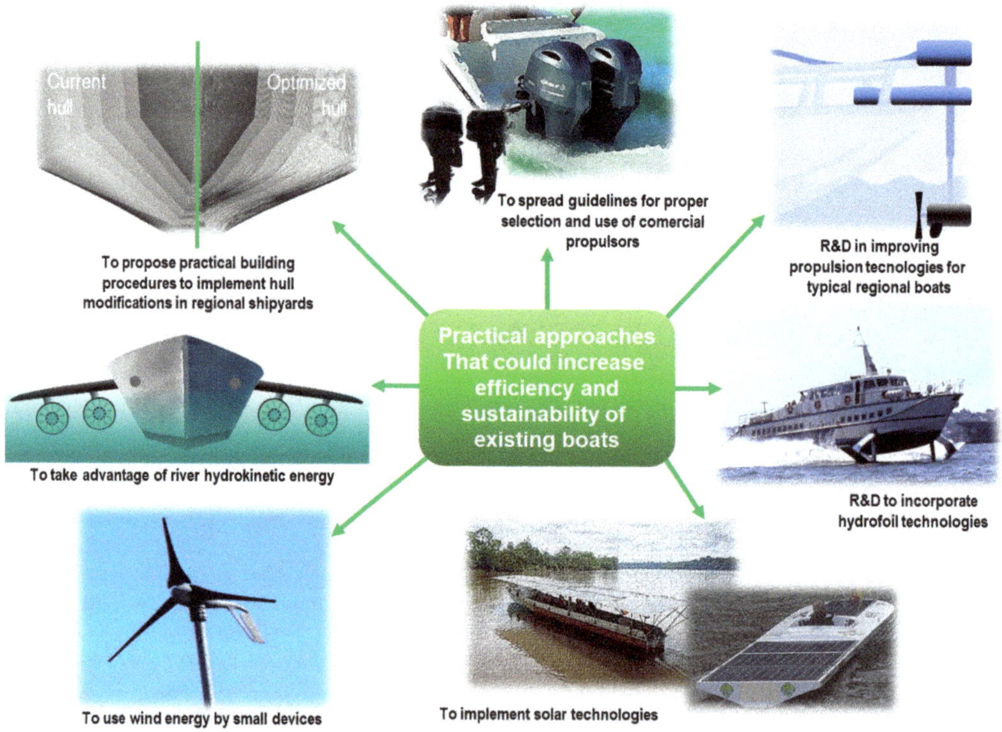


**Figure 6.** Possible improvements in the efficiency and sustainability of the existing boats used for daily activities in remote regions of the Amazon and similar regions elsewhere. Adapted image credits: commercial propulsors [89]; hydrofoil technologies [90]; solar technologies (left: [80], right: [83]); wind energy [91].

For all these innovations, a feasibility analysis should always be carried out to evaluate possible environmental impacts in the Amazonian environment.

## 5. Conclusions

Some alternatives for increasing the efficiency and sustainability of river transport in remote regions using small boats were explored in this paper. The focus was on improving the performance of existing boats, thereby contributing to the sustainable development of remote communities. Several means of improving the performance of small boats used for school transport in an Amazon riverine region were described. Improving propulsion and reducing the hydrodynamic resistance of vessels are proposed in the literature to improve their performance, whereas permanent modifications to the boat hulls may provide long-term benefits in terms of efficiency. Hull modifications to existing small boats used for river transport should be feasible. Perhaps these modifications could be carried out by small-scale, local shipbuilders, adding further value to these recommendations through the generation of local employment. Assuming that an improved school boat in the Amazon may have a ~5% reduction in fuel consumption, it was estimated that it would save thousands of liters of fuel per year. This would have socioeconomic and environmental benefits for the communities through the resulting reduction in greenhouse gas emissions.

Other means that could contribute to more sustainable boats in the future were also described. R&D are suggested to develop guidelines for the selection and development of improved propulsion devices, the improvement of hulls in existing boats, and the implementation of hydrofoil technologies. Renewable energy sources available in the region, such as hydrokinetic, solar, and wind energies, were also identified as a means of improving the sustainability of river transport in remote areas.

**Author Contributions:** Conceptualization, J.V.H.-F. and R.S.; methodology, J.V.H.-F., H.W.S.M., and R.S.; software, J.V.H.-F.; validation, J.V.H.-F.; formal analysis, J.V.H.-F., R.S., H.W.S.M., and V.C.; investigation, J.V.H.-F., H.W.S.M., V.C., and R.S.; resources, V.C. and R.S.; data curation, H.W.S.M., V.C., and J.V.H.-F.; writing—original draft preparation, J.V.H.-F., H.W.S.M., and R.S.; writing—review and editing, H.W.S.M., J.V.H.-F., V.C., and R.S.; visualization, H.W.S.M., J.V.H.-F., and V.C.; supervision, R.S.; project administration, J.V.H.-F.; funding acquisition, V.C. and R.S. All authors have read and agreed to the published version of the manuscript.

**Funding:** CONACYT-SENER-Sustentabilidad Energética, CEMIE-Océano project, Grant Agreement No. FSE-2014-06- 249795.

**Data Availability Statement:** Not applicable.

**Acknowledgments:** The authors wish to thank the CEMIE-Océano (project FSE-2014-06- 249795). The help provided by Jill Taylor for the revision of the manuscript is gratefully acknowledged.

**Conflicts of Interest:** The authors declare that they have no conflict of interest.

## References

1. Hilling, D. *Transport and Developing Countries*; Routledge: London, UK, 2003.
2. Joia, L.A.; dos Santos, R.P. ICT-Equipped Bank Boat and the Financial Inclusion of the Riverine Population of Marajó Island in the Brazilian Amazon. *Inf. Syst. J.* **2019**, *29*, 842–887. [CrossRef]
3. Globo Criança do AM Dribla Barreiras e Acha em Canoa "Cantinho" Para Estudar. Available online: http://g1.globo.com/am/amazonas/noticia/2013/10/crianca-do-am-dribla-barreiras-e-acha-em-canoa-cantinho-para-estudar.html (accessed on 23 June 2020). (In Portuguese).
4. Silva, J.G.D.R. Saberes e Práticas Tradicionais: As Condições Do Trabalho Nos Estaleiros Navais à Beira-Rio Da Cidade de Manaus. Ph.D. Thesis, Universidade Federal do Amazonas, Manaus, Brasil, 2016. (In Portuguese).
5. UN United Nations—Department of Economig and Social Affairs, Sustainable Development. The 17 Goals. Available online: https://sdgs.un.org/goals (accessed on 2 January 2021).
6. Ali, S.; Hussain, T.; Zhang, G.; Nurunnabi, M.; Li, B. The Implementation of Sustainable Development Goals in "BRICS" Countries. *Sustainability* **2018**, *10*, 2513. [CrossRef]
7. Crespo, B.; Míguez-Álvarez, C.; Arce, M.E.; Cuevas, M.; Míguez, J.L. The Sustainable Development Goals: An Experience on Higher Education. *Sustainability* **2017**, *9*, 1353. [CrossRef]
8. Centro de Referências Em Educação Integral. A Escola Dos Povos Ribeirinhos: Entre a Potência e Os Desafios. Available online: https://educacaointegral.org.br/reportagens/a-escola-dos-povos-ribeirinhos-entre-a-potencia-e-os-desafios/ (accessed on 2 January 2021).

9. Schneider, R.R.; Ariam, E.; Veríssimo, A.; Souza, C.J.; Barreto, P. *Sustainable Amazon: Limitations and Opportunities for Rural Development*; The World Bank: Washington, DC, USA, 2002.
10. Greene, D.L.; Wegener, M. Sustainable Transport. *J. Transp. Geogr.* **1997**, *5*, 177–190. [CrossRef]
11. Pojani, D.; Stead, D. Sustainable Urban Transport in the Developing World: Beyond Megacities. *Sustainability* **2015**, *7*, 7784–7805. [CrossRef]
12. Lins, N.V.M.; Rodrigues, L.R.Q.; Barreiros, N.R.; Machado, W.V. Construção Naval No Amazonas: Proposições Para o Mercado. In Proceedings of the Copinaval, Congreso Panamericano de Ingenieria Naval, Motevideo, Uruguay, 18–22 October 2009.
13. Canoagem Com Quantos Paus Se Faz Uma Canoa Tradicional Ribeirinha? Available online: http://www.canoagem.org.br/arquivos/ckfinder/files/artigosobreaconstrucaodecanoas.pdf (accessed on 4 November 2020).
14. Lameira, P.; Loureiro, E.; Moraes, H.; Figueiredo, N.; Benjamin, C. Optimization Design of Planning and Production Control in a Shipyard—Case Study: The Amazon Region. *Marit. Technol. Eng.* **2014**, *1*, 373.
15. Andrade, C.E.R.; Santos, M.F. A Carpintaria Naval Do Nordeste Paraense: "Do Ontem Ao Hoje". *Acta Fish. Aquat. Resour.* **2017**, *5*, 28–36.
16. Medeiros, J.T.D.S. O Transporte Fluvial e o Direito à Dignidade Da Pessoa Humana Na Amazônia. Ph.D. Thesis, Universidade do Estado do Amazonas, Manaus, Brasil, 2011.
17. Marinha do Brasil. *NORMAM—Normas Da Autoridade Marítima*; Marinha do Brasil: Brasilia, Brasil, 2020. (In Portuguese)
18. George, V.; Pillai, S.A. Energy Consumption and Conservation in Indian Fisheries. *Infofish* **1993**, *2*, 62.
19. Németh, P.S. *O Feitio Da Canoa Caiçara de Um Só Tronco: A Cultura Imaterial de Uma Nação, Em 25 Linhas. Dossiê Para Instrução de Processo de Registro de Bem Cultural de Natureza Imaterial Junto Ao IPHAN*; IPHAN: Sao Paulo, Brasil, 2018.
20. Nepstad, D.; Carvalho, G.; Barros, A.C.; Alencar, A.; Capobianco, J.P.; Bishop, J.; Moutinho, P.; Lefebvre, P.; Silva, U.L., Jr.; Prins, E. Road Paving, Fire Regime Feedbacks, and the Future of Amazon Forests. *For. Ecol. Manag.* **2001**, *154*, 395–407. [CrossRef]
21. Silva, S.B.; De Oliveira, M.A.; Severino, M.M. Economic Evaluation and Optimization of a Photovoltaic-Fuel Cell-Batteries Hybrid System for Use in the Brazilian Amazon. *Energy Policy* **2010**, *38*, 6713–6723. [CrossRef]
22. Hochkirch, K.; Bertram, V. Engineering Options for More Fuel Efficient Ships. In Proceedings of the First International Symposium on Fishing Vessel Energy Efficiency, Vigo, Spain, 20–22 May 2010.
23. Ghose, J. *Basic Ship Propulsion*; Allied Publishers: New Delhi, India, 2004.
24. Molland, A.F.; Turnock, S.R.; Hudson, D.A. *Ship Resistance and Propulsion*; Cambridge University Press: Cambridge, UK, 2017.
25. Isaac, V.; Almeida, M.; Cruz, R.; Nunes, L. Artisanal Fisheries of the Xingu River Basin in Brazilian Amazon. *Braz. J. Biol.* **2015**, *75*, 125–137. [CrossRef] [PubMed]
26. Papanikolaou, A.; Xing-Kaeding, Y.; Strobel, J.; Kanellopoulou, A.; Zaraphonitis, G.; Tolo, E. Numerical and Experimental Optimization Study on a Fast, Zero Emission Catamaran. *J. Mar. Sci. Eng.* **2020**, *8*, 657. [CrossRef]
27. Huang, F.; Chi, Y. Hull Form Optimization of a Cargo Ship for Reduced Drag. *J. Hydrodyn. Ser. B* **2016**, *28*, 173–183. [CrossRef]
28. Percival, S.; Hendrix, D.; Noblesse, F. Hydrodynamic Optimization of Ship Hull Forms. *Appl. Ocean Res.* **2001**, *23*, 337–355. [CrossRef]
29. Saha, G.K.; Suzuki, K.; Kai, H. Hydrodynamic Optimization of Ship Hull Forms in Shallow Water. *J. Mar. Sci. Technol.* **2004**, *9*, 51–62. [CrossRef]
30. Wilson, W.; Hendrix, D.; Gorski, J. Hull Form Optimization for Early Stage Ship Design. *Nav. Eng. J.* **2010**, *122*, 53–65. [CrossRef]
31. Kim, H.; Yang, C. A New Surface Modification Approach for CFD-Based Hull Form Optimization. *J. Hydrodyn.* **2010**, *22*, 503–508. [CrossRef]
32. Hollenbach, U.; Friesch, J. Efficient Hull Forms—What Can Be Gained. In Proceedings of the 1st International Conference on Ship Efficiency, Hamburg, Germany, 8–9 October 2007.
33. Spagnolo, G.S.; Papalillo, D.; Martocchia, A.; Makary, G. Solar-Electric Boat. *J. Transp. Technol.* **2012**, *2*, 144–149. [CrossRef]
34. Ozden, M.C.; Demir, E. The Successful Design and Construction of Solar/Electric Boats Nusrat and Muavenet: An Overview. In Proceedings of the Ever Monaco 2009 Conferences on Ecological Vehicles and Renewable Energies, Monte Carlo, Monaco, 26–29 March 2009; pp. 27–29.
35. Guamán, F.; Ordoñez, J.; Espinoza, J.; Jara-Alvear, J. Electric-Solar Boats: An Option for Sustainable River Transportation in the Ecuadorian Amazon. In Proceedings of the 6th International Conference on Energy and Sustainability, WIT Transactions on Ecology and The Environment, Medelin, Columbia, 2–4 September 2015; Volume 195, pp. 439–448.
36. Jiang, F.; Xie, H.; Ellen, O. Hybrid Energy System with Optimized Storage for Improvement of Sustainability in a Small Town. *Sustainability* **2018**, *10*, 2034. [CrossRef]
37. Zhang, B.; Ma, K.; Ji, Z. The Optimization of the Hull Form with the Minimum Wave Making Resistance Based on Rankine Source Method. *J. Hydrodyn.* **2009**, *21*, 277–284. [CrossRef]
38. Park, J.-H.; Choi, J.-E.; Chun, H.-H. Hull-Form Optimization of KSUEZMAX to Enhance Resistance Performance. *Int. J. Nav. Archit. Ocean Eng.* **2015**, *7*, 100–114. [CrossRef]
39. Grigoropoulos, G.J. Hull Form Optimization for Hydrodynamic Performance. *Mar. Technol.* **2004**, *41*, 167–182.
40. Jeong, K.-L.; Jeong, S.-M. A Mesh Deformation Method for CFD-Based Hull Form Optimization. *J. Mar. Sci. Eng.* **2020**, *8*, 473. [CrossRef]
41. Seok, W.; Kim, G.H.; Seo, J.; Rhee, S.H. Application of the Design of Experiments and Computational Fluid Dynamics to Bow Design Improvement. *J. Mar. Sci. Eng.* **2019**, *7*, 226. [CrossRef]

42. Cordonnier, J.; Gorintin, F.; De Cagny, A.; Clément, A.; Babarit, A. SEAREV: Case Study of the Development of a Wave Energy Converter. *Renew. Energy* **2015**, *80*, 40–52. [CrossRef]
43. Garcia-Teruel, A.; Forehand, D. Optimal Wave Energy Converter Geometry for Different Modes of Motion. In Proceedings of the 3rd International Conference on Renewable Energies Offshore, Lisbon, Portugal, 8–10 October 2018; pp. 299–305.
44. Goggins, J.; Finnegan, W. Shape Optimisation of Floating Wave Energy Converters for a Specified Wave Energy Spectrum. *Renew. Energy* **2014**, *71*, 208–220. [CrossRef]
45. Han, S.; Lee, Y.-S.; Choi, Y.B. Hydrodynamic Hull Form Optimization Using Parametric Models. *J. Mar. Sci. Technol.* **2012**, *17*, 1–17. [CrossRef]
46. Yu, J.-W.; Lee, C.-M.; Lee, I.; Choi, J.-E. Bow Hull-Form Optimization in Waves of a 66,000 DWT Bulk Carrier. *Int. J. Nav. Archit. Ocean Eng.* **2017**, *9*, 499–508. [CrossRef]
47. Brizzolara, S.; Bruzzone, D. Optimising the Steady Hydrodynamic Performance of Two High-Speed Trimaran Hull Forms. In Proceedings of the International Conference on Ship and Shipping Research, NAV 2006, Genova, Italy, 21–23 June 2006; Volume 3, pp. 1–3.
48. Kükner, A.; Sariöz, K. High Speed Hull Form Optimisation for Seakeeping. *Adv. Eng. Softw.* **1995**, *22*, 179–189. [CrossRef]
49. Song, S.; Demirel, Y.K.; Muscat-Fenech, C.D.M.; Tezdogan, T.; Atlar, M. Fouling Effect on the Resistance of Different Ship Types. *Ocean Eng.* **2020**, *216*, 107736. [CrossRef]
50. El Moctar, O.; Sigmund, S.; Ley, J.; Schellin, T.E. Numerical and Experimental Analysis of Added Resistance of Ships in Waves. *J. Offshore Mech. Arct. Eng.* **2017**, *139*. [CrossRef]
51. Sigmund, S.; El Moctar, O. Numerical and Experimental Investigation of Added Resistance of Different Ship Types in Short and Long Waves. *Ocean Eng.* **2018**, *147*, 51–67. [CrossRef]
52. Gerritsma, J.; Beukelman, W. Analysis of the Resistance Increase in Waves of a Fast Cargo Ship. *Int. Shipbuild. Prog.* **1972**, *19*, 285–293. [CrossRef]
53. Fundo Nacional de Desenvolvimento Da Educação (FNDE). FNDE Doa Primeiras Lanchas Escolares a Municípios Do Amazonas. Available online: https://www.fnde.gov.br/index.php/acesso-a-informacao/institucional/area-de-imprensa/noticias/item/1978-fnde-doa-primeiras-lanchas-escolares-a-munic%C3%ADpios-do-amazonas (accessed on 7 November 2020).
54. Critica Oito Lanchas de Escolas Públicas de Manaus São Abandonadas. Available online: https://www.acritica.com/channels/cotidiano/news/oito-lanchas-de-escolhas-publicas-sao-abandonadas (accessed on 7 November 2020). (In Portuguese).
55. Globo. Ribeirinhos Na Amazônia Viajam Mais de 3h de Barco Para Ir à Escola. Available online: http://g1.globo.com/am/amazonas/noticia/2012/04/ribeirinhos-na-amazonia-viajam-mais-de-3h-de-barco-para-ir-escola.html (accessed on 7 November 2020).
56. Fundo Nacional de Desenvolvimento Da Educação (FNDE). *National Fund For Development of Education. Relatório de Gestão*; Ministry of Education: Brasília, Brasil, 2017.
57. Correio-da-Amazonia Correio Da Amazônia. Crianças Do Amazonas Enfrentam Embarcações Para Chegar Na Escola. Available online: https://correiodaamazonia.com/criancas-amazonas-enfrentam-embarcacoes-para-chegar-escola/ (accessed on 1 July 2020).
58. BetterBoat. How Much Is a Boat Actually Going to Cost You? That All Depends. Available online: https://betterboat.com/how-much-is-a-boat/ (accessed on 25 June 2020).
59. Global-Petrol-Prices. Gasoline Prices, Liter. Available online: https://www.globalpetrolprices.com/gasoline_prices/ (accessed on 25 June 2020).
60. Vinh, B.T.; Truong, N.H. Erosion Mechanism of Nga Bay Riverbanks, Ho Chi Minh City, Vietnam. *Asean Eng. J.* **2014**, *3*, 132–141.
61. Holanda, F.S.R.; Santos, L.G.D.C.; Santos, C.M.D.; Casado, A.P.B.; Pedrotti, A.; Ribeiro, G.T. Riparian Vegetation Affected by Bank Erosion in the Lower São Francisco River, Northeastern Brazil. *Rev. Árvore* **2005**, *29*, 327–336. [CrossRef]
62. Jun, P.; Gillenwater, M.; Barbour, W. $CO_2$, $CH_4$, and $N_2O$ Emissions from Transportation-Water-Borne-Navigation. In *Good Practice Guidance and Uncertainty Management in National Greenhouse Gas Inventories*; IPCC: Geneva, Switzerland, 2002; pp. 71–92.
63. 2006 IPCC Guidelines for National Greenhouse Gas Inventories. Available online: https://www.ipcc-nggip.iges.or.jp/public/2006gl/ (accessed on 1 March 2020).
64. Wu, Y.-H.; Shaw, H.-J. Document Based Knowledge Base Engineering Method for Ship Basic Design. *Ocean Eng.* **2011**, *38*, 1508–1521. [CrossRef]
65. Eyres, D.J. *Ship Construction*; Elsevier: Amsterdam, The Netherlands, 2006.
66. Favacho, B.I.; Vaz, J.R.P.; Mesquita, A.L.A.; Lopes, F.; Moreira, A.L.S.; Soeiro, N.S.; Rocha, O.F.L.D. Contribution to the Marine Propeller Hydrodynamic Design for Small Boats in the Amazon Region. *Acta Amaz.* **2016**, *46*, 37–46. [CrossRef]
67. Pinheiro, K.; Neto, G.; Filho, S.; Morais, S.; Mesquita, A. Análise de Alternativas Para Novo Sistema de Propulsão de Lanchas Escolares. In Proceedings of the CONEM 2016—Congresso Nacional de Engenharia Mecânica, Fortaleza, CE, Brasil, 21–25 August 2016. (In Portuguese).
68. Tronstad, T.; Astrand, H.H.; Haugom, G.P.; Langfeldt, L. *Study on the Use of Fuel Cells in Shipping*; European Maritime Safety Agency: Lisbon, Portugal, 2017.
69. BBC. The Fuel That Could Transform Shipping. Available online: https://www.bbc.com/future/article/20201127-how-hydrogen-fuel-could-decarbonise-shipping (accessed on 4 February 2021).
70. ShipTechnology. HySHIP: Inside Europe's Flagship Hydrogen Ship Demonstrator Project. Available online: https://www.ship-technology.com/features/hydrogen-vessel/ (accessed on 4 February 2021).

71. Reuters. First Wave of Ships Explore Green Hydrogen as Route to Net Zero. Available online: https://www.reuters.com/article/shipping-energy-hydrogen-focus-int-idUSKBN27F18U (accessed on 12 December 2020).
72. Craft, U.N.H. Balancing Mission Requirements and Hydrofoil Design Characteristics. *J. Hydronautics* **1967**, *1*. [CrossRef]
73. Ellsworth, W. Hydrofoil Development—Issues and Answers. In Proceedings of the AIAA/SNAME Advanced Marine Vehicles Conference, San Diego, CA, USA, 25–27 February 1974.
74. Dezeen. Cas Dahmen Creates Zero-Emission Hydrofoil Concept. Available online: https://www.dezeen.com/2018/08/06/cas-dahmen-zero-emission-hydrofoil-concept-hydropace-design/ (accessed on 25 January 2021).
75. DNV-GL. A New Era for Yacht Design. Available online: https://www.dnvgl.com/expert-story/maritime-impact/a-new-era-of-yacht-design.html (accessed on 13 January 2021).
76. Aziz, A.S.; Tajuddin, M.F.N.; Adzman, M.R.; Ramli, M.A.; Mekhilef, S. Energy Management and Optimization of a PV/Diesel/Battery Hybrid Energy System Using a Combined Dispatch Strategy. *Sustainability* **2019**, *11*, 683. [CrossRef]
77. Matos, F.B.; Camacho, J.R.; Rodrigues, P.; Guimarães, S.C., Jr. A Research on the Use of Energy Resources in the Amazon. *Renew. Sustain. Energy Rev.* **2011**, *15*, 3196–3206. [CrossRef]
78. Martins, F.R.; Rüther, R.; Pereira, E.B.; Abreu, S. Solar Energy Scenarios in Brazil. Part Two: Photovoltaics Applications. *Energy Policy* **2008**, *36*, 2865–2877. [CrossRef]
79. Canal Solar. Barco Hotel No Amazonas Inova e Opera Com 24 KWp de Energia Solar e Baterias. Available online: https://canalsolar.com.br/noticias/item/1214-barco-hotel-no-amazonas-inova-e-opera-com-24-kwp-de-energia-solar-e-baterias (accessed on 1 January 2021).
80. Plitt, L.; BBC News Brasil. A Canoa Solar Na Amazônia Que Ajuda Comunidades a Navegar Sem Gasolina Pela Selva, by Laura Plitt, BBC Mundo, Enviada Especial à Amazônia Equatoriana. Available online: https://www.bbc.com/portuguese/brasil-43824464 (accessed on 3 January 2021).
81. EST. Projeto Leviatã. Available online: https://est.uea.edu.br/#/projeto/1/leviata (accessed on 4 January 2021).
82. UFSC. Estudantes Da UFSC e UFRJ Participam de Copa Do Mundo de Barcos Solares. Available online: https://noticias.ufsc.br/2014/07/estudantes-da-ufsc-e-ufrj-participam-de-copa-do-mundo-de-barcos-solares/ (accessed on 4 January 2020).
83. Gorter, T. Design Considerations of a Solar Racing Boat: Propeller Design Parameters as a Result of PV System Power. *Energy Procedia* **2015**, *75*, 1901–1906. [CrossRef]
84. Volkswagen, AG. With the MEB and CUPRA on the High Seas. The "Silent" Yatch with MEB and CUPRA Design. Available online: https://www.volkswagenag.com/en/news/stories/2021/01/with-the-meb-and-cupra-on-the-high-seas.html# (accessed on 1 February 2021).
85. SkyWind Wind Turbines: Powerful, Reliable & Easy To Set Up! Available online: https://myskywind.shop/en (accessed on 29 December 2020).
86. da Silva Cruz, J.; Blanco, C.J.C.; Junior, A.C.P.B. Flow-Velocity Model for Hydrokinetic Energy Availability Assessment in the Amazon. *Acta Sci. Technol.* **2020**, *42*, e45703. [CrossRef]
87. Waterotor. Waterotor: Low Cost Power from Slow Moving Water. Available online: https://waterotor.com/ (accessed on 4 January 2021).
88. Waterlily. WaterLily: Rivers Flow 24/7—So Does Your Power. Available online: https://www.waterlilyturbine.com/ (accessed on 4 January 2020).
89. Hendricks, J. Boats: Selecting the Best Outboard Motor; How to Choose between Four- and Six-Cylinder Outboard Motors for Your Boat. Available online: https://www.sportfishingmag.com/selecting-best-outboard-motor/ (accessed on 3 January 2021).
90. express000 00297 (348) 30-12-1970 Sydney Ferries Hydrofoil Fairlight Passing Garden Island on Sydney Harbour, Sydney, New South Wales, Australia. by Express000 Is Licensed with CC BY-NC-SA 2.0. 1970. Available online: https://search.creativecommons.org/photos/61a782ff-0ada-4cb5-8c0c-da00e57de59d (accessed on 1 February 2020).
91. Somma, R. "Small Wind Turbines" by Ryan Somma Is Licensed with CC BY-SA 2.0. 2021. Available online: https://search.creativecommons.org/photos/fafb492c-64b0-4837-b29f-22f570bb0296 (accessed on 1 February 2020).

applied sciences

MDPI

Article

# Assessment of Cold Ironing and LNG as Mitigation Tools of Short Sea Shipping Emissions in Port: A Spanish Case Study

**Alba Martínez-López [1,*], Alejandro Romero [2] and José A. Orosa [3]**

1 Department of Mechanical Engineering, University of Las Palmas de Gran Canaria, 35014 Las Palmas de Gran Canaria, Spain
2 Oceanic Platform of the Canary Islands (PLOCAN), 35214 Telde, Spain; alejandro.romero@plocan.eu
3 Department of Nautical Science and Marine Engineering, E.T.S.N.M., University of A Coruña, 15011 Corunna, Spain; jose.antonio.orosa@udc.es
* Correspondence: alba.martinez@ulpgc.es

**Abstract:** By the end of 2025 European ports are required to provide (Directive 2014/94/EU) facilities to ensure the Liquefied Natural Gas (LNG) use and on-shore electricity supply for vessels (Cold Ironing—CI). Even though this involves considerable port investment, many uncertainties about CI and LNG performance exist because their application depends on vessel operators' willingness. Additionally, lag times for CI connection/disconnection along with methane emissions from LNG undermine their feasibility for Short Sea Shipping (SSS). Since, among the SSS aims are the reduction in berthing times and its effectiveness for-inter-islands' traffic where, land electricity grids are frequently dependent on the fuel burning generation by penalizing the CI performance. This paper introduces a calculation method to evaluate the pollution savings in monetary terms by CI and LNG use in SSS. The method is applied to three European routes by testing the environmental performance of two fleets: feeder and Ro-Pax vessels. The results show that feeders reach higher environmental improvements by using port mitigation than Ro-Pax vessels. Additionally, the need for ensuring the sustainability of on-shore grids before the CI implementation was evinced, especially in insularity frameworks, where the environmental benefits from LNG use proved to be more effective.

**Keywords:** short sea shipping; cold ironing; on-shore power supply; LNG; port sustainability

**Citation:** Martínez-López, A.; Romero, A.; Orosa, J.A. Assessment of Cold Ironing and LNG as Mitigation Tools of Short Sea Shipping Emissions in Port: A Spanish Case Study. *Appl. Sci.* **2021**, *11*, 2050. https://doi.org/10.3390/app11052050

Academic Editor: Thomas Brewer

Received: 19 January 2021
Accepted: 22 February 2021
Published: 25 February 2021



## 1. Introduction

Directive 2014/94/EU requires European Union (EU) ports that are part of the core Trans-European Network for Transport (TEN-T) to provide the facilities to enable alternative fuel use by the end of 2025. Significant port investment is needed to accomplish this Directive; especially to provide on-shore electricity supply (Cold Ironing (CI)) for vessels during berthing and access to Liquefied Natural Gas (LNG) refueling points. Despite the investment required, the current European regulation does not oblige vessel operators to use them (Directive 1999/32/EC, Directive 2005/33/EC). Thus, it can be affirmed that CI and LNG might not be used, even if the facilities exist. Since their capacity to abate port pollution relies on the willingness of vessel operators (the penetration rate) the concern about "guesstimates" for the ports [1] is shared by all geographical zones where the use of sustainable solutions in port is governed by less strict legislation.

The uncertainties about the CI and LNG penetration rates for vessel operators open up an interesting discussion, not only about the expected effectiveness of these measures to mitigate port pollution [2–4] but also about who should pay for it. The answer is not self-evident [1,5] since the benefits of their use are related to society and the environment. Thereby, some of key investors (the vessel operators) do not necessarily obtain economic benefits from them.

This attitude contrasts with the EU treatment of other transport modes, where the "polluter pays" and "user pays" principles are consistently applied through the internalization of external costs via taxes and charges collected in European Regulation (Directive 1999/62/EC amended by Directive 2011/76/EU, and currently affected by the proposal to amend COM (2017) 275 for road transport). From an environmental perspective, the European regulation has been more restrictive and efficient for load transport by road than maritime transport [6]. Thus, the traditional green label for Short Sea Shipping (SSS) in the domestic traffic is currently under discussion [6,7], since its current technical limitations are far from meeting the emission standards required by the EU for the land transport (Euro VI standards) [8,9].

In light of the foregoing, the performance of port mitigation alternatives is particularly interesting for SSS [7,10]. The intensity of this type of traffic (high number of calls and sailing close to the coast), along with its energy requirements during berthing, have led to the identification of SSS vessels as the largest beneficiaries of CI [3,5,11] and LNG [12]. In this regard, Christodoulou and Cullinane, 2020 [10] highlighted the evidence of the environmental benefits obtained in SSS from the combination of technical and operational initiatives of shipping companies adopted as complementary measures to the normative efforts of the authorities.

However, a number of authors have raised concerns about this type of traffic's mitigation practices in port: additional times are required for CI connection that could make SSS competitiveness unfeasible for short calls [13,14]; the islands are frequently involved in SSS activity and their electricity grids are mainly supplied by burning oil, leading CI to be a poor sustainable alternative [5].

Beyond the Sulfur Emission Control Areas (SECA) and Emission Control Area (ECA) zones, defined in Annex VI to MARPOL where $SO_x$ and $NO_x$ emissions are restricted during sailing, regulatory pressure has been increasing in ports. Hence, in the EU, since 2010 Community ports have demanded a maximum limit of 0.1% sulfur by weight for marine fuels used by inland waterway vessels and ships at berth (Directive 2005/33/EC; amending Directive 1999/32/EC). However, emission —equivalent to those that would be achieved through the limits on sulfur in fuel—can be authorized through the use of abatement technologies. Therefore, the sulfur limitations collected in the Directive were met through different operational solutions (e.g., MGO fuel types 0.1% sulfur, HFO fuel with scrubber use or LNG as fuel, etc.). Despite this restriction, the Commission in 2006 recognized that the regulation established up to that date had been insufficient to maintain port air quality. Consequently, it recommended shore-side electricity (CI) for use by ships at berth in Community ports (Recommendation 2006/339/EC), since it provides additional benefits such as noise reduction, which is especially important for ports situated near residential areas. The COM 2013(295) highlighted the need for stricter requirements on environmental performance in ports. In a further step, Directive 2014/94/EU forces Member States to ensure not only shore-side electricity supply (CI) for vessels but also available refueling points for LNG by 31 December 2025 in TEN-T Core Network ports. In parallel, EU Regulation (2017/352) seeks to define the common criteria (European Commission and Member States jointly) for voluntary environmental charges in port. Additionally, this Regulation gives ports the freedom to make changes in order to promote the sustainable use of their own infrastructures.

The provision of LNG and CI services involve significant port investment; nonetheless, CI is often presented as a feasible solution for emissions mitigation [1–4]. This is so, even when CI's high dependence on on-shore energy (the vessels' electricity consumption is met by plugging them into the on-shore electricity network during berthing time). The main advantage of CI is the relocation of the emission source far from ports where the existence of significant population centers in the port's hinterlands makes $NO_x$ and $SO_x$ emissions from vessels especially harmful, since these pollutants have a direct impact on citizens. The additional advantages of CI in contrast to other mitigation solutions are the elimination of port waste, noise, vibrations etc. Nevertheless, Winkel et al. [5] warned about the risk of CI use in regions that are highly dependent on fossil fuels for electricity generation (many of the islands), while Zis et al. [14] highlighted the possible lack of feasibility of CI for short call traffic (lag connection times). In turn, LNG has proved to be the most suitable fuel alternative for SSS when cost, time and sustainability are jointly evaluated from the vessel operator's point of view [12]. However, methane emissions remain a weak point that penalizes the reliability of LNG as a mitigation port solution.

In order to reduce the uncertainties about CI and LNG performance on the mitigation of SSS port pollutant emissions, this paper introduces a mathematical method to calculate emission savings by LNG and CI use versus traditional on-board electricity supply. The method quantifies in monetary terms the emissions savings for SSS in such a way that it permits us to assess in several routes, not only the performance of one port abatement alternative, but also to compare several systems' performance with diverse types of suitable vessels to SSS. The method has been tested through its application on three SSS routes, where ports with different electricity grids are involved: the Atlantic coast between France and Spain; the South of Spain and the Canary Islands; and interisland traffic (Canary Islands). The application case considered two different kinds of vessels for the routes: feeder and Ro-Pax vessels.

The paper offers quantitative results that are useful not only for the port authorities and stakeholders involved in the case studies, but also for policy makers and ship operators who can evaluate the influence of on-shore electricity grids and vessel characteristics on the performance of the mitigation measures in port.

## 2. Methods

In order to accurately assess the impact of port mitigation solutions on SSS activity, the following paragraphs introduce a calculation model able to quantify in monetary terms the performance of these measures, with special attention being paid to CI and LNG alternatives. Appendix A provides the detailed nomenclature used in this section for the variables involved.

The calculation procedure assumes two circumstances: on-board electricity supply and off-shore electricity supply. The former offers two possibilities: the conventional generation of electricity on board through Low Sulfur Marine Gas Oil (LMGO) (0.5% sulfur content) and the use of on-board generating sets driven by LNG dual engines; while the latter involves CI use (see Figure 1). Even though CI is just an alternative for the berthing time of the vessels (see Figure 1), in order to offer greater knowledge of port pollution, the air emissions generated during vessel maneuvers are also evaluated.

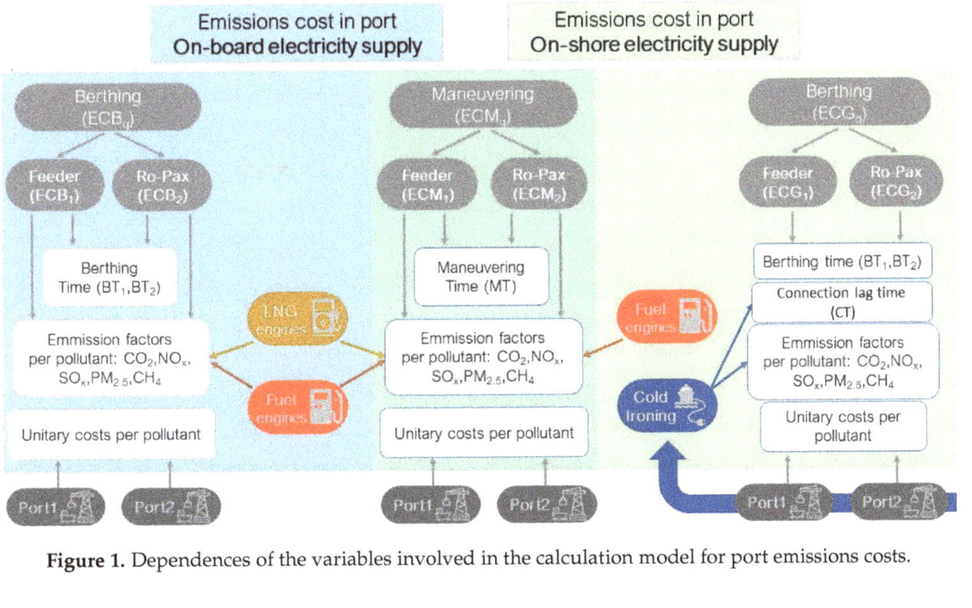


**Figure 1.** Dependences of the variables involved in the calculation model for port emissions costs.

Thus, Equations (1) and (3) provide the yearly environmental cost (EUR/year) for on-board electricity generation, during the maneuvering stage ($ECM_q$, $\forall q \in Q$) and berthing stage ($ECB_q$, $\forall q \in Q$) on a SSS line.

$$ECM_q = N \cdot \sum_{k=1}^{2} (ECMqk);\ \forall q \in Q \tag{1}$$

$$ECM_{q,k} = \sum_{u=1}^{5} (EFMuq \cdot (MT) \cdot UCukv) + IPBq \cdot EFe \cdot LFMq \cdot MT \cdot UC6kv; \\ \forall q \in Q \wedge \forall k \in K \wedge \forall v \in V \wedge \forall e \in EE \tag{2}$$

$$ECB_q = N \cdot \sum_{k=1}^{2} (ECBqk);\ \forall q \in Q \tag{3}$$

$$ECB_{q,k} = \sum_{u=1}^{5} (EFBuq \cdot (BTq) \cdot UCukv) + IPBq \cdot EFe \cdot LFBq \cdot BTq \cdot UC6kv; \\ \forall q \in Q \wedge \forall k \in K \wedge \forall v \in V \wedge \forall e \in EE \tag{4}$$

Likewise, Equation (5) ($ECG_q$, $\forall q \in Q$) offers information about the yearly environmental costs during berthing when the vessel is plugged into the electricity grid (CI option).

$$ECG_q = N \cdot \sum_{k=1}^{2} (ECGqk);\ \forall q \in Q \tag{5}$$

$$ECG_{q,k} = \sum_{u=1}^{5} (EFGuk \cdot (BTq + CT) \cdot PBq \cdot UCukv) + PBq \cdot EFG6k \cdot (BTq + CT) \cdot UC6kv; \\ \forall q \in Q \wedge \forall k \in K \wedge \forall v \in V \tag{6}$$

These Equations (1)–(6) quantify in monetary terms the impact of the pollutants (U = {1, ... , $u$}): $NO_x$ (ozone precursors), $SO_2$ (acidifying substances), $PM_{2.5}$, $PM_{10}$ (particular matter mass), and the greenhouse gases ($CO_2$) that are emitted to the air by vessels that are suitable for SSS (Q = {1, ... , $q$}, container and roll-on, roll-off vessels) by operating in several ports (K = {1, ... , $k$}). Additionally, the methane impact ($CH_4$ pollutant) is also integrated into the assessment by considering its emission factor ($EF_e$; $\forall e \in EE$) for the different kinds of engines (EE = {1, ... , $e$}) and their load factor ($LFM_q$-maneuvering-$LFB_q$-berthing-, $\forall q \in Q$).

The calculation of the environmental costs (Equations (1)–(6)) also considers the yearly trips of the shipping lines ($N$), the times (hours) invested in the stages ($MT$-maneuvering time-, $BTq$; $\forall q \in Q$ -berthing time-), emission factors, unitary costs for the pollutants and the installed power-MCR- for operations during every operational stage ($IPB_q$; $\forall q \in Q$). This power is only the auxiliary engines' power for the berthing whereas, in the maneuvering stage propulsion power (main engine power) is also considered. Due to the features of the electricity grid, equation 6 additionally integrates the connection lag time ($CT$) and the required power for the berthing operations ($PB_q$; $\forall q \in Q$).

The unitary cost of pollutants ($UC_{ukv}$; $\forall u \in U \wedge \forall k \in K \wedge \forall v \in V$) involves the average "damage cost" for transport emissions (EUR/kg) by taking into account the geographical location of the port (K = {1, ... , $k$}) and the population density in its hinterlands (V = {1, ... , $v$}). For European cases this information is regularly published by the European Commission in the "*Handbook on the external costs of transport* (last updated in 2019 [15]) for every country. Successive updates can be estimated with the national Consumer Price Index (CPI) for every country (National Institute of Statistics and Economic Studies of France [16], National Statistics Institute of Spain [17], etc.).

The emission factors for the vessels during the port operations—that is, during maneuvering ($EFM_{uq}$; $\forall u \in U \wedge \forall q \in Q$ see Equation (2)) and berthing ($EFB_{uq}$; $\forall u \in U \wedge \forall q \in Q$ see Equation (4))—are highly dependent on the kinds of engines and power required at each operational stage. These emission factors (kg/h) can be obtained through the calculation tools, whereas the emission factors for $CH_4$ ($EF_e$; $\forall e \in EE$) can be taken from previous publications for each kind of engine (i.e., dual fuel engines operating with LNG [8,18] and for fuel-based engines [19]).

In turn, the emission factors related to the electricity grid ($EFG_{uk}$; $\forall u \in U \wedge \forall k \in K$, see Equation (6)) are a direct consequence of the dominant sources in the land generating plants. Consequently, the localization of the port and the relative weight of renewable sources on the land-based energy mix are the main drivers of these emission factors (kg/kWh).

*Practical Case Study*

The selected case study attempt to extract significant findings through comparing CI and LNG performance not only in different frameworks but also through different fleets. Thus, the selected examples are application cases where the expected benefits from mitigation port systems use would reach maximum levels. This aim involves the analysis of the most demanding SSS lines: high frequency transport services (5 calls/week) through the operation of vessels with high electricity requirements in port: Ro-Pax vessels [3,5] and feeder vessels with reefer containers [11,14].

The analysis considers the fleets shown in Table 1; optimized container vessel (capacity to 185 reefer container) [20] and conventional Ro-Pax vessels by operating both of them under SSS conditions. The focus is on the vessels' electricity balance, with the most demanding operational stage being maneuvering. This is so because this stage demands additional electrical power to supply, besides the steering and mooring systems, the bow thrusters (see Table 1). This fact, along with the significance of the time invested in maneuvering over the whole time in SSS, forces us to include this operational stage in the analysis.

**Table 1.** Technical features of the fleets considered for the application cases.

| | | Feeder Vessel (reefer) ** | Ro-Pax Vessel *** |
|---|---|---|---|
| | Cargo units | 184 (reefer containers) | Pax:1000<br>Lane length for trailers: 1600 m<br>Cars: 160<br>Reefer cargo's plugs: 100 |
| **Technical Features** | Lbp (m) | 78.15 | 153.25 |
| | B (m) | 14.46 | 28.65 |
| | D (m) * | 7.41 | 13.85 |
| | GT | 2456 | 26,916 |
| | Service Speed (kn) | 19.49 | 23.00 |
| | Main Engine (kW) | 7000 | 2 × 15,600 |
| | Type of Main Engine | Tier III (MGO) | Tier III (MGO) |
| | Auxiliary engines (kW) | 2 × 662 | 3 × 1254 |
| | Power take off (PTO) (kW) | 1500 | 2 × 1000 |
| | Bow thruster (kW) | 350 | 2 × 1000 |
| **Electricity demand** | Maneuvering (kW) | 1620 | 4936 |
| | Berthing (kW)-*PBq*- | 1200 | 1880 |
| **Electricity support** | Maneuvering | PTO + 1 auxiliary engine | 2 PTO + 3 auxiliary engines |
| | Berthing | 2 auxiliary engines | 2 auxiliary engines |
| **Maritime-Route 1** | | ***Vigo(Spain)–St. Nazaire (France)*** | |
| **Maritime-Route 2** | | ***Las Palmas (Gran Canaria island)–Huelva (Spain)*** | |
| **Maritime-Route 3** | | ***Las Palmas (Gran Canaria island)–Sta. Cruz de Tenerife (Tenerife island)*** | |

* Depth to upper continuous deck for Ro-Pax vessel. ** More technical features in Martínez-López et al. [20]. *** More technical features on the base vessels: "Sorolla" (http://www.hjbarreras.es/?page=lis-ferries&idp=10) and "Martí i Soler" (http://www.hjbarreras.es/?page=lis-ferries.2&idp=35).

Container vessels' electricity plants are designed to supply all cargo units' demand as reefer containers (which needs most electricity power). Likewise, the engine room arrangement for the Ro-Pax vessels is designed on the basis of two main engines, (with two main engine driven alternators; Power take off (PTO) in their gearboxes) by moving two propellers. This propulsion configuration, together with the installation of two bow thrusters, significantly improves the maneuvering capacity of these vessels, but also increases their electricity demands (see Table 1). Finally, the Ro-Pax vessels assumed in the application cases require a sufficient energy to supply in port: a plugging capacity for 100 reefer cargo units, high capacity for the cold storage warehouses, air conditioning systems, two stern amp doors and one movable car deck in the garage.

It should be borne in mind that the vessels' boiler emissions have not been considered in the analysis, since shore power is not an alternative for their activity [4,14]. Consequently, even though the vessel is plugged into the land grid during the berthing stage, the vessels' boilers are working to provide the steam required for the fuel tanks' heating coils (among other uses).

Table 1 also shows the maritime routes selected for the case studies. These routes involve ports that belong to electricity grids with large differences in terms of sustainability. The first, between Vigo in the northwest of Spain and St. Nazaire in Brittany (France), was selected mainly due to the different environmental footprint of the electricity grids of both countries (see Section 3.1). Additionally, this route currently operates as a European Motorway of the Sea (MoS), having been a frequently studied route for optimizing fleets in the past [20,21]. This permits us to widen the analysis by comparing unconventional vessels (e.g., optimized feeder vessels) to the conventional ones.

Case study 2, between the Canary Islands (Las Palmas port) and the Iberian Peninsula (Huelva, Spain), represents a habitual kind of shipping between continental Europe and its islands. Each island is supplied by its own electricity plants and therefore the sustainability of its grid is completely different from that on the continent (see Section 3.1). Finally, case study 3, between two islands of the Canary archipelago (Las Palmas and Sta. Cruz de Tenerife), introduces one of the most frequented maritime routes in Europe (interisland), but with the singularity of a scarce share of renewable energy sources in their electricity generation.

The operation times in port were estimated through expressions published by Martínez-López et al. [21] for SSS vessels (feeder and Ro-Pax vessels). The results were tested with real values from Las Palmas port (an average loading/unloading time of 3.5 and 2.3 h for feeder and Ro-Pax, respectively, and 30 minutes for the maneuvering stage). Additionally, a hoteling time of 8 h per day (from 23 p.m. to 7 a.m.) was assumed for each shipping line. Despite some of the study ports being involved in CI projects (like the Core LNGas hive project, 2020 [22]) none of them has currently permanent on-shore supply facilities for SSS traffic. Thus, 1 hour as a connection/disconnection lag time ($CT$) was assumed for all cases.

The emission factors for the vessels ($EFM_{uq}$; $\forall u \in U \wedge \forall q \in Q$ see Equation (2) and $EFB_{uq}$; $\forall u \in U \wedge \forall q \in Q$ see Equation (4)) have been obtained through the calculation tools developed by Kristensen and Bingham [23] for container vessels and Ro-Pax vessels [24]. Moreover, the emission factors for $CH_4$ ($EF_e$; $\forall e \in EE$) have been assumed with the following values $EF_1$ = 0.040 g/kWh for fuel-based engines and $EF_2$ = 5.79 g/kWh dual engines operating with LNG [12].

The electricity grids' emissions ($EFG_{uk}$; $\forall u \in U \wedge \forall k \in K$) for the European application cases can be obtained, per country and year, through the European Pollutant Release and Transfer Register -E-PRTR- (Regulation (EC) No 166/2006) along with the Energy Statistics published by EUROSTAT (EU Commission, DG Energy, Unit A4,2020) [25]. Despite the E-PTRT providing a register for 91 different pollutants, the particulate matters $PM_{2.5}$ are not specifically given. Due to the significant impact of this pollutant on human health (there were an estimated 412,000 premature deaths in Europe through long-term exposure to this pollutant in 2016 [26]), the $PM_{2.5}$ amount can be estimated through its relationship with the $PM_{10}$ (see Table 2), since the latter is offered by E-PTRT [27].

**Table 2.** Emission factors $PM_{2.5}/PM_{10}$ for source category in energy industries (g/GJ).

| *Hard Coal* | *Brown Coal* | *Natural Gas* | *Derived Gases* | *Heavy fuel Oil* | *Other Liquefied fuels* | *Biomass* |
|---|---|---|---|---|---|---|
| '9/20' | '9/20' | '0.9/0.9' | '5/5' | '18/13' | '2/1' | '38/33' |

Source: [28].

EUROSTAT (EU Commission, DG Energy, Unit A4, 2020) [25] collects information about gross electricity generation, by fuel or product, as well as by type of generation. In such a case, the calculation of the weighting factor of the source in an electricity grid can be calculated for each country and year by making it possible to estimate the $PM_{2.5}$ amount from $PM_{10}$ information. However, this information is not available for smaller geographical levels in EUROSTAT; for these application cases the official data from regional institutions must be consulted. The European Pollutant Release and Transfer Register (E-PRTR) (Regulation (EC) No 166/2006) not only offers information per kind of economic sector (NACE) and industrial activity, but also desegregates information beyond the national level [27].

Finally, Table 3 provides the air pollutant costs where metropolitan or urban areas are related to the port hinterland population: larger than 0.5 million or less than this population, respectively.

**Table 3.** The unitary cost of air pollutants for France and Spain in 2017.

| Pollutants (EUR/kg) | France | | Spain | |
|---|---|---|---|---|
| | Urban | Metropolitan | Urban | Metropolitan |
| $NO_x$ | 28.56 | 28.56 | 8.64 | 8.64 |
| $SO_2$ | 14.59 | 14.59 | 6.91 | 6.91 |
| $PM_{2.5}$ | 137.55 | 427.35 | 113.79 | 353.57 |
| $PM_{10}$ | 6.19 | 6.19 | 10.36 | 10.36 |
| $CO_2$ | 0.10 | 0.15 | 0.10 | 0.10 |
| $CH_4$ | 2.62 | 2.62 | 2.54 | 2.54 |

Data source: [15–17].

## 3. Results

Despite the varying maritime distances, a standard operational number (five calls per week) for transport services has been taken for all case study routes, in order to simplify the analysis.

### *3.1. Estimation of Emission Factors for the Electricity Grids in the CI Alternative*

EUROSTAT (EU Commission, DG Energy, Unit A4, 2020) [25] information about gross electricity generation per type of generation was used for the electricity grid of France and continental Spain, while the required information for the islands was taken from the Canary Islands' Energy yearbook [29]. This information, reflected in Table 4, along with the relationship between $PM_{2.5}$ and $PM_{10}$ offered by E-PTRT (see Table 2) permitted us to obtain the emission factors of these pollutants, as reflected in Table 5.

**Table 4.** Gross electricity generation by fuel in 2017 (%).

| | Hard Coal | Brown Coal | Oil and Petroleum Products | Natural Gas and Manufactured Gas | Solid Biofuels and Renewable Wastes | Renewable | Nuclear | Main Activity Electricity Only Plants [TWh] |
|---|---|---|---|---|---|---|---|---|
| SPAIN | 15.44 | 0.93 | 5.72 | 23.68 | 1.86 | 31.31 | 21.06 | 239.59 |
| FRANCE | 2.28 | 0.00 | 1.24 | 7.64 | 1.01 | 16.84 | 70.98 | 530.76 |
| GRAN CANARIA ISLAND | 0.00 | 0.00 | 91.86 | 0.00 | 0.00 | 8.14 | 0.00 | 3.48 |
| TENERIFE ISLAND | 0.00 | 0.00 | 92.27 | 0.00 | 0.24 | 7.49 | 0.00 | 3.53 |

Data source: [25,29].

Finally, the pollutant emissions of the electric power generating plants (main activity-NACE: 35.1 (Electric power generation, transmission, and distribution)) not just for the countries (France and Spain) but also for the islands, has been taken from the European Pollutant Release and Transfer Register -E-PRTR- (Regulation (EC) No 166/2006). This procedure is valid for all pollutants except for $CH_4$ emissions on the islands [29], due to the absence of information in E-PRTR [27] in this regard.

In light of the foregoing, Table 5 shows the estimated emission factors for the European grids considered in the application cases: Spain, France, Gran Canaria island and Tenerife island ($EFG_{uk}$; $\forall u \in U \wedge \forall k \in K$).

**Table 5.** Average emission factor for European grids in 2017.

| | $NO_x$ (g/kWh) | $SO_2$ (g/kWh) | $PM_{2.5}$ (g/kWh) | $PM_{10}$ (g/kWh) | $CO_2$ (g/kWh) | $CH_4$ (g/kWh) |
|---|---|---|---|---|---|---|
| **SPAIN (continental)** | 0.567 | 0.402 | 0.012 | 0.013 | 291.369 | 0.018 |
| **FRANCE** | 0.066 | 0.042 | 0.001 | 0.001 | 4.301 | 0.001 |
| **GRAN CANARIA ISLAND** | 1.796 | 0.785 | 0.030 | 0.015 | 656.034 | 0.026 |
| **TENERIFE ISLAND** | 2.344 | 0.762 | 0.033 | 0.016 | 645.609 | 0.026 |

Data source: [25,27–29].

## 3.2. *Performance of the Port Mitigation Alternatives*

Tables 6 and 7 show the "saved emissions" by onshore power supply and by dual engine use regarding fuel-based engines (conventional engines, see Table 1). This saving is measured in monetary terms per year (EUR) and in the percentage of saving emissions regarding fuel-based engines on-board.

**Table 6.** Emissions reduction in feeder vessels by port mitigation alternatives use.

| | | | **Vigo-St. Nazaire** | | **Las Palmas -Huelva** | | **Las Palmas -Sta. Cruz de Tenerife** | |
|---|---|---|---|---|---|---|---|---|
| N (yearly trips) | | | 476 | | 476 | | 476 | |
| Calls per week and direction | | | 5 | | 5 | | 5 | |
| Number of vessels (Nb) | | | 3 | | 4 | | 1 | |
| Maritime Distance (Nautical Miles) | | | 464 | | 702 | | 53 | |
| **Ports** | | | **Vigo** | **St. Nazaire** | **Las Palmas** | **Huelva** | **Las Palmas** | **Sta. Cruz Tenerife** |
| **On-board power supply (Tier III-MGO-engines)** | Maneuvering emissions | Per trip and port-$ECM_{1k}$-(EUR/trip) | 112.03 | 246.51 | 157.49 | 112.03 | 157.49 | 112.03 |
| | | Per trip (EUR/trip) | 358.54 | | 269.52 | | 269.52 | |
| | | Per year-$ECM_1$-(EUR/year) | 170,663.98 | | 128,291.96 | | 128,291.96 | |
| | Berthing emissions | Per trip and port-$ECB_{1k}$-(EUR/trip) | 436.01 | 883.19 | 611.92 | 436.01 | 611.92 | 436.01 |
| | | Per trip (EUR/trip) | 1319.20 | | 1047.94 | | 1047.94 | |
| | | Per year-$ECB_1$-(EUR/year) | 1,135,935.53 | | 902,932.98 | | 902,932.98 | |
| | Port Emissions | Per trip (EUR/trip) | 1677.74 | | 1317.46 | | 1317.46 | |
| | | **Per year (EUR/year)** | **1,307,329.62** | | **1,031,224.94** | | **1,031,224.94** | |
| **On-board power supply (dual-LNG-engines)** | Maneuvering emissions | Per trip and port-$ECM_{1k}$-(EUR/trip) | 95.00 | 184.10 | 107.67 | 95.00 | 107.67 | 95.00 |
| | | Per trip (EUR/trip) | 279.10 | | 202.66 | | 202.66 EUR | |
| | | Per year-$ECM_1$-(EUR/year) | 132,850.77 | | 96,468.27 | | 96,468.27 | |
| | Berthing emissions | Per trip and port-$ECB_{1k}$-(EUR/trip) | 355.50 | 624.81 | 404.52 | 355.50 | 404.52 | 355.50 |
| | | Per trip (EUR/trip) | 980.31 | | 760.02 | | 760.02 | |
| | | Per year-$ECB_1$-(EUR/year) | 844,661.29 | | 654,855.01 | | 654,855.01 | |
| | Port emissions | Per trip (EUR/trip) | 1,259.40 | | 962.68 | | 962.68 | |
| | | **Per year (EUR/year)** | **977,512.06** | | **751,323.28** | | **751,323.28** | |

**Table 6.** *Cont.*

| | | | | | | | | |
|---|---|---|---|---|---|---|---|---|
| **On-shore power supply** | Maneuvering emissions | Per trip and port-$ECM_{1k}$-(EUR/trip) | 112.03 | 246.51 | 157.49 | 112.03 | 157.49 | 112.03 |
| | | Per trip (EUR/trip) | 358.54 | | 269.52 | | 269.52 | |
| | | Per year-$ECM_1$-(EUR/year) | 170,663.98 | | 128,291.96 | | 128,291.96 | |
| | Berthing emissions | Per trip and port-$ECG_{1k}$-(EUR/trip) | 201.58 | 79.63 | 475.07 | 201.58 | 475.07 | 447.59 |
| | | Per trip (EUR/trip) | 281.21 | | 676.65 | | 922.66 | |
| | | Per year-$ECG_1$-(EUR/year) | 200,133.81 | | 537,585.04 | | 743,067.54 | |
| | Port emissions | Per trip (EUR/trip) | 639.75 | | 946.17 | | 1.192.18 | |
| | | **Per year (EUR/year)** | **370,797.79** | | **665,877.00** | | **871,359.50** | |
| ***Save emissions per year*** | | *On-shore power supply (EUR)* | *936,531.82* | | *365,347.94* | | *159,865.44* | |
| | | *Dual LNG engines use (EUR)* | *329,817.56* | | *279,901.67* | | *279,901.67* | |
| ***Emissions reduction per year*** | | **On-shore power supply (%)** | **71.64%** | | **35.43%** | | **15.50%** | |
| | | **Dual LNG engines use (%)** | **25.23%** | | **27.14%** | | **27.14%** | |

Tables 6 and 7 results show that the emissions reduction by CI use is strongly conditioned by the sustainability of the electricity grids in port. Since the French grids are the most sustainable (from the air emissions standpoint, see Table 4), the routes where St. Nazaire is involved are those most benefited by on-shore power supply.

**Table 7.** Emissions reduction in Ro-Pax vessels by port mitigation alternatives use.

| | | | **Vigo-St. Nazaire** | | **Las Palmas -Huelva** | | **Las Palmas -Sta. Cruz de Tenerife** | |
|---|---|---|---|---|---|---|---|---|
| N (yearly_trips) | | | 476 | | 476 | | 476 | |
| Calls per week and direction | | | 5 | | 5 | | 5 | |
| Number of vessels (Nb) | | | 3 | | 4 | | 1 | |
| Maritime Distance (Nautical Miles) | | | 464 | | 702 | | 53 | |
| **Ports** | | | **Vigo** | **St. Nazaire** | **Las Palmas** | **Huelva** | **Las Palmas** | **Sta. Cruz Tenerife** |
| **On-board power supply (Tier III-MGO-engines)** | Maneuvering emissions | Per trip and port-$ECM_{2k}$-(EUR/trip) | 286.95 | 605.54 | 403.15 | 286.95 | 403.15 | 286.95 |
| | | Per trip (EUR/trip) | 892.49 | | 690.10 | | 690.10 | |
| | | Per year-$ECM_2$-(EUR/year) | 424,826.90 | | 328,488.10 | | 128,291.96 | |
| | Berthing emissions | Per trip and port-$ECB_{1k}$-(EUR/trip) | 407.11 | 845.36 | 592.52 | 407.11 | 592.52 | 407.11 |
| | | Per trip (EUR/trip) | 1,252.47 | | 999.63 | | 999.63 | |
| | | Per year-$ECB_2$-(EUR/year) | 1,178,840.38 | | 1,049,115.55 | | 1,049,115.55 | |
| | Port emissions | Per trip (EUR/trip) | 2,144.97 | | 1,689.73 | | 1,689.73 | |
| | | **Per year (EUR/year)** | **1,603,667.28** | | **1,377,603.66** | | **1,377,603.66** | |
| **On-board power supply (dual-LNG-engines)** | Maneuvering emissions | Per trip and port-$ECM_{2k}$-(EUR/trip) | 232.18 | 433.82 | 264.56 | 232.18 | 264.56 | 232.18 |
| | | Per trip (EUR/trip) | 665.99 | | 496.74 | | 496.74 | |
| | | Per year-$ECM_2$-(EUR/year) | 317,013.32 | | 236,447.41 | | 236,447.41 | |
| | Berthing emissions | Per trip and port-$ECB_{2k}$-(EUR/trip) | 356.23 | 626.03 | 407.90 | 356.23 | 407.90 | 356.23 |
| | | Per trip (EUR/trip) | 982.26 | | 764.12 | | 764.12 | |
| | | Per year-$ECB_2$-(EUR/year) | 1,030,882.51 | | 801,948.83 | | 801,948.83 | |
| | Port emissions | Per trip (EUR/trip) | 1648.25 | | 1260.86 | | 1260.86 | |
| | | **Per year (EUR/year)** | **1,347,895.84** | | **1,038,396.24** | | **1,038,396.24** | |

**Table 7.** *Cont.*

| | | | | | | | | |
|---|---|---|---|---|---|---|---|---|
| On-shore power supply | Maneuvering emissions | Per trip and port-$ECM_{1k}$-(EUR/trip) | 286.95 | 605.54 | 403.15 | 286.95 | 403.15 | 286.95 |
| | | Per trip (EUR/trip) | 892.49 | | 690.10 | | 690.10 | |
| | | Per year-$ECM_2$-(EUR/year) | 424,826.90 | | 328,488.10 | | 328,488.10 | |
| | Berthing emissions | Per trip and port-$ECG_{1k}$-(EUR/trip) | 229.11 | 108.37 | 533.14 | 229.11 | 533.14 | 497.04 |
| | | Per trip (EUR/trip) | 337.48 | | 762.25 | | 1030.18 | |
| | | Per year-$ECG_2$-(EUR/year) | 264,534.38 | | 700,627.20 | | 966,701.54 | |
| | Port emissions | Per trip (EUR/trip) | 1,229.98 | | 1,452.35 | | 1,720.28 | |
| | | **Per year (EUR/year)** | **689,361.28** | | **1,029,115.30** | | **1,295,189.64** | |
| *Emissions saved per year* | | *On-shore power supply (EUR)* | *914,305.99* | | *348,488.35* | | *82,414.01* | |
| | | *Dual LNG engines use (EUR)* | *255,771.44* | | *339,207.41* | | *339,207.41* | |
| *Emissions reduction per year* | | **On-shore power supply (%)** | **57.01** | | **25.30** | | **5.98** | |
| | | **Dual LNG engines use (%)** | **15.95** | | **24.62** | | **24.62** | |

The progressive reduction of the advantage provided by CI on the, on the routes between the Iberian Peninsula and inter-island, can be also explained by the nature of the electricity generation sources in those regions (see Tables 3 and 4). Contrary to CI performance occurs when LNG is used as a fuel for on-board engines. In this latter case, the unitary costs of air pollutants per country (see Table 3) are the main drivers of the results. Consequently, the routes through French ports achieve the least advantage from LNG use.

This behavior can also be seen in Figures 2 and 3 whereas the best environmental performance is achieved through CI for the routes involving French ports from peninsular Spain, SSS traffic among the Canary Islands provides the greatest sustainability by operating with LNG in dual engines (see Tables 6 and 7). In this case, not only the berthing stage ($ECB_q$, $\forall q \in Q$, see Figures 2 and 3) but also the maneuvering stage ($ECM_q$, $\forall q \in Q$) benefit from using sustainable fuel. These results are true regardless of the kind of vessel (see Figures 2 and 3); however, when attention is focused on the kind of fleet, feeder vessels prove to be more sensitive to the mitigation alternatives. Thus, feeder fleets achieve greater improvements in terms of sustainability than Ro-Pax vessels (see Tables 6 and 7), on all routes. This is mainly due to the feeder vessels' longer berthing times in comparison to the Ro-Pax vessels.

Berthing time also determines sensitivity to the connection/disconnection lag times of CI use; the longer the berthing time is, the lower this sensitivity is. Therefore Ro-Pax is not only more influenced by this lag time but also, this additional time can make the CI alternative unfeasible. This is the case of Sta. Cruz de Tenerife port (see Table 7) where the CI alternative is less sustainable per trip than the on-board supply (Tier III-MGO-engines): 497.04 versus 407.11 EUR/trip (see Table 7); only when hoteling is considered in the analysis (inactive time in port from 23 p.m. to 7 a.m.) does the CI alternative become more favorable (1,377,603.66 EUR/year versus 1,295,189.64 EUR/year).

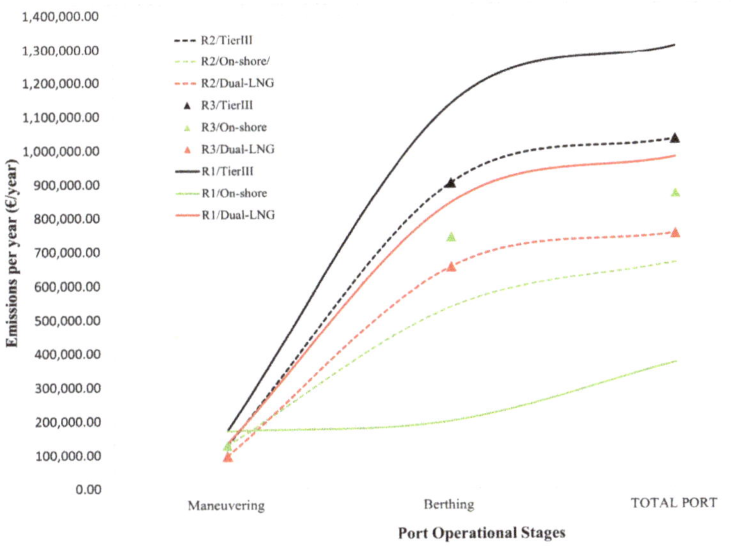


**Figure 2.** Emissions per year of feeder vessels in port. R1: Vigo (Spain)–St. Nazaire (France) R2: Las Palmas (Gran Canaria island)–Huelva (Spain) R3: Las Palmas (Gran Canaria island–Sta. Cruz de Tenerife (Tenerife island).

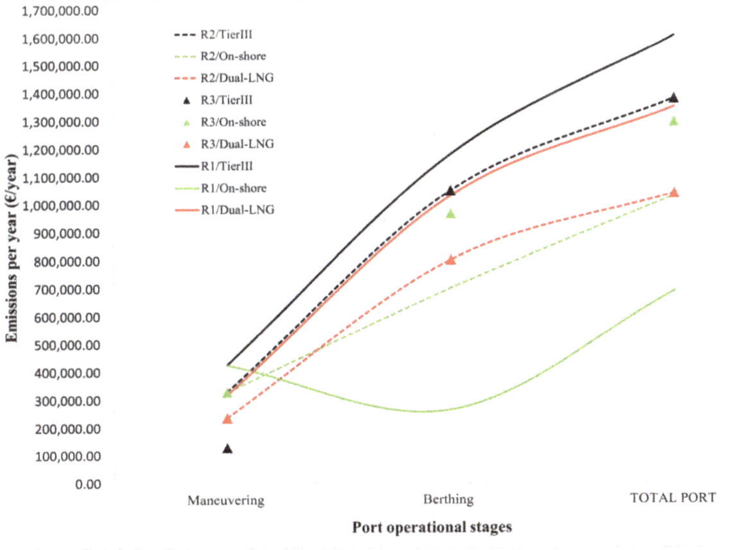


**Figure 3.** Emissions per year of Ro-Pax vessels in port. R1: Vigo (Spain)–St. Nazaire (France) R2: Las Palmas (Gran Canaria island)–Huelva (Spain) R3: Las Palmas (Gran Canaria island)–Sta. Cruz de Tenerife (Tenerife island).

## 4. Discussion

Even though real measures for the case studies are not available to test the effectiveness of the method proposed, the first approach of its reliability can be made from the data published by other authors.

Spengler and Tovar [30] published external costs at berth (EUR/h) related to 2016 for several Spanish ports. Despite the authors' advice about the nature of this indicator, it is a homogeneous value (an aggregated value that includes whole fleet activity and all the operational stages in port), that collects the same pollutants as this paper, excepting the $CH_4$, and therefore this value can be taken as a reference point to test the reliability of the results obtained by the proposed method. For this aim, two assumptions are considered: firstly, most of the 2016 fleet operated with conventional on-board electricity supply in Spanish ports and secondly, since time invested in berthing is significantly higher than in maneuvering, the former has been predominant in the Spengler and Tovar (2020) indicators.

Spengler and Tovar [30] found that in 2016, the Las Palmas port had an external cost at berth of 215 EUR/h. Tables 6 and 7 show that for Las Palmas port the values obtained for the berthing stage per trip (3.42 h for feeders and 2.3 h for Ro-Pax) are 178.92 EUR/h for feeder vessels (611.92 EUR/3.42 h see Table 6) and 257.61 EUR/h for Ro-Pax vessels (592.52 EUR/2.3 h see Table 7). Therefore, despite the simplifications assumed, the values obtained are close to those previously published by other authors, which suggest an acceptable level of reliability in the calculation model.

On the other hand, the results achieved in this paper confirm the need for increasing the energy environmental sources in electricity insular generation to enhance the desired effects from the application of Directive 2014/94/EU. To this aim, self-generated local renewable energy in port [5] has proved to be one effective solution (in Gothenburg, through local surplus wind generated power, among others). Nevertheless, due to the absence of a continental grid connection and the uncertainty about on-shore renewable resource availability, a challenge arises for the islands. Off-shore wind energy can reduce that uncertainty since the further away from the coast, the higher and more constant the wind energy source is. In fact, floating off-shore wind energy is currently able to overcome the inherent difficulties of traditional "bottom-fixed" foundations for larger depths (such a case of the Canary Islands) by avoiding interferences with port activity [31]. According to the FLOTANT project [32], two 600 MW floating off-shore wind farms (a turbine of 12 MW is supported by each floating substructure) could provide 2.8 TWh per year [32]. This is Gran Canaria island's required renewable power to reach almost the same share of air pollutant sources in its energy mix as in France (see Table 4). In other words, in such a scenario CI performance in the Las Palmas de Gran Canaria port would be equivalent to that of St. Nazaire.

As has been shown in this paper, in the meantime, LNG use could be a more effective option in insular ports. Despite the fact that its mitigation potential has proved to be lower than CI (methane effect), LNG provides an additional benefit by also acting during the maneuvering stage. However, LNG feasibility is again conditioned by the availability of LNG refueling points in ports (Directive 2014/94/EU) and dual engines installed in vessels. An intermediate solution is currently being studied by some European ports—through the Core LNGas hive project [22]—where the shore-side electricity supply is based on off-grid power production through quayside generating sets. This solution does not eliminate local emissions and is not suitable for supplying the most demanding vessels but, by means of burning LNG, the emissions of port generating sets are cleaner, the system is more flexible and, in the short term, the cost–benefit analysis is favorable.

## 5. Conclusions

The obligation to provide CI facilities and refueling points for LNG in EU ports by 2025 (Directive 2014/94/EU) involves institutional support to boost these technologies to advance port sustainability. However, their effectiveness is highly conditioned on port features that include the hinterland population size and the share of renewable sources in the energy mix of the regions where the ports are located. This concern is especially justified when SSS is considered, since this type of traffic is characterized by short call times, shipping close to the coast, and frequent operation in insular scenarios with an electricity generation mix that offers few renewable sources.

In light of the foregoing, this paper contributes to knowledge about the port performance of LNG and CI alternatives for SSS traffic, by introducing a calculation method that is able to evaluate the environmental advantages provided by their use. The method was applied to three routes with ports located in regions with varying sustainability in their land electricity grids: Peninsular Spain–France, South Spain–Canary Islands and interisland (Canary Islands). Simultaneously, two different SSS fleets were analyzed: feeder and Ro-Pax vessels.

From the application cases, a number of findings can be drawn. One important influence is the connection/disconnection lag time on CI performance for SSS. Specifically, this is inversely proportionate to berthing time. Thus, the feeder vessels with longer loading times than Ro-Pax vessels show lower sensitivity to this variable by achieving greater environmental benefits by CI use. Likewise, SSS routes with ports located in regions where low-emission sources predominate in electricity generation clearly benefit more from CI use than the LNG alternative in port. On the other hand, on those routes whose ports offer an electricity supply with high dependence on fossil fuel generation, the convenience of CI versus LNG use is highly conditioned on the circumstances of the regions involved. The latter has been exemplified in this paper through the analysis of interisland traffic (Las Palmas de Gran Canaria–Sta. Cruz de Tenerife); the results show LNG performance is clearly greater in environmental terms, regardless of the fleet type.

Thus, the results obtained in this paper quantitatively confirm prior warnings about the benefits of CI in the insular framework, especially with short call traffic. This highlights the need for additional measures to ensure the sustainability of the on-shore electricity supply for the application of Directive 2014/94/EU in the insular territories. Otherwise, Directive 2014/94/EU will not only involve considerable port investment but also will not bring about the expected health benefits. These measures necessarily require a prior increase in the share of renewable sources in the port electricity supply.

**Author Contributions:** Conceptualization, A.M.-L., A.R. and J.A.O.; methodology, A.M.-L. and A.R.; validation, A.M.-L. and J.A.O.; formal analysis, A.M.-L. and J.A.O.; resources, A.R. and J.A.O.; data curation, A.R.; writing—original draft preparation, A.M.-L.; writing—review and editing, J.A.O.; visualization, A.R. and J.A.O.; supervision, J.A.O. All authors have read and agreed to the published version of the manuscript.

**Funding:** This publication has been supported by the H2020 project FLOTANT, which is coordinated by the Oceanic Platform of the Canary Islands (PLOCAN). The project has received funding from the European Union's Horizon 2020 research and innovation program under grant agreement No. 815289.

**Institutional Review Board Statement:** Not applicable.

**Informed Consent Statement:** Not applicable.

**Data Availability Statement:** Not applicable.

**Conflicts of Interest:** The authors declare no conflict of interest.

## Appendix A. Nomenclature

Subscripts:

| | |
|---|---|
| EE = {1. ... . *e*}: | Kinds of engines involved. Medium speed four-stroke diesel engine with MGO (Tier III) and four-stroke dual-fuel engines (LNG plant). |
| K = {1. ... . *k*} | Ports for linear shipping line: Vigo. St. Nazaire. Las Palmas de Gran Canaria and Sta. Cruz de Tenerife. These ports articulate the lines: Vigo (Spain)–St. Nazaire (France); Las Palmas de Gran Canaria (Gran Canaria island)–Huelva (Spain) and Las Palmas de Gran Canaria (Gran Canaria island)–Sta. Cruz de Tenerife (Tenerife island). |
| Q = {1. ... . *q*} | Kind of vessel: feeder vessel and Ro-Pax |
| U = {1. ... .*u*} | Kind of emission pollutants: $NO_x$, $SO_2$, $PM_{2.5}$, $PM_{10}$, $CO_2$ and $CH_4$ |
| V = {1. ... .*v*} | Classification of ports according to the population of their hinterlands: metropolitan zone (over 0.5 million inhabitants) and urban zone. |

Variables:

| | |
|---|---|
| $BT_q$ | Berthing time (h); this is the loading/unloading time. ∀q∈Q |
| $CT$ | Lag connection time (h). |
| $ECB_{qk}$ | Environmental costs during berthing for every kind of vessel and port (EUR); ∀q∈Q ∧ ∀k∈K |
| $ECG_{qk}$ | Environmental costs by on-shore power use during berthing for every kind of vessel and port (EUR); ∀q∈Q ∧ ∀k∈K |
| $ECM_{qk}$ | Environmental costs during maneuvering for every kind of vessel and port (EUR); ∀q∈Q ∧ ∀k∈K |
| $EF_e$ | Emission factor of methane for every kind of engine (g/kWh) |
| $EFB_{uq}$ | Emission factors for every kind of vessel and pollutant during berthing.(kg/h) ∀u∈U ∧ ∀q∈Q |
| $EFG_{uk}$ | Emissions from the electricity grid for every kind of pollutant and port during berthing (kg/kW.h) ∀u∈U ∧ ∀k∈K |
| $EFM_{uq}$ | Emissions from every kind of vessel and pollutant during maneuvering (kg/h) ∀u∈U ∧ ∀q∈Q |
| $IPB_q$ | The installed power-MCR-of the engines involved in every operational stage and for every kind of vessel (*IPBq*; ∀q∈Q) |
| $LFB_q$ | Load factor of the engines involved in the berthing stage (%) ∀q∈Q |
| $LFM_q$ | Load factor of the engines involved in the maneuvering stage (%) ∀q∈Q |
| $MT$ | Maneuvering time (h). This time collects the pilot and towing time when they are necessary. |
| $N$ | Number of yearly trips of a shipping line. |
| $N_b$ | Number of vessels in a particular SSS line. |
| $PB_q$ | Required power for berthing operations for every kind of vessel (kW) ∀q∈Q |
| $UC_{ukv}$ | Unitary costs for every kind of pollutant and port (EUR/kg) ∀u∈U ∧ ∀k∈K ∧ ∀v∈V |

## References


1. Tseng, P.H.; Pilcher, N. A study of the potential of shore power for the port of Kaohsiung. Taiwan: To introduce or not to introduce? *Res. Transp. Bus. Manag.* **2015**, *17*, 83–91. [CrossRef]
2. Ballini, F.; Rozzo, R. Air pollution from ships in ports: The socio-economic benefit of cold-ironing technology. *Res. Transp. Bus. Manag.* **2015**, *17*, 92–98. [CrossRef]
3. Innes, A.; Monios, J. Identifying the unique challenges of installing cold ironing at small and medium ports—The case of Aberdeen. *Transp. Res. D Transp. Environ.* **2018**, *62*, 298–313. [CrossRef]
4. Zis, T.P.V. Prospects of cold ironing as an emissions reduction option. *Transp. Res. Part A Policy Pract.* **2019**, *119*, 82–95. [CrossRef]
5. Winkel, R.U.; Weddige, U.; Johnsen, D.; Hoen, V.; Papaefthimiou, S. Shore Side Electricity in Europe: Potential and environmental benefits. *Energy Policy* **2016**, *88*, 584–593. [CrossRef]
6. Hjelle, H.M. Short Sea Shipping's Green Label at Risk. *Transp. Rev.* **2020**, *30*, 617–640. [CrossRef]
7. Martínez-López, A.; Caamaño Sobrino, P.; Míguez González, M. Influence of external costs on the optimisation of container fleets by operating under motorways of the sea conditions. *Int. J. Shipp. Transp. Logist.* **2016**, *8*, 653–686. [CrossRef]
8. Brynolf, S.; Magnusson, M.; Fridell, E.; Andersson, K. Compliance possibilities for the future ECA regulations through the use of abatement technologies or change of fuels. *Transp. Res. D Transp. Environ.* **2014**, *28*, 6–18. [CrossRef]
9. Bengtsson, S.; Fridell, E.; Andersson, K. Fuels for short sea shipping: A comparative assessment with focus on environment impact. *Proc. Inst. Mech. Eng.* **2014**, *228*, 44–54. [CrossRef]
10. Christodoulou, A.; Cullinane, K. Potential of, and drivers for, private voluntary initiatives for the decarbonisation of short sea shipping: Evidence from a Swedish ferry line. *Marit. Econ. Logist.* **2020**. [CrossRef]
11. Winkel, R.; Weddige, U.; Johnsen, D.; Hoen, V.; Papaefthymiou, G. Project number: TRANL14441: Potential for Shore Side Electricity in Europe. January 2015. Ecofys. Available online: https://es.scribd.com/document/366724117/Ecofys-2014-Potential-for-Shore-Side-Electricity-in-Europe (accessed on 11 June 2020).
12. Martínez-López, A.; Caamaño, P.; Chica, M.; Trujillo, L. Choice of propulsion plants for container vessels operating under Short Sea Shipping conditions in the European Union: An assessment focused on the environmental impact on the intermodal chains. *Proc. Inst. Mech. Eng.* **2019**, *233*, 653–669. [CrossRef]
13. Khersonsky, Y.; Islam, M.; Peterson, K. Challenges of connecting shipboard marine systems to medium voltage shoreside electrical power. *IEEE Trans. Ind. Appl.* **2007**, *43*, 838–844. [CrossRef]
14. Zis, T.; North, R.J.; Angeloudisa, P.; Ochieng, W.; Bell, M.G.H. Evaluation of cold ironing and speed reduction policies to reduce ship emissions near and at ports. *Logist. Marit. Econ. Logist.* **2014**, *16*, 371–398. [CrossRef]
15. European Commission. *4.K83 -Handbook on the External Costs of Transport*; January 2019–V1.1; European Commission: Brussels, Belgium, 2019; ISBN 978-92-76-18184-2.

16. National Institute of Statistics and Economic Studies of France. 2017. Available online: https://www.insee.fr/en/statistiques/4771934#tableau-ipc-flash-g1-en (accessed on 30 September 2020).
17. National Statistics Institute of Spain. 2017. Available online: https://www.ine.es/en/prensa/ipc_tabla_en.htm (accessed on 30 September 2020).
18. Contessi, C. Wartsila Ship Power 4 Stroke Applications Development. Gas Engine Emissions Wartsila Dual. December 2013. Available online: http://www.lngnordwest.de/files/lng_downloads/WS_131216/20131216%20Wartsila%20Leer_DF%20emission%20conference.pdf (accessed on 19 January 2016).
19. Nielsen, J.; Stenersen, D. *Emission factors for CH4. NOx. Particulates and Black Carbon for Domestic Shipping in Norway. Revision 1*; Marintek. Klima og forurensningsdirektoratet; Norwegian Pollution Agency: Trondheim, Norway, 2010.
20. Martínez-López, A.; Caamaño Sobrino, P.; Chica González, M.; Trujillo, L. Optimization of a container vessel fleet and its propulsion plant to articulate sustainable intermodal chains versus road transport. *Transp. Res. D Transp. Environ.* **2018**, *59*, 134–147. [CrossRef]
21. Martínez-López, A.; Kronbak, J.; Jiang, L. Cost and time models for the evaluation of intermodal chains by using short sea shipping in the North Sea Region: The Rosyth-Zeebrugge route *Int. J. Shipp. Transp. Logist.* **2015**, *7*, 494–520. [CrossRef]
22. CORE LNGas hive.2020; European Commission. Trans-European Transport Networks (Call 2014 Connect Europe Facility CEF). Available online: http://corelngashive.eu/en/ (accessed on 11 January 2020).
23. Kristensen, H.O.; Bingham, H. *Project no. 2016-108: Update of Decision Support System for Calculation of Exhaust Gas Emissions*; Report no. 06; HOK Marineconsult ApS and Technical University of Denmark: Lyngby, Denmark, 2020.
24. Kristensen, H.O.; Psaraftis, H. *Project no. 2014-122: Mitigating and Reversing the Side-Effects of Environmental Legislation on Ro-Ro Shipping in Northern Europe*; Work Package 2.3; Report no. 07; HOK Marineconsult ApS and Technical University of Denmark: Lyngby, Denmark, 2016.
25. EU Commission DG Energy. Unit A4. EUROSTAT: Energy Statistics. Energy Datasheets: EU Countries. 2020. Available online: https://data.europa.eu/euodp/en/data/dataset/information-on-energy-markets-in-eu-countries-with-national-energy-profiles/resource/24184068-8ec3-470a-ba28-5ca2317c6f6f (accessed on 30 September 2020).
26. European Environment Agency. EEA Report No 10/2019: Air Quality in Europe-2019 Report. 2019. Available online: https://www.eea.europa.eu/publications/air-quality-in-europe-2019 (accessed on 30 September 2020).
27. European Environment Agency. European Pollutant Release and Transfer Register. 2020. Available online: https://prtr.eea.europa.eu/#/industrialactivity (accessed on 30 September 2020).
28. European Environment Agency. 1.A.1 Combustion in energy industries GB2009 update June 2010. In *EMEP/EEA Air Pollutant Emission Inventory Guidebook—2009*; EEA Technical Report No 9/2009; European Environment Agency: Copenhagen, Denmark, 2009; p. 96.
29. Government of the Canary Islands. Economy, Industry and Trade Counseling. Anuario Energético de Canarias 2017. 2018. Available online: http://www.gobiernodecanarias.org/istac/jaxi-istac/menu.do?uripub=urn:uuid:131cf873-66a9-408d-8cfa-537d6be05067 (accessed on 30 September 2020).
30. Spengler, T.; Tovar, B. Potential of cold-ironing for the reduction of externalities from in-port shipping emissions: The state-owned Spanish port system case. *J. Environ. Manag.* **2021**, *279*, 111807. [CrossRef]
31. FLOTANT Project. Innovative, Low Cost, Low Weight and Safe Floating Wind Technology Optimized for Deep Water Wind Sites. European Union's Horizon 2020 Programme-No815289-Deliverable D4.1: Structural and Naval Architecture Design Basis. 2019. Available online: https://flotantproject.eu/wp-content/uploads/2020/12/191025-FLT_WP4_D4.1_DesignBasis_v3.pdf (accessed on 31 December 2020).
32. FLOTANT Project Innovative, Low Cost, Low Weight and Safe Floating Wind Technology Optimized for Deep Water Wind Sites. European Union's Horizon 2020 Programme-No815289-Deliverable D4.2: Specifications of a Generic Wind Turbine. 2019. Available online: https://flotantproject.eu/wp-content/uploads/2020/04/190927_FLT-D.4.2-Specification_Turbine_V2.pdf (accessed on 31 December 2020).

*applied sciences*

MDPI

*Article*

# Truck Appointment System for Cooperation between the Transport Companies and the Terminal Operator at Container Terminals

**Hyeonu Im [1], Jiwon Yu [1] and Chulung Lee [2,*]**

1 Department of Industrial and Management Engineering, Korea University, 145, Anam-ro, Seongbuk-gu, Seoul 02841, Korea; gusdndla@korea.ac.kr (H.I.); vermouth28@korea.ac.kr (J.Y.)
2 School of Industrial and Management Engineering, Korea University, 145, Anam-ro, Seongbuk-gu, Seoul 02841, Korea
* Correspondence: leecu@korea.ac.kr; Tel.: +82-2-3290-3395

**Abstract:** Despite the number of sailings canceled in the past few months, as demand has increased, the utilization of ships has become very high, resulting in sudden peaks of activity at the import container terminals. Ship-to-ship operations and yard activity at the container terminals are at their peak and starting to affect land operations on truck arrivals and departures. In response, a Truck Appointment System (TAS) has been developed to mitigate truck congestion that occurs between the gate and the yard of the container terminal. The vehicle booking system is developed and operated in-house at large-scale container terminals, but efficiency is low due to frequent truck schedule changes by the transport companies (forwarders). In this paper, we propose a new form of TAS in which the transport companies and the terminal operator cooperate. Numerical experiments show that the efficiency of the cooperation model is better by comparing the case where the transport company (forwarder) and the terminal operator make their own decision and the case where they cooperate. The cooperation model shows higher efficiency as there are more competing transport companies (forwarders) and more segmented tasks a truck can reserve.

**Keywords:** truck congestion problem; truck appointment system; cooperation model; scheduling of truck arrivals

**Citation:** Im, H.; Yu, J.; Lee, C. Truck Appointment System for Cooperation between the Transport Companies and the Terminal Operator at Container Terminals. *Appl. Sci.* **2021**, *11*, 168. https://dx.doi.org/10.3390/app11010168

Received: 27 November 2020
Accepted: 23 December 2020
Published: 27 December 2020



## 1. Introduction

The global cargo operations, which had been steadily increasing every year since the global financial crisis in 2008, have been hit hard by COVID-19. Sea transportation regulations such as temporary suspension and cancellation of operations occurred, and 11% of ship operations were canceled for six months since December 2019 when the first case of COVID-19 appeared [1]. For example, 120 out of 126 countries have had restrictions on crew rotation, 92 of these countries have banned crew rotation, and 28 of these countries have allowed crew rotation through the search and approval of the authorities [2]. These restrictions will prevent ships from entering the container terminal until it confirms that the crew has not been infected with the virus (mostly 14 days), impeding the smooth operation of maritime transport.

According to the International Association of Ports and Harbors' Port Economic Impact Barometer report [3], major container terminals in Europe and North America need more moves per ship than ever due to the wave of blank sailings. Despite the number of sailings canceled in the past few months, as demand has increased, the utilization of ships has become very high, resulting in sudden peaks of activity at the import container terminals. As a result, ship-to-ship operations and yard activity at the container terminals are at their peak, especially starting to affect land operations on truck arrivals and departures. With several days off duty, the pressure on the workforce in some ports has increased. Increasing levels of congestion at port access roads exacerbates these issues. Therefore, it is necessary to discuss the truck congestion problem that occurs between the gate and the yard of the container terminal (see Figure 1).

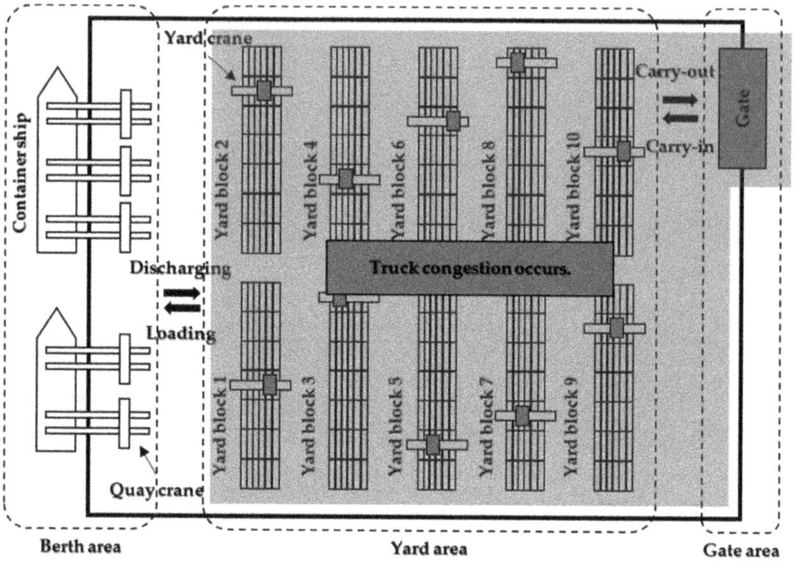


**Figure 1.** Truck congestion at the container terminal.

The truck congestion between the gate and the yard of a container terminal is the main topic of discussion in large-scale import and export container terminals around the world as it causes many trucks to wait and emit more $CO_2$ than usual [4]. The reasons for congestion of trucks between the gate and the yard vary, and are mainly due to uncertainty in the truck arrivals, an increase in container ship size, an increase in container volume, failure of logistics equipment or IT systems, and changes in work rules. The basic strategy for mitigating the truck congestion between the gate and the yard is to expand yard spaces and gate spaces. However, space expansion requires a lot of time and cost, and we cannot apply a space expansion strategy in a narrow area where it is impossible to secure additional space. Therefore, it is necessary to solve the truck congestion problem not only for the profitability of stakeholders but also for the reduction of $CO_2$ emissions.

If the terminal operator can predict the truck arrivals a few hours in advance, they can adjust the yard volume accordingly. The vehicle booking system is developed and operated in-house at large-scale container terminals. However, the efficiency of the system is low due to frequent truck schedule changes by the transport companies (forwarders). Therefore, it is possible to solve the truck congestion problem by developing a Truck Appointment System (TAS) that allows the transport companies (forwarders) and the terminal operator to cooperate to achieve their goals.

The TAS is a practical way for the transport company (forwarder) and the terminal operator to communicate. Some container terminals such as Los Angeles, Long Beach, Hong Kong, Jebel Ali, Antwerp Gateway, and Southampton are applying a TAS [5,6].

In a traditional TAS, the terminal operator pre-sets the maximum number of trucks that can arrive at the gate for each time window, and the transport company (forwarder) books appropriately so as not to exceed the maximum number of trucks per time window. The terminal operator then rejects reservations for trucks from transport companies (forwarders) that exceed the maximum number of trucks per time window [4]. The traditional TAS allows the terminal operator to control the truck congestion between the gate and the yard by limiting the maximum number of truck arrivals. However, transport companies (forwarders) have difficulty meeting the different situations and the transport requirements of individual containers.

In this paper, we propose a new TAS that helps improve the profitability of stakeholders by considering the positions of the transport companies (forwarders) and the terminal operator, respectively, and comparing them with cooperation cases. To this end, we develop a mathematical model from the perspective of (1) the transport company (forwarder), (2) the terminal operator, and (3) cooperation between the transport companies (forwarders) and the terminal operator.

The new TAS reflects each stakeholder's penalties as the objective function of the mathematical model. We define these penalties in terms of collectively expressing the time, number, and capacity that should be optimized to reduce truck congestion. In a mathematical model for the transport company (forwarder), the penalties are the waiting time for the truck and the number of rehandling for carry-out containers. In a mathematical model for the terminal operator, the penalties are the unassigned and unreserved slot capacity (the tasks a truck can reserve).

Figure 2 shows a framework of the new truck appointment system. From the perspective of the transport company (forwarder), it is necessary to reduce the truck turnaround time. The turnaround time consists of a fixed service time, a variable waiting time, and a variable rehandling time. Therefore, the transport companies (forwarders) aim to reduce the truck waiting time and the container rehandling time. From the perspective of the terminal operator, it is necessary to balance the workload of the yard crane. The more tasks that can be reserved and booked, the better the productivity of the container terminal. Therefore, the terminal operator aims to increase the number of slots that can be reserved. From the perspective of cooperation between the transport companies (forwarders) and the terminal operator, it is necessary to consider the purpose of the transport company (forwarder) and the terminal operator at the same time. We multiply each objective function by its weight and add them. Weight refers to the position of the stakeholders in cooperation. In this study, the transport company (forwarder) and the terminal operator have the same weight (transport company's weight = terminal operator's weight = 1).

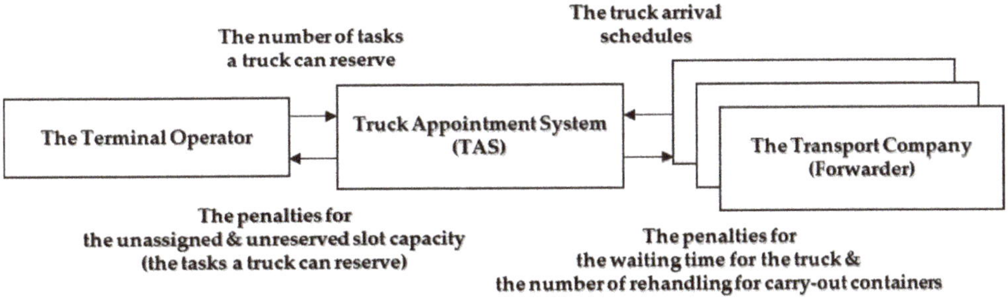


**Figure 2.** A framework of the new Truck Appointment System (TAS).

By comparing the results of numerical experiments for cases where transport companies (forwarders) and terminal operator make their own decisions and cooperate, we propose a model that is beneficial to all stakeholders.

The rest of this paper consists of the following topics. Section 2 discusses the literature related to this paper. Section 3 defines the TAS problem and proposes a mathematical model for the perspective of the transport company (forwarder), the terminal operator, and cooperation between the transport companies (forwarders) and the terminal operator. Section 4 shows the results of numerical experiments on the mathematical model in three cases. Section 5 provides conclusions and implications for all stakeholders.

## 2. Literature Review

The transmission of COVID-19 has destroyed the economy and immediate action is required. A study proposes the implementation of new green infrastructure along with the maintenance of the existing infrastructure for economic growth while protecting the environment [7]. However, this approach incurs additional costs because it requires rebuilding the existing infrastructure. In this paper, we improve the productivity of the container terminal by adjusting the work schedule through TAS while using only the existing facilities.

There have been several optimization studies that apply a TAS to the container terminal. One study developed a model that supports the decision making of transport companies (forwarders). A truck schedule was assigned to each time window to mitigate congestion occurring at the container terminal [8]. However, this study considered only export containers, not import containers arriving at the container terminal during peak times when truck congestion mainly occurs. This study also distributed the workload during peak times to each yard block but did not account for the increased waiting times for trucks at other times. Other studies developed a model that assigns truck schedules to each time window and optimized the time window by estimating the queue length [9,10]. However, these studies only considered the waiting time when calculating the truck turnaround time and did not take into account other factors (e.g., the number of rehandling).

Similar to general TAS, which considers the gate and the yard at the same time, there are studies conducted separately in the gate system and the yard system. A study in the gate system minimized the truck waiting cost and the gate operating cost in the queuing problem reflecting truck arrivals and gate processing [11]. Another study in the yard system calculated the expected number of rehandlings through a simulation of the yard crane handling import containers and applied it to the truck queuing model to improve the crane productivity and minimize the truck transaction time [12]. Also, there is a study that managed truck arrivals based on the truck-vessel service relationship in the congestion of the container terminal [13]. This study shows that the congestion of the container terminal is also affected by berth operations.

A study that solved the congestion of the container terminal by simulation determines the container arrival sequence through the gate simulation model reflecting the truck arrival distribution and calculates the container rehandling efficiency according to the heuristic procedure [14]. However, the system that includes both a gate system and a yard system has too many considerations, so it is more difficult to implement in one simulation. Likewise, in this study, only the gate system (not the entire system) was implemented as a simulation. Some studies improved the traditional TAS and developed a new TAS through a negotiation process that considers the needs of both the transport companies (forwarders) and the terminal operator [4,15]. When the transport companies (forwarders) inform the terminal operator of the truck arrivals, the terminal operator calculates an estimated turnaround time for each time window and provides it to the transport companies (forwarders). Then the transport companies (forwarders) reschedule truck arrivals according to the estimated turnaround time per time window. This sequential decision making process has a problem that requires a lot of procedures and time.

One of the methodologies to support the decision making process is to use cluster analysis that takes into account numerical or categorical data. Clustering means dividing data into meaningful groups, and one study reviewed cluster analysis techniques that support the decision-making process [16]. In particular, it provided technical details for cluster analysis dealing with mixed data consisting of numerical and categorical attributes. Meanwhile, in this paper, we propose an integrated decision making process that shares information (e.g., the purpose of each stakeholder) through cooperation between stakeholders and benefits all stakeholders.

In the broad context of competition between transport chains, a dry port is an extended version of a seaport [17,18]. In particular, since ports affect the competitive advantage of the hinterland [19], inland distribution is an important factor [20]. The dry ports are far from typical borders but have access to major metropolitan areas, highways, and labor bases [21]. From a functional perspective, the dry ports consist of close, mid-range, and distant dry ports [18]. Therefore, the TAS of this paper is available to a dry port as well as a seaport.

Most previous studies have focused only on strategies for determining truck schedules that can reduce the turnaround time from the perspective of the transport company (forwarder). From the perspective of the terminal operator, they have focused only on strategies for determining yard crane operations that can improve the productivity of the container terminal.

This paper differs from previous studies in the following aspects:

1. As a factor in reducing truck turnaround time from the perspective of the transport company (forwarder), most of the previous studies only considered truck waiting time, which did not reflect container rehandling time. However, this paper covers not only truck waiting time but container rehandling time as well. In this way, this paper can express truck turnaround time in more detail.
2. As a factor for improving terminal productivity from the perspective of the terminal operator, most previous studies only considered workload balancing and distribution containers to each yard block. However, this paper not only considers workload balancing but also reduces the waste of resources by increasing the number of available and reserved slots together. In this way, this paper can more directly express terminal productivity.
3. Previous studies of the negotiation process between the transport companies (forwarders) and the terminal operator required a sequential decision making process of exchanging their own decisions. However, in this paper we develop a mathematical model for integrated decision making that considers the transport companies (forwarders) and the terminal operator at the same time. Furthermore, it compares the results of numerical experiments for cases where transport companies (forwarders) and terminal operator make their own decisions and cooperate. Then we propose a model that benefits all stakeholders.

For example, we suppose there are one-yard block and two-time windows ($t_1$ and $t_2$) for two transport companies (forwarders).

In the sequential decision making process, each company allocates trucks by time window without information from each other. (Company A: five trucks for time window; Company B: four trucks for time window $t_1$ and two trucks for time window $t_2$). There are nine trucks for time window $t_1$ and two trucks for time window $t_2$. Then, the terminal operator informs each company that congestion will occur due to a longer turnaround time in time window $t_1$. Therefore, each company decides to move the two trucks from the time window $t_1$ to time window $t_2$. (Company A: three trucks for time window $t_1$ and two trucks for time window $t_2$; Company B: two trucks for time window $t_1$ and four trucks for time window $t_2$). Finally, the terminal operator informs each company that congestion is unlikely to occur.

On the other hand, the integrated decision making process proposed in this study considers cooperation between transport companies (forwarders) and the terminal operator so they share their information. Company A should allocate trucks to time window $t_1$ as much as possible, and company B does not have a problem with either time window $t_1$ or time window $t_2$. Therefore, through a mathematical model provided by the terminal operator, company A moves only one truck to time window $t_2$, and company B moves two trucks to time window $t_2$. (Company A: four trucks for time window $t_1$ and one truck for time window $t_2$; Company B: two trucks for time window $t_1$ and four trucks for time window $t_2$).

In summary, compared to the sequential decision making process, the integrated decision making process can make decisions in a short time, and company A can reduce the cost of changing work by moving only one truck.

## 3. Model Description and Formulation

### 3.1. Definition of a New Truck Appointment System(TAS)

In this paper, we consider a new TAS between the transport companies (forwarders) and the terminal operator at the gate and the yard of the container terminal. There is variability in truck arrivals due to the operational complexity of the container terminal and the uncertainty of inland flows [12]. The terminal operating system assigns tasks continuously, but to reduce the complexity of the continuity problem, the TAS is expressed in a discrete form (see Figure 3).

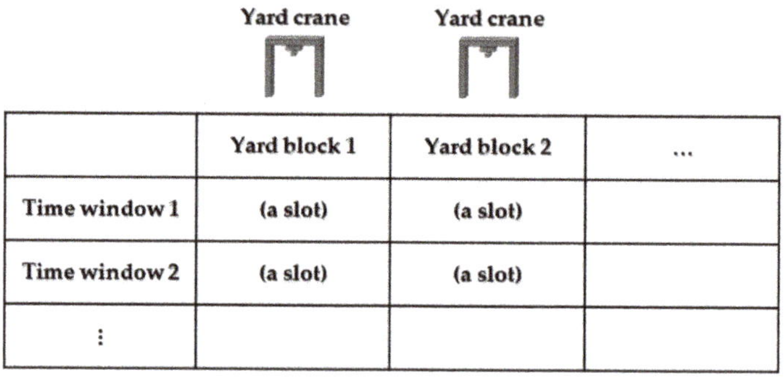


**Figure 3.** The Truck Appointment System (TAS) expressed in a discrete form.

The TAS consists of yard blocks (space concept, see Figure 4) and time windows (time concept) in a two-dimensional matrix, and each area is called a slot, and the slot capacity refers to the workload of the yard crane. The terminal operator determines the number of tasks a truck can reserve per slot, and the transport companies (forwarders) assign trucks per slot. Then, the yard crane assigned to each block performs the reserved tasks.

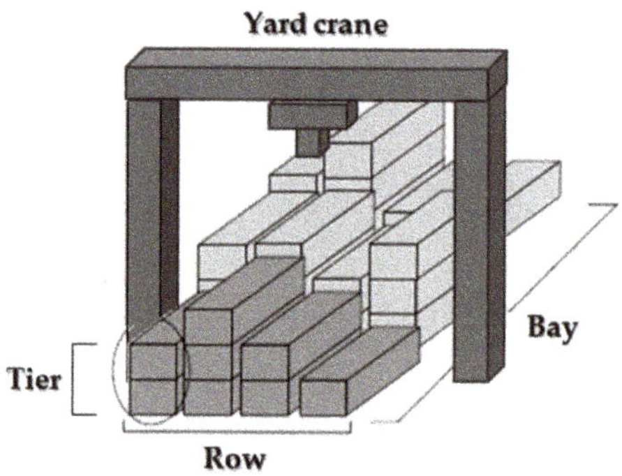


**Figure 4.** The configuration of the yard block.

From the perspective of the transport company (forwarder), the mathematical model assigns a truck schedule to the slots that can be reserved and set by the terminal operator for each yard block and time window. To do this, we minimize the waiting time for the truck and the number of rehandling for carry-out containers. From the perspective of the terminal operator, the mathematical model optimizes the number of tasks a truck can reserve according to the truck allocations set by the transport companies (forwarders) for each yard block and time window. To do this, we consider whether there is unassigned slot capacity other than the number of tasks a truck can reserve and whether there is unreserved slot capacity. From the perspective of cooperation between the transport companies (forwarders) and the terminal operator, the mathematical model simultaneously determines the number of tasks a truck can reserve and the truck allocations.

We consider the following assumptions to formulate a mathematical model:

1. The container handling process only includes loading/discharging, carry-in/out, and rehandling, not re-marshaling.
2. The priority of loading/discharging is highest and carry-in/out proceeds when there is no loading/discharging. (The truck arrives at the yard block according to the carry-in/out schedule).
3. All export containers arrive at the yard before loading, all import containers leave the container terminal after discharging, and we do not consider other exceptions.
4. We consider rehandling for carry-out, but not rehandling for loading.
5. We apply different stacking rules according to the task type of the container. (In the case of import containers, the carry-out times are different, so they are spread low to reduce rehandling. In the case of export containers, the loading times are similar, so they are stacked high with a small number of rows to facilitate shipment).
6. Carry-in/out or rehandling operations belonging to the same yard block and time window start with the truck that arrives first.

### 3.2. *A Mathematical Formulation for the New Truck Appointment System*

We use the following notations to formulate a mathematical model:

**Indices and parameters**

| | |
|---|---|
| $i$ | Index for a task type, $i \in \{out(bound), in(bound)\}$ |
| $j$ | Index for a yard block, $j = 1, 2, \cdots, J$ |
| $k$ | Index for a time window, $k = 1, 2, \cdots, K$ |
| $n$ | Index for a transport company (forwarder), $n = 1, 2, \cdots, N$ |
| $r^n$ | The rehandling rate required during inbound operations by the transport company (forwarder) $n, 0 \leq r^n \leq 1$ |
| $t$ | The tier where an inbound container is stored |
| $T$ | Maximum number of storage tiers of a yard block |
| $w^n_{ij}$ | The number of tasks for the transport company (forwarder) $n$ of task type $i$ assigned to the yard block $j$ |
| $v^n_k$ | The number of trucks from the transport company (forwarder) $n$ available in the time window $k$ |
| $B_{jk}$ | Maximum number of tasks for a yard crane (TC) in the yard block $j$ and the time window $k$ |
| $s_{ijk}$ | The number of loading/discharging operations of a container of task type $i$ assigned to the yard block $j$ and the time window $k$ |
| $g_k$ | Maximum number of trucks that can pass through the gate in time window $k$ |
| $\alpha$ | Transport company's weight |
| $\beta$ | Terminal operator's weight |

**Decision variables**

| | |
|---|---|
| $X^n_{ijk}$ | The number of trucks of task type $i$ assigned by the transport company (forwarder) $n$ to the yard block $j$ and the time window $k$ (from the perspective of the terminal operator we apply this variable as a parameter.) |
| $Y_{jk}$ | The number of tasks a truck can reserve allocated by the terminal operator for the yard block $j$ and the time window $k$ (from the perspective of the transport company (forwarder), we apply this variable as a parameter). |

#### 3.2.1. A Mathematical Model from the Perspective of the Transport Company (Forwarder)

The mathematical model (Model 1) for the transport company (forwarder) takes into account the penalties for the waiting time for the truck and the number of rehandling for carry-out containers.

$$\text{Minimize } \sum_n \sum_j \sum_k \left\{ \frac{\sum_n \sum_i X^n_{ijk} - 1}{2} \sum_i X^n_{ijk} + r^n \frac{T+1}{2} \sum_{i \in \{in\}} X^n_{ijk} \right\}. \quad (1)$$

Subject to,

$$\sum_k X^n_{ijk} \geq w^n_{ij} \text{ for all } i,\ j,\text{ and } n; \tag{2}$$

$$\sum_i \sum_j X^n_{ijk} \leq v^n_k \text{ for all } k \text{ and } n; \tag{3}$$

$$\sum_n \sum_i X^n_{ijk} \leq Y_{jk} \text{ for all } j \text{ and } k; \tag{4}$$

$$X^n_{ijk} \geq 0 \text{ for all } i,\ j,\ k, \text{and } n. \tag{5}$$

Equation (1) is an objective function that minimizes the sum of the penalties for each transport company's truck waiting and rehandling of carry-out containers. Equation (2) is a constraint, and the number of trucks of task type assigned by the transport company (forwarder) to the yard block must satisfy the number of tasks for the transport company (forwarder) of task type assigned to the yard block. Equation (3) is a constraint, and the number of trucks from the transport company (forwarder) available in time window must be satisfied. Equation (4) is a constraint, and the number of tasks a truck can reserve allocated by the terminal operator for the yard block and the time window must be satisfied. Equation (5) represents the non-negative condition of the decision variable $X^n_{ijk}$. In the mathematical model from the perspective of the transport company (forwarder), we apply $Y_{jk}$ as a parameter.

The penalty for the waiting time for the truck is the value calculated by multiplying the number of trucks assigned to the same yard block and time window by the expected time waiting for the task to start. We calculate the expected time waiting for the operation to begin like (6)–(8). $P_1$ is the probability that a truck arrives at the yard block $j$ and the time window $k$. $W$ is the number of tasks for trucks arriving earlier (the waiting time of the target truck), and $\lambda$ the sequence in which the trucks arrived in the same yard block and time window. $E_1$ is the expected time waiting for the operation to begin.

$$P_1 = 1/\sum_n \sum_i X^n_{ijk}, \tag{6}$$

$$W = \lambda - 1,\ for\ \lambda = 1, 2, \ldots, \sum_n \sum_i X^n_{ijk}, \tag{7}$$

$$E_1 = \sum_\lambda W \times P_1 = \sum_{\lambda=1}^{\sum_n \sum_i X^n_{ijk}} (\lambda - 1)/\sum_n \sum_i X^n_{ijk} = \left(\sum_n \sum_i X^n_{ijk} - 1\right)/2. \tag{8}$$

The penalty for the number of rehandling for carry-out containers is the value calculated by multiplying the number of carry-out tasks, the expected number of rehandlings required to take out one, and the rate that requires rehandling during the carry-out operation. We calculate the expected number of rehandlings to take out one container like (9)–(11) [22]. $P_2$ is the probability of taking a container out of tier $t$. $R$ is the number of rehandlings required to take it out of a specific location. $E_2$ is the expected number of rehandlings to take a container out of a bay.

$$P_2 = 1/T; \tag{9}$$

$$R = T - t + 1,\ for\ 1 \leq t \leq T; \tag{10}$$

$$E_2 = \sum_t R \times P_2 = \sum_{t=1}^{T} (T - t + 1)/T = (T + 1)/2. \tag{11}$$

3.2.2. A Mathematical Model from the Perspective of the Terminal Operator

The mathematical model (Model 2) for the terminal operator takes into account the penalties for the unassigned and unreserved slot capacity (the tasks a truck can reserve).

$$\text{Minimize} \sum_j \sum_k \left\{ \frac{B_{jk}}{\sum_j \sum_k B_{jk}} \left(B_{jk} - \sum_i s_{ijk} - Y_{jk}\right) + \left(Y_{jk} - \sum_n \sum_i X^n_{ijk}\right) \right\}. \tag{12}$$

Subject to,

$$\sum_n \sum_i X^n_{ijk} \le Y_{jk} \text{ for all } j \text{ and } k; \tag{13}$$

$$Y_{jk} \le B_{jk} - \sum_i s_{ijk} \text{ for all } j \text{ and } k; \tag{14}$$

$$\sum_j Y_{jk} \le g_k \text{ for all } k; \tag{15}$$

$$Y_{jk} \ge 0 \text{ for all } j \text{ and } k. \tag{16}$$

Equation (12) is an objective function that minimizes the sum of the penalties for the unassigned and unreserved slot capacity (the tasks a truck can reserve). The unassigned slot penalty is proportional to the maximum number of tasks for a yard crane (TC) in each yard block and time window. Equation (13) and Equation (14) are constraints, and each equation means a limit on the number of tasks a truck can reserve allocated by the terminal operator and the number of unassigned slots for the yard block and the time window. Equation (15) is a constraint, and the maximum number of trucks that can pass through the gate in each time window must be satisfied. Equation (16) represents the non-negative condition of the decision variable $Y_{jk}$. In the mathematical model from the perspective of the terminal operator, we apply $X^n_{ijk}$ as a parameter.

#### 3.2.3. A Mathematical Model from the Perspective of Cooperation between the Transport Companies (Forwarders) and the Terminal Operator

The mathematical model (Model 3) for cooperation between the transport companies (forwarders) and the terminal operator is considered simultaneously by combining their respective models.

$$\text{Minimize } \sum_j \sum_k \left[ \begin{array}{c} \alpha \sum_n \left\{ \frac{\sum_n \sum_i X^n_{ijk} - 1}{2} \sum_i X^n_{ijk} + r^n \frac{T+1}{2} \sum_{i \in \{in\}} X^n_{ijk} \right\} \\ +\beta \left\{ \frac{B_{jk}}{\sum_j \sum_k B_{jk}} \left( B_{jk} - \sum_i s_{ijk} - Y_{jk} \right) + \left( Y_{jk} - \sum_n \sum_i X^n_{ijk} \right) \right\} \end{array} \right]. \tag{17}$$

Subject to,

$$\sum_k X^n_{ijk} \ge w^n_{ij} \text{ for all } i,\ j, \text{ and } n, \tag{18}$$

$$\sum_i \sum_j X^n_{ijk} \le v^n_k \text{ for all } k \text{ and } n, \tag{19}$$

$$\sum_n \sum_i X^n_{ijk} \le Y_{jk} \text{ for all } j \text{ and } k, \tag{20}$$

$$Y_{jk} \le B_{jk} - \sum_i s_{ijk} \text{ for all } j \text{ and } k, \tag{21}$$

$$\sum_j Y_{jk} \le g_k \text{ for all } k, \tag{22}$$

$$X^n_{ijk}, Y_{jk} \ge 0 \text{ for all } i,\ j,\ k, \text{ and } n. \tag{23}$$

Equation (17) is an objective function and reflects the weighted sum of the perspectives of the transport company (forwarder) (Equation (1)) and the terminal operator (Equation (12)). Equation (18) is a constraint, and the number of trucks of task type assigned by the transport company (forwarder) to the yard block must satisfy the number of tasks for the transport company (forwarder) of task type assigned to the yard block. Equation (19) is a constraint, and the number of trucks from the transport company (forwarder) available in time window must be satisfied. Equation (20) and Equation (21) are constraints, and each equation means a limit on the number of tasks a truck can reserve allocated by the terminal operator and the number of unassigned slots for the yard block and the time window. Equation (22) is a constraint, and the maximum number of trucks that can pass through the gate in each time window must be satisfied. Equation (23) represents the non-negative condition of the decision variable $X^n_{ijk}$ and $Y_{jk}$.

### 3.3. Analysis Procedure for the New Truck Appointment System

We carry out numerical experiments according to the following analysis procedure to propose a model that is beneficial to all stakeholders and draw implications. For numerical experiments, we apply the Monte Carlo approximation, a method of approximating the expected value using sampling. The greater the number of samples extracted by the Monte Carlo approximation, the higher the accuracy of the approximate expected value. This method utilizes the central limit theorem, which states that if you have a population with mean $\mu$ and standard deviation $\sigma$ and take sufficiently large random samples from the population, then the distribution of the sample means will be approximately normally distributed. Therefore, in this paper, the expected value is approximated by sampling more than 100 times.

Step 1-1: Solve the mathematical model from the perspective of the transport company (forwarder) (Section 3.2.1). We reflect an arbitrary $Y_{jk}$ value that follows a uniform distribution as an input variable and find the decision variable $X^n_{ijk}$. We derive the average of $X^n_{ijk}$ by sampling more than 100 times and reflect it as an input variable in step 1-2.

Step 1-2: Solve the mathematical model from the perspective of the terminal operator (Section 3.2.2). We reflect the average of $X^n_{ijk}$ derived in step 1-1 as an input variable and find the decision variable $Y_{jk}$. We derive the average of $Y_{jk}$ by sampling more than 100 times.

Step 2: Solve the mathematical model from the perspective of cooperation between the transport companies (forwarders) and the terminal operator (Section 3.2.3). We derive the averages of the decision variable $X^n_{ijk}$ and $Y_{jk}$ by sampling more than 100 times.

Step 3: Compare the objective values derived in step 1-1 and step 1-2 with the value derived in step 2. We compare the mathematical model solved by the transport companies (forwarders) and the terminal operator from their respective perspectives and the one that reflects their goals at the same time. We propose a model that benefits all stakeholders.

## 4. Numerical Experiments and Results

We conducted numerical experiments using IBM ILOG CPLEX 12.8, and we used a personal laptop with Intel ® Core ™ i7-9750H CPU and 16GB memory specification.

The mathematical model in this paper considers task type, yard block, time window, and transport company (forwarder) as indices. Excluding the task type, the remaining three factors affect the objective function of both the transport company (forwarder) and the terminal operator. Therefore, the numerical experiments consider three factors of yard block, time window, and transport company (forwarder).

Through the sensitivity analysis, we examined the change of the objective value according to the change in these three factors. Since the yard block and time window are common elements constituting the two-dimensional matrix of TAS, they are considered together in the sensitivity analysis.

We first conducted a sensitivity analysis according to the change in the number of yard blocks and time windows under the control of the number of transport companies (forwarders). Next, we conducted a sensitivity analysis according to the change in the number of transport companies (forwarders) under the control of the number of yard blocks and time windows. Finally, we analyzed how the changes in these three factors affect the objective value of the mathematical model.

### 4.1. The Change in the Number of Yard Blocks and Time Windows

As shown in Table 1, we experimented with three cases of yard blocks and time windows based on two transport companies (forwarders). The input parameters for each case reflect random values generated according to the uniform distribution (as shown in Table 2). Table 3 and Figure 5 show the experimental results for the three mathematical models.

**Table 1.** Cases for the change in the number of yard blocks and time windows.

| Cases | No. of Transport Companies (Forwarders) | No. of Yard Blocks | No. of Time Windows |
|---|---|---|---|
| Case 1 | 2 | 2 | 2 |
| Case 2 | 2 | 3 | 3 |
| Case 3 | 2 | 4 | 4 |

**Table 2.** Input parameters for the change in the number of yard blocks and time windows.

| Input Parameters | Values |
|---|---|
| $r^n$ | u (0,1.0) |
| $w_{ij}^n$ | u (0,10) |
| $v_k^n$ | u (10,30) |
| $B_{jk}$ | u (20,40) |
| $s_{ijk}$ | u (0,5) |
| $g_k$ | u (40,60) |
| $Y_{jk}$ | u (5,15) |

**Table 3.** The experimental results for the change in the number of yard blocks and time windows.

| Cases | The Objective Value of Model 1 (A) | The Objective Value of Model 2 (B) | The Objective Value of Model 3 (C) | Gap (%) ((A+B) − (C))/(C) |
|---|---|---|---|---|
| Case 1 | 210.19 | 16.56 | 225.19 | 0.69 |
| Case 2 | 229.41 | 19.40 | 244.79 | 1.64 |
| Case 3 | 251.78 | 21.59 | 266.80 | 2.46 |

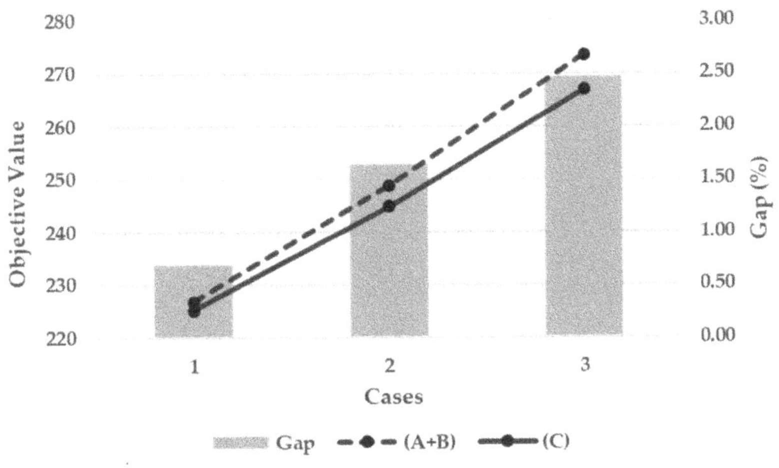


**Figure 5.** Comparison of experimental results for the change in the number of yard blocks and time windows.

According to the experimental results, in all cases, the objective value when the transport companies (forwarders) and the terminal operator cooperated was lower than when the transport company (forwarder) and the terminal operator decided independently. As the number of yard blocks and time windows increased, the gap in the objective value gradually increased.

### 4.2. The Change in the Number of Transport Companies (Forwarders)

As shown in Table 4, we experimented with three cases of transport companies (forwarders) based on two-yard blocks and two-time windows. The input parameters for each case reflect random values generated according to the uniform distribution (as shown in Table 5). Compared with the input parameters in Table 2, since the number of transport companies (forwarders) is large, the range of $Y_{jk}$ is higher, but other input parameters are the same. Table 6 and Figure 6 show the experimental results for the three mathematical models.

**Table 4.** Cases for the change in the number of transport companies (forwarders).

| Cases | No. of Transport Companies (Forwarders) | No. of Yard Blocks | No. of Time Windows |
|---|---|---|---|
| Case 1 | 2 | 2 | 2 |
| Case 2 | 3 | 2 | 2 |
| Case 3 | 4 | 2 | 2 |

**Table 5.** Input parameters for the change in the number of transport companies (forwarders).

| Input Parameters | Values |
|---|---|
| $r^n$ | u (0,1.0) |
| $w^n_{ij}$ | u (0,10) |
| $v^n_k$ | u (10,30) |
| $B_{jk}$ | u (20,40) |
| $s_{ijk}$ | u (0,5) |
| $g_k$ | u (40,60) |
| $Y_{jk}$ | u (10,25) |

**Table 6.** The experimental results for the change in the number of transport companies (forwarders).

| Cases | The Objective Value of Model 1 (A) | The Objective Value of Model 2 (B) | The Objective Value of Model 3 (C) | Gap (%) ((A+B) − (C))/(C) |
|---|---|---|---|---|
| case 1 | 210.19 | 16.56 | 225.19 | 0.69 |
| case 2 | 407.26 | 12.39 | 397.15 | 5.67 |
| case 3 | 770.86 | 8.84 | 697.22 | 11.83 |

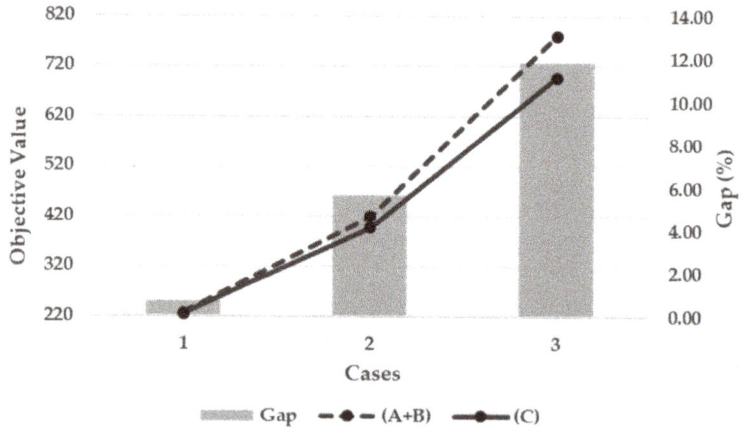


**Figure 6.** Comparison of experimental results for the change in the number of transport companies (forwarders).

According to the experimental results, in all cases, the objective value when the transport companies (forwarders) and the terminal operator cooperated was lower than when the transport company (forwarder) and the terminal operator decided independently. Also, as the number of transport companies (forwarders) increased, the gap in the objective value increased rapidly.

### 4.3. Analysis of the Results from the Perspective of Each Transport Company (Forwarder) and Terminal Operator

Based on the results of the sensitivity analysis performed in Sections 4.1 and 4.2, we analyzed how the changes in the number of yard blocks, time windows, and transport companies (forwarders) affected each transport company (forwarder) and terminal operator. Figure 7 shows the experimental results of the impact of the three factors on the transport company (forwarder). Figure 8 shows the experimental results of the effect of these changes on the terminal operator.

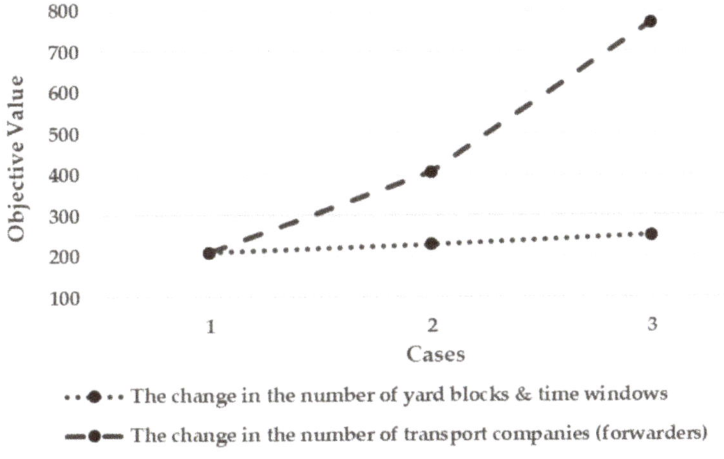


**Figure 7.** The relationship between the transport company (forwarder) and the change in the number of yard blocks, time windows, and transport companies (forwarders).

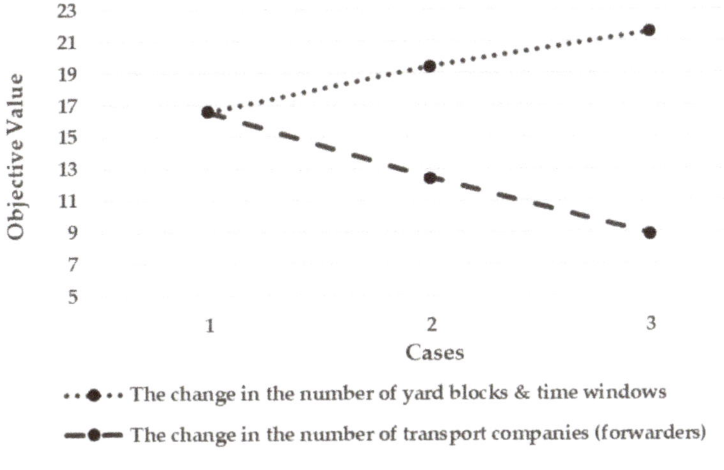


**Figure 8.** The relationship between the terminal operator and the change in the number of yard blocks, time windows, and transport companies (forwarders).

Both the number of yard blocks and time windows and the number of transport companies (forwarders) have a positive effect on the transport company (forwarder). In other words, as the number of yard blocks and time windows and the number of transport companies (forwarders) increased, the objective value of the transport company (forwarder) increased.

The terminal operator has a positive relationship with the number of yard blocks and time windows. However, it has the opposite relationship to the number of transport companies (forwarders). In other words, as the number of yard blocks and time windows increased, the terminal operator's objective value increased. On the other hand, as the number of transport companies (forwarders) increased, the terminal operator's value decreased.

## 5. Discussion

The results of the sensitivity analysis on the number of yard blocks, time windows, and transport companies (forwarders) clearly show the impact of maximizing the benefits of stakeholders.

When looking at the case of cooperation, the efficiency of the cooperation model was higher as the number of yard blocks and time windows increased. In case 3, the cooperation model showed an efficiency (gap) of 2.46%. This means that the more segmented tasks a truck can reserve, the more cooperation is required. The more detailed the reservation, the easier it is to plan the gate and the yard operations. As the number of transport companies (forwarders) increased, the efficiency of the cooperation model rapidly increased. In case 3, the cooperation model showed an efficiency (gap) of 11.83%. This proves that cooperation is necessary as the competitive relationship intensifies. Transport company (forwarder) can share information with other transport companies (forwarders) beyond communication with the terminal operator, enabling efficient truck allocation.

Looking at the perspective of the transport company (forwarder), the penalty of the transport company (forwarder) proportionally increased as the number of yard blocks, time windows, and transport companies (forwarders) increased. The increase in the number of tasks a truck can reserve has a minor impact on the transport company (forwarder). However, increasing the number of transport companies (forwarders) means increasing competitors, so the penalty for each transport company (forwarder) increases dramatically.

From the perspective of the terminal operator, results were quite different from those of the transport company (forwarder). The terminal operator's penalty is proportional to the number of yard blocks and time windows. However, as the number of transport companies (forwarders) increased, the terminal operator's penalty decreased in inverse proportion. The fact that a small number of transport companies (forwarders) participate in the TAS means that the management efficiency of the terminal operator is low. It is because one transport company's allocation schedule is a big part of it. On the other hand, as the number of participating transport companies (forwarders) increases, the terminal operator's management efficiency increases. This is because as the number of transport companies (forwarders) increases, the number of unreserved slots decreases.

## 6. Conclusions

Despite the number of sailings canceled in the past few months, as demand has increased, the utilization of ships has become very high. As a result, ship-to-ship operations and yard activity at the container terminals are at their peak, especially starting to affect land operations on truck arrivals and departures. The truck congestion causes many trucks to wait and emit more $CO_2$ than usual. Therefore, we solved the truck congestion problem by developing a new TAS that allows the transport companies (forwarders) and the terminal operator to cooperate to achieve their goals. The TAS in this paper helps reduce the truck congestion by considering the positions of the transport companies (forwarders) and the terminal operator, respectively, and comparing them with cooperation cases. To this end, we developed a mathematical model from the perspective of (1) the transport company (forwarder), (2) the terminal operator, and (3) cooperation between the transport companies (forwarders) and the terminal operator.

We reflected each stakeholder's penalties as the objective function of the mathematical model. In a mathematical model for the transport company (forwarder), the penalties are the waiting time for the truck and the number of rehandling for carry-out containers. In a mathematical model for the terminal operator, the penalties are the unassigned and unreserved slot capacity (the tasks a truck can reserve). From the perspective of cooperation between the transport companies (forwarders) and the terminal operator, we multiplied each objective function by its weight and added them. The transport company (forwarder) and the terminal operator have the same weight (transport company's weight = terminal operator's weight = 1).

For numerical experiments, we applied the Monte Carlo approximation, a method of approximating the expected value using sampling. The greater the number of samples extracted by the Monte Carlo approximation, the higher the accuracy of the approximate expected value. We sampled more than 100 times to approximate the expected value. Through the sensitivity analysis, we examined the change of the objective value according to the change in three factors of yard block, time window, and transport company (forwarder). Since the yard block and time window are common elements constituting the two-dimensional matrix of TAS, they are considered together in the sensitivity analysis.

As a result of the experiments, the cooperation model shows higher efficiency as the number of competing transport companies (forwarders) increases. Also, the more segmented tasks a truck can reserve in TAS, the easier it is to plan the gate and the yard operations. From the perspectives of the transport company (forwarder) and the terminal operator, a large number of yard blocks and time windows benefits both the transport company (forwarder) and the terminal operator. On the other hand, as the number of competitors increases, the penalty of the transport company (forwarder) increases, whereas the management efficiency of the terminal operator tends to improve.

This study attempted to see the efficiency of the cooperation model from the assumption that the transport companies (forwarders) and the terminal operator cooperate equally. However, in reality, each container terminal has a subordinate relationship between stakeholders and there is a limit that cannot reflect this legal and political environment.

**Author Contributions:** Conceptualization, H.I. and C.L.; methodology, H.I. and C.L.; software, H.I.; validation, H.I.; formal analysis, H.I.; investigation, H.I. and J.Y.; data curation, H.I. and J.Y.; writing-original draft preparation, H.I.; writing-review and editing, H.I. and C.L.; visualization, H.I.; supervision, C.L.; project administration, C.L.; funding acquisition, C.L. All authors have read and agreed to the published version of the manuscript.

**Funding:** This work was supported by the National Research Foundation of Korea (NRF) grant funded by the Korea government (MSIT) (NRF-2020R1F1A1076812). This research was also supported by Brain Korea 21 FOUR.

**Institutional Review Board Statement:** Not applicable.

**Informed Consent Statement:** Not applicable.

**Data Availability Statement:** Not applicable.

**Conflicts of Interest:** The authors declare no conflict of interest.

## References

1. COVID-19 Impacts Global Container Ship Trade. Available online: https://safety4sea.com/covid-19-impacts-global-container-ship-trade/ (accessed on 24 April 2020).
2. Coronavirus (COVID-19) Port/Country Implications. Available online: https://www.iss-shipping.com/pages/coronavirus-port-country-implications (accessed on 17 May 2020).
3. IAPH-WPSP Port Economic Impact Barometer. Available online: https://sustainableworldports.org/wp-content/uploads/2020-07-06-COVID19-Barometer-Report.pdf (accessed on 6 July 2020).
4. Phan, M.H.; Kim, K.H. Negotiating truck arrival times among trucking companies and a container terminal. *Transp. Res. Part E Logist. Transp. Rev.* **2015**, *75*, 132–144. [CrossRef]

5. Giuliano, G.; O'Brien, T. Reducing port-related truck emissions: The terminal gate appointment system at the Ports of Los Angeles and Long Beach. *Transp. Res. Part D Transp. Environ.* **2007**, *12*, 460–473. [CrossRef]
6. Yi, S.; Scholz-Reiter, B.; Kim, T.; Kim, K.H. Scheduling appointments for container truck arrivals considering their effects on congestion. *Flex. Serv. Manuf. J.* **2019**, *31*, 730–762. [CrossRef]
7. D'Adamo, I.; Rosa, P. How Do You See Infrastructure? Green Energy to Provide Economic Growth after COVID-19. *Sustainability* **2020**, *12*, 4738. [CrossRef]
8. Murty, K.G.; Wan, Y.W.; Liu, J.; Tseng, M.M.; Leung, E.; Lai, K.K.; Chiu, H.W. Hongkong International Terminals gains elastic capacity using a data-intensive decision-support system. *Interfaces* **2005**, *35*, 61–75. [CrossRef]
9. Zhang, X.; Zeng, Q.; Chen, W. Optimization model for truck appointment in container terminals. *Procedia Soc. Behav. Sci.* **2013**, *96*, 1938–1947. [CrossRef]
10. Chen, G.; Govindan, K.; Yang, Z. Managing truck arrivals with time windows to alleviate gate congestion at container terminals. *Int. J. Prod. Econ.* **2013**, *141*, 179–188. [CrossRef]
11. Guan, C.; Liu, R.R. Container terminal gate appointment system optimization. *Marit. Econ. Logist.* **2009**, *11*, 378–398. [CrossRef]
12. Zhao, W.; Goodchild, A.V. Using the truck appointment system to improve yard efficiency in container terminals. *Marit. Econ. Logist.* **2013**, *15*, 101–119. [CrossRef]
13. Chen, G.; Jiang, L. Managing customer arrivals with time windows: A case of truck arrivals at a congested container terminal. *Annals Oper. Res.* **2016**, *244*, 349–365. [CrossRef]
14. Ramírez-Nafarrate, A.; González-Ramírez, R.G.; Smith, N.R.; Guerra-Olivares, R.; Voß, S. Impact on yard efficiency of a truck appointment system for a port terminal. *Annals Oper. Res.* **2017**, *258*, 195–216. [CrossRef]
15. Phan, M.H.; Kim, K.H. Collaborative truck scheduling and appointments for trucking companies and container terminals. *Transp. Res. Part B Methodol.* **2016**, *86*, 37–50. [CrossRef]
16. Caruso, G.; Gattone, S.A.; Fortuna, F.; Di Battista, T. Cluster Analysis as a Decision-Making Tool: A Methodological Review. In *Decision Economics: In the Tradition of Herbert A. Simon's Heritage*; Bucciarelli, E., Chen, S.H., Corchado, J., Eds.; Springer: Cham, Switzerland, 2018; Volume 618, pp. 48–55.
17. Van Der Horst, M.R.; De Langen, P.W. Coordination in hinterland transport chains: A major challenge for the seaport community. *Marit. Econ. Logist.* **2008**, *10*, 108–129. [CrossRef]
18. Roso, V.; Woxenius, J.; Lumsden, K. The dry port concept: Connecting container seaports with the hinterland. *J. Transp. Geogr.* **2009**, *17*, 338–345. [CrossRef]
19. Lam, J.S.L.; Yap, W.Y. Container port competition and complementarity in supply chain systems: Evidence from the Pearl River Delta. *Marit. Econ. Logist.* **2011**, *13*, 102–120. [CrossRef]
20. Notteboom, T.E.; Rodrigue, J.P. Port regionalization: Towards a new phase in port development. *Marit. Econ. Logist.* **2005**, *32*, 297–313. [CrossRef]
21. Allen, R.S. Inland port savings. *J. Commer.* **2008**, *10*, 27.
22. Lee, H.; Kim, K.H. Comparing Expected Numbers of Re-Handles for Empty Containers During Gate-Out Operation. *J. Navig. Port Res.* **2018**, *42*, 207–216.

MDPI
St. Alban-Anlage 66
4052 Basel
Switzerland
Tel. +41 61 683 77 34
Fax +41 61 302 89 18
www.mdpi.com

*Applied Sciences* Editorial Office
E-mail: applsci@mdpi.com
www.mdpi.com/journal/applsci